On Disasters In India

Anu Kapur

FOUNDATION® BOOKS

Delhi • Bangalore • Mumbai • Kolkata • Chennai • Hyderabad • Pune

Published by

Cambridge University Press India Pvt. Ltd.
under the Foundation Books imprint
Cambridge House, 4381/4 Ansari Road, Daryaganj,
New Delhi 110 002

Cambridge University Press India Pvt. Ltd.
C-22, C-Block, Brigade M.M., K.R. Road, Jayanagar,
Bangalore 560 070
Plot No. 80, Service Industries, Shirvane, Sector-1, Nerul,
Navi Mumbai 400 706
10 Raja Subodh Mullick Square, 2nd Floor, **Kolkata** 700 013
21/1 (New No. 49), 1st Floor, Model School Road, Thousand Lights,
Chennai 600 006
House No. 3-5-874/6/4, (Near Apollo Hospital), Hyderguda,
Hyderabad 500 029
Agarwal Pride, 'A' Wing, 1308 Kasba Peth, Near Surya Hospital,
Pune 411 011

First Published 2009

ISBN 978-81-7596-622-2

Published by Manas Saikia for Cambridge University Press India Pvt. Ltd. and printed at Sanat Printers, Kundli, Haryana.

Contents

List of Figures

List of Tables

Acknowledgements

To plant a seed in a sterile landscape is difficult. Yet all forests evolved from barren rock, when soil was formed step by step. Such has been the evolution of this book. The idea for this book was sown while my students were pursuing research on disasters in India. Overcoming difficult circumstances, as they searched and brought home material and documents, there emerged the possibility of this research on disasters in India. *On Disasters in India*, therefore, is for my students. It is they who took it to completion.

As an infant it only had Deeptima. Soon Deebanjali, Shabana and Rimi joined in. Through its childhood, Roshini, Meeta and Neeti provided tender help. A fresh breeze from Purnima and Suman and occasional yet timely help from Anuradha and Amrita put it on the path to maturity. As Pooja Yadav deftly applied her skills of pruning, there arrived Punam. With her usual smile it was Punam Tripathi who ensured that not a single of the 4004 was lost or neglected. Devoting a full calendar year, her deep attachment to 4004 took the book to its flowering. But the path was not without its share of weeds and thorns.

It was the keen eye and constant concern of Gopal Krishan, Emeritus Professor, Panjab University, that kept the work abreast and abloom. Aware that plants wither out of harshness, my parents were wise to wait, watch and heal.

Providing a friendly space, all my colleagues at the Department of Geography, Delhi School of Economics, University of Delhi, were a sure support. The editors at Cambridge University Press India Pvt. Ltd. were a delight to work with. Jacketing the book with finesse they made it possible for it to reach out to one and all. The circle of energy and support fills me with deep gratitude.

Anu Kapur

I

Introduction

India has witnessed countless disasters in the last hundred years. This book is based on India's 4004 disaster research writings. Towards this goal, the number was neither predetermined nor an end in itself. It is an auto-evolved number. As a destination is to the seeker, so 4004 is to this research. This book is a research on India's disaster research using 4004.

A Veritable Niche

Why be concerned with the research already accomplished in the field? Why not simply get on with research itself? This too is a valid query. Research is perceived and perhaps truly is an intellectual engagement of a higher order. In the realm of academics, one is judged by individual research, not by the work of others. It could also be added that in a world of specialisation, the compilation of research is the forte of another group of professionals. Here, there is need to draw a fine line. Firstly, before knowledge was caged into cubbyholes, it was academia that compiled and interpreted the research in a given field. It remains to be seen what would have been the shades of research had this role sustained itself. Further, research on research is not equivalent to compilation. It also differs from a review of literature. The investigation of any idea or theme demands a reading of the relevant literature which functions as a background to evaluate a specific research objective. This is certainly important, but when the research itself is researched, it builds the very contours of the researchscape.

Secondly, in a review it is rarely possible to cull more than a few hundred references. The constraint of time and the drive to achieve the goals that have been set deters a wide search. Generally a research article barely carries more than fifty references. In a book or doctoral thesis, the fact that they rarely cross the 500 mark is an attestation of this. While a feeble count lends a worm's eye to the review, its small size mars a bird's eye view. A larger number of reference points are needed to sketch a scape.

Finally, the matter is not just about numbers. It is also about content. Each reference is a store of different bytes of information. The caption of each carries details like the theme, approach, methodology, area of study and the focus of inquiry. Depending upon the framework, a reference may carry some or all of these attributes. These bits of information can be made to behave like a set of data. As this diversity of information is pooled and analysed, it helps build an understanding of the landscape of research. It is of course difficult to define the critical number of references that are needed to build a quantum field of understanding. But there is little doubt that it is in the midst of such a scenario that a research on research carves its own veritable niche. This is true for India, and especially for the subject of its disasters.

Disasters: A Subject for India

Disasters scar the face of any country. India frequently heads the list as a country with the largest number of people dead and affected by disasters. During 1982–2001, as many as 59 million people were affected by disasters. This equals the total population of the United Kingdom. The average annual deaths in India due to disasters was four times the average for Asia, and eight times the world average during the years 1982–2001.

India has not only faced the maximum number of disasters, but to her discredit are disasters that have beaten world records. Evidence to this is found in the Guinness World Records – an institution that collects and presents data with investment in authenticity. In its records, India is listed under three entries on account of disasters; the worst toll a flood disaster could take of homes, a train disaster that has killed the maximum number of people, and the worst industrial-chemical disaster.

The 1978 flood that swept West Bengal and rendered 15 million homeless out of the state's total population of 44 million is in its log. In this flood, 13,000 people were killed and 1.3 million dwellings were destroyed. Also listed is the train that fell off a bridge into the River Baghmati at Samastipur, Bihar in 1981. The death of 800 passengers is the highest casualty in the history of train disasters in the world. Almost 4000 people were killed and thousands were left with permanent disabilities in 1984 when the toxic methyl isocyanate gas leaked from the Union Carbide Corporation pesticide plant in Bhopal.

In addition, India makes it to the top ten world disasters in case of landslide and avalanche, railway and aviation accidents, and tropical cyclones (Table 1.1).

Indian rail accidents hold the first, fourth and fifth positions among the top ten rail disasters during the 1980s and 1990s. India ranked sixth among the ten worst aviation disasters. In the number of people killed in an avalanche and in a landslide, India held the seventh rank, and for fatalities in a cyclone it stood fifth in 1882.

Disasters cost India nearly two billion rupees each year. The economic losses in a single disaster are staggering. An estimated US$ 2.1 billion was lost in assets in the Bhuj earthquake, which occurred on 26 January 2001 in Gujarat. A financial loss of over US$ 1.2 billion, equalling one per cent of the total government expenditure, is estimated from the devastation on the east coast of India from the tsunami on 26 December 2004 (*Business India,* 2005). Losses of this scale cut into economic growth rates. While it is difficult to estimate the total cost of damage, the constant cry for funds by the states to the Finance Commission confirms the plight.

A high pitched wail for aid arises from different parts of the country. In 1982–86, when the states sought an assistance of 251 billion rupees, the actual assistance given was 40 billion rupees, only 16 per cent of the total requirement. The turndown was in the vicinity of one seventh to one eighth times of what Haryana, Himachal Pradesh, Kerala and

Table 1.1 India's Rank in the World's Top Ten Fatal Disasters

Disaster type	Year	Place	Killed	India's rank
Railway accident	1981	Samastipur, Bihar	800	1
	1995	Firozabad, Uttar Pradesh	> 300	4
Tropical cyclone	1737	West Bengal	> 300,000	2
	1876	West Bengal	200,000	4
	1882	Mumbai	> 100,000	5
	1864	Kolkata, West Bengal	> 50,000	6
Aviation accident	1985	Ireland (Air India crash)	329	6
Avalanche	1979	Lahaul valley, Himachal Pradesh	> 200	7
Landslide	1980	Darjeeling, West Bengal	250	7

Based on *Department of Earth and Atmospheric Sciences Hazards and Disasters, 10 'Worst' Natural Disasters, Last Update 1997.*

Andhra Pradesh had called for. The release of funds was just one tenth in the case of Uttar Pradesh. So acute is the crunch that when the Union Government announced a National Calamity Fund in 1995–2000 of 70 billion rupees to deal with 'serious disasters', as many as seventy memoranda from 23 states forwarded a combined request for 240 billion rupees. Given the difference between the demand and supply, one is compelled to conclude that first aid is all that the Finance Commission seems to pay for. A poor country can just not afford to waste its scarce resources.

Fresh to human memory is the cyclone that hit Orissa on the east coast on 29 October 1999; the earthquake on the west coast in Gujarat on 26 January 2001; and the tsunami along both coasts of India on 26 December 2004. While the scale of all such disasters brought them to notice, there are countless instances where a house caved in, a farm was wiped out, and a livelihood came to an end but no one cared or took note.

Big or small, known or unknown, disasters are a subject of concern for India. Decisions about disasters can impact the lives of millions. Research provides reliable data and concrete solutions instead of opinions and hypothetical conjectures. Any decisions about managing disasters, after all, cannot be based on guesswork, chance or whimsy feeling. The thousands of students and learners enrolling for courses and training programmes in disaster management cannot be educated without an understanding soaked in research. Knowledge of what to, how to, and why to select a particular *modus operandi* of management is important. One only has to recall the mass deaths at the hands of epidemics and the success of health and medical sciences to acknowledge that serious research is what has extended the longevity of human life. Research on disasters is an essential need. As a prelude, the theme of research on research is an urgent call.

Call for Research

In India, time runs short. The victims, service providers, journalists, scientists and even the general public generate an outcry for information when confronted with frequent disasters. Policy-makers, administrators, media, educators and decision-makers are all in seeming hurry and for obvious reasons. India can just not afford to wait for millions more to die before it begins to act. A wisdom honed in one's own territory is the only way that light can be shed on the needs of the people. While one can learn from the culture of other countries, the staple for research has to be Indian. A quick access to the available research is thus a necessity.

Stock taking has many merits. It paves the way for an easier and less time consuming search. Insight about the past and the present builds foresight about the future. A research on research helps unfold the socio-historical context of research. The changing face of the subject can also be gleaned from such an inquiry. By ferreting conclusions about the coverage of themes, topics and regions, it provides a glimpse of current knowledge. A spin off from such an investigation is that it reveals the key contributors, unfolds the central debates and helps identify the paradigm of research. While much depends on the skill of the researcher, if aptly handled, the data from a research on research allows one to peek at the intellectual landscape of a subject. At the same time, such a research acts as a springboard that expedites and stimulates interest towards the frontiers of research. It helps identify and retrieve what is required with greater ease. The collection, classification, description and analysis of research is thus of utmost value. While one can beg for aid and borrow modern technologies, research has not only to be grown but also to be known on home turf. But this 'home' of India is vast and the call for research resounds from many a disaster. Before aspiring to reach any number, let alone 4004, there is a need to understand the terrain, chart the best possible route and spellout some of the ground rules.

II

Before 4004

Before setting out to search for the research many prior decisions had to be taken. There was a need to fix the territory, identify the language of the literature to be surveyed, define the framework of time, and decide the sources that would be tapped. Decisions had to be taken about what to collect, how much to look for, and where to explore. While a roster can be drawn and an itinerary customised, some decisions could only be taken while in the field.

Delimiting the Territory

India was picked as the spatial unit for study. The political map of the world defines the territorial extent of this country as covering an area of 3.29 million square kilometres. Lying entirely in the northern hemisphere, the mainland measures 3214 kilometres from north to south between the latitudes and about 2933 kilometres from east to west between the longitudes. The Indian subcontinent was partitioned in 1947. For ensuring compatibility, the boundaries of present day India were superimposed on the pre-partition map and references to Pakistan or Bangladesh were excluded. Any research that made peripheral or incidental reference to India was omitted. Therefore, works at the global scale or even concerning Asia or South Asia were set aside. The selection was strictly confined to references centred on India.

Selecting the Language

Research can be written and communicated in many languages. Of these, English was singled out as the language for this research. It indeed is ironical that although the research is bound to India, a language not native to it has been adopted for an inquiry. The possibility of literature on disasters in the twenty-two regional languages of India cannot be ruled out, yet English was selected as the language for this research. This was not because it is an adjunct to the main official language of Hindi, or the largest spoken business exchange language in the world, or the major medium for international publications, but mainly because the author is monolingual. Since translations create their own share of problems, English was adhered to as the language for research. A document in Hindi, Sanskrit, Urdu or any of the regional language was omitted. Such a decision had implications on the time and nature of outlets that were researched. Therefore, while a historical benchmark was not strictly defined, yet manuscripts, diaries, inscriptions, letters, and travelogues that were not in English were excluded. The research was confined largely to publications that belong to colonial and independent India except of course if there were works in English on disasters of medieval or ancient times. While some hazy lines could be drawn for the past, it was more difficult to mark the present.

Marking the Years

Research is never-ending. It is not easy to bring it to a halt. Today, there is active interest and research on disasters in India. This compounds the situation. Whereas the decision about English being the language for consideration set the reference frame to the colonial era, the date for closing was selected as the second Wednesday of October 2006. The choice of this date was not arbitrary. This day and month do share a concern for disasters. On 22 December 1989, the General Assembly of the United Nations designated the second Wednesday of October as the international day for natural disaster reduction. Across the world, themes like women, children, hospitals, schools, and the role of media in disasters were recalled and discussed on this day. As a vehicle to promote a global culture of disaster reduction, this date became a marker of time for ten years from 1990 to 1999. Even when the baton of the international decade of disaster reduction was handed over to the office of the International Strategy for Disaster Reduction at Geneva, the relevance of second Wednesday of October continued. But the more significant question is why end the investigation in 2006? This was done because the haul of references began to carry far too many duplications rather than provide fresh additions. This large proportion of repetitions was a definitive clue that time had come to close. While the spatio-temporal domains were easier to identify, embanking the theme was a bit more perplexing.

Embanking the Theme

What is to be included in the canopy of a disaster? This question is not simple to answer and riddles even some well-established disciplines like geography, sociology and environmental studies. To avoid getting lost in the maze of definitions or add another to the growing lot, a few guidelines were used to identify the research on disasters. It was decided to hand pick the research if it:

- pondered on the fundamentals
- depicted the temporal and spatial occurrence
- described the characteristics
- explained the cause
- identified the impact
- examined the response and
- suggested the mitigation and management of disasters.

Even though the above parameters were spelt out, the diversity in the design of research on disasters made the selection difficult. A research on disasters could range from being the study of an individual event like the Orissa cyclone or a lone force like earthquakes, drought or floods. It could also be one that investigates a series of events, multiple processes or a range of effects. There are those which deal with the source, some that select the cause, while others that focus on the force of a disaster. A research may take a holistic perspective or primarily adopt a sectional view. Wide is the mix and numerous are the ways to research disasters. Depending on the goals, some may prefer to labour on disasters that arrive suddenly while others could work on the slow onset types.

Following a different approach, another way would be to group them into the human and the natural. The High Powered Committee Report on Disaster Management in India, 2002, lists thirty-one types of disasters. They are classified into five categories, of which three belong to the natural realm. These include ten disasters under the heading of water and climate, four under the geological group and a similar number under the ones that are biologically related.

There is variety in each case. The human disasters range from fire and riots to terrorism and even include the group of technological disasters. There are still further subdivisions. Take the example of technology – it includes disasters that belong to industry, transport and other infrastructure. Transport includes the disasters of different modes of surface, air and

Figure 2.1 **Separating Natural from Human Disasters**

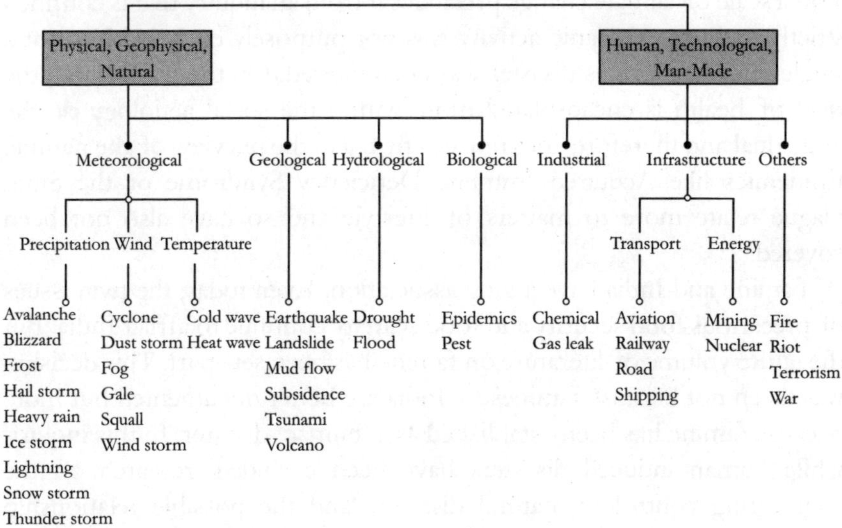

water carriage. This is not the end. Rail and road disasters are forms of surface transport while on water there are disasters of inland and ocean liners (Figure 2.1). Surrounded by this astonishing variety, this research opted to confine itself to the research on India's natural disasters.

Opting for Natural Disasters

The natural disaster too is not a single or simple entity. Therefore, before looking for readings on this theme, it was essential to draw boundaries around the content of this word. This was because in order to use this work effectively, an understanding of the topical coverage is required. In this research, nine type of disasters that cover the major meteorological, hydrological and geological types, namely cyclone, heat and cold wave, heavy rainfall, storm, flood, drought and earthquake, landslide and tsunami have been included. To grasp the entire range, a tenth category was added. This consisted of an assortment of references on hazards, risks or multiple natural disasters. In spite of the relaxed rules of inclusion, there were some which did not fit the frame of inquiry and were thus eliminated.

Soil erosion, forest fires and climate change were left out. This was not without reason. Soil erosion is ascribed to a human cause and therefore fits better within the realm of environmental problems. Forest fires have been excluded for the reason that they affect flora and fauna more than human life. The theme of climate change was also eliminated because the global scale of climate change precludes it from an inquiry that is confined strictly to India. Volcanic activity was not purposely excluded but not a single reference to this disaster was encountered. On the other hand, the field of health is encapsulated more within the social aetiology of the individual and therefore does not strictly fall in the purview of the natural. Epidemics like Acquired Immune Deficiency Syndrome or the Surat plague relate more to matters of lifestyle and so have also not been covered.

Famine and India have a long association. Even today, the twin issues of precarious food security and food scarcity continue to afflict India. But the entire volume of literature on famine has been set apart. This decision was taken not because famines of India are better documented but more because famine has been established as a 'human' disaster. In this inquiry, while human induced disasters have been excluded, research on the engineering control of natural disasters and the possible relationship between dams and floods, and dams and seismicity have been taken into account. This is because it is the technology that perpetuated the very disasters that it was designed to control. The inclusion of these was also intended to shed light on the demerits of modern technology as a fix for natural disasters.

Rules for insertion were applied with greater discretion in areas with an abundance of literature like the geologically studies on earthquakes and landslides. By comparison, the limits of eligibility were stretched for themes that have been poorly covered, like heat wave or cold wave, or studies on deaths and the social and psychological impact of disasters.

With these as the broad frames of inclusion and exclusion, the material covered a wide array. The studies on the causes, and the solutions sought to control, direct or manage natural disaster are part and parcel of this theme. Similarly, the entire range of impact from life to livelihood, physical to psychological, social to political are dimensions of this study. Response and mitigation also fall directly in the lap of studies of natural disaster. Even though the two, natural and human, share a lot in common, it primarily is the concern with nature that is used to distinguish natural from human disasters. Having laid the broad parameters, a question that could be raised is why select natural and not human disasters? Many reasons could be forwarded in defence of this.

Deciding for the Natural Disaster

A plausible reason that could be offered is that the range of disasters and the constraints of time do not allow the possibility of doing justice to both human and natural disasters. One could have asserted that dealing with all the disasters would be unwieldy and a confinement to natural disasters does provide a measure of coherence and manageability. There could also be an alternative explanation for this. A laggard would add that any search is like a relay and some other researcher could well follow with a series on human disasters. An optimist would say that a sequel to this, a volume two, that devotes itself to human disasters is on the cards. However each of these arguments could be forwarded in support of human disasters. If natural and human are divisions, why give the former priority over the latter? Here of course one could promote the selection on the grounds of the ecology of India.

It is commonly claimed that India's size, location and disposition breed natural disasters. With an equal number of latitudes and longitudes (29 degree), India is endowed with a shapely body. While the Tropic of Cancer divides India into two halves, the physiographic character gives it an altitudinal sweep of 8598 metres. Including the islands, it has a 7516 kilometre long coastline. Its geography is often quoted as a cause for natural disasters. Take the case of the Himalayas. Their height and geometry breed earthquakes, landslides, avalanches and cold waves. Swept on all three sides by the Indian Ocean, the coastal peninsular region of India has to shoulder the wrath of heavy rain, cyclones and tsunamis. Etched with innumerable rivers, the extensive basin of the Ganga-Brahmaputra system is infested with floods. Situated in the rain shadow of the hills, semi-arid conditions typify the interior plateau, while the northwest of India suffers from extreme drought. Since two-thirds of India continues to eke a living off the physical environment, it is the natural disaster that becomes more prominent and glaring.

A vast majority of the people in India endure extremes. When there are droughts, they hope and when floods arrive, they cope. When they get sick, they die. Millions of families in India live in shacks with porous roofs and walls through which wind blows and water drips. It is only for a handful of elites that weather is incidental; sheltered in air-conditioned homes and cocooned in their 24 degree centigrade offices, they dash from the home to the car buffered from cold, heat or rain. When it rains or snows, they have an umbrella and an overcoat. But for the teeming majority in India, this protection has simply not been created. The result is that much of India lives on the brink of a natural disaster. While the world over people's anxieties about human disasters like terrorism and nuclear bombs are growing, for the millions in India it is the 'natural' that is still daunting and real. The devastating disasters in 2004, 2001 and 1999 are also ascribed to three different naturals – the tsunami, earthquake and cyclone. This scenario could well be a convenient and convincing justification for selecting natural disasters as the inquiry for research.

However neither the land, the people of India, the recent disasters, nor lack of time or search for manageability dictated the decision to centre on natural disasters. The real reason was different.

While analysing the research on natural disasters, one wanted to question: Are there in essence 'natural disasters' or is it the focus and method of research that has catapulted 'nature' into a cause of disaster. How and why has the researcher held nature responsible for the disaster? Would the continuation of this approach yield benefits for managing disasters in India? These and other such questions led to the selection of work on natural disasters. A need was felt to collect the research done on natural disaster and then explore the possible entrenchment of this perception.

In the case of human disasters, there is no such controversy. War, terrorism, riots are man-made. Although industrial and transportation disasters are labelled as accidents (a position which also would be worth its share of investigation), yet there is uncontroversial acceptance of these within the realm of human disasters. Given this premise, one opted to research on natural disasters of India.

Picking the Package

Research expresses itself in varied forms and hues. It could be communicated as a spoken word, rhyme as a poem, set to a song, motioned in a dance, staged as a drama, captured in a photograph, rolled on a film, laid out in a map or chronicled in a text. Research has a place in a seminar, cinema, theatre, exhibition, or written word. The word could be penned on a mural, diary or note pad. It could be circulated through the newsprint or the Internet. Depending upon the depth of work, the venues could be a pamphlet, newspaper, magazine, thesis, proceedings, report, article in a journal, chapter in a book, or a book itself. The avenues are many and desire insatiable. Yet, to know what to wall in and also out is critical. Ephemeral sources like items in newspapers, pamphlets, short notices of publication and mimeographs were left out. Similarly, the diffused discussion of disasters in gazetteers and guidebooks led to their elimination The scarcity of material and data in statistical publications and census bulletins on the theme of disasters reduced their value in the search. Since websites and e-material are readily and easily accessible to audiences, a listing of these too has been left out.

This research is confined to publications in the shape of books, articles in journals, reports, and theses, along with features in a few relevant magazines. As channels for research, these are the more easily available and accessible. Further, to decipher the pattern of research, the source has to be large and continuous. Whereas paintings are sporadic, photographs and newspapers ephemeral and films on disaster a once in a while venture, publications enjoy greater continuity, durability and credibility. They were preferred and therefore picked.

Selections are difficult yet necessary. Identifying the country, period of work, theme and outlets that were to be searched set the road to be traversed. Be it a landmark, the North Star, the global positioning system, or a *guru*, direction backs search.

III

The Search

Where does one begin? What is the first step? The library undoubtedly is the finest institution of civilisation. An ideal library is an ocean of published knowledge. World-class libraries exemplify this description. The Library of Congress is the oldest federal cultural institution of the United States of America. It has a collection of more than 130 million items, which includes 29 million books and other printed materials. Taking a walk along its bookshelves spanning over 850 kilometres would be like traversing four times the distance between Delhi and Agra. The library of Princeton University includes nearly 6.2 million books, 36,000 linear feet of manuscripts, and impressive holdings of rare books, prints and archives. The University of Cambridge has a collection of 4.4 million while the Oxford University Library Service holds 11 million items.

Compared to these world-class centres, the libraries in India seem like ponds. The Parliament Library, the largest in Delhi and the second largest in India after the National Library at Kolkata, holds 1.27 million volumes of printed books, reports, government publications, reports, debates, gazettes and other documents. While it would be unfair to compare an ocean with a pond, what is of help is the fact that the two are not isolated. As a hydrological cycle connects a pond, lake and an ocean, such are the currents of globalisation. Today, trade and the Internet allow a larger access to publications than was ever possible. But unlike a hydrological cycle which circulates within a finite resource of water, the number of publications that are added afresh each year are many. Nearly one million books were published in the year 1996. Despite the boom, there is always a dearth of what one is looking for. This is specially so in the sphere of research on disasters of India. Yet one hopes, begins, pursues and does locate (Annexure 1).

To help read through the stock of publications, a library offers three different forms of compilation. There is the bibliography, the state of the art handbook and the catalogue. These therefore topped the list of search.

Not in the Records of Publications

The world has an amazing number and type of bibliographies which function as records of publications. There are ones on religion, war, plant species, computers and even a bibliography of articles on dog bites.

A search did not bring forward even a single bibliographical category on natural disasters of India. What did emerge were seventeen bibliographies on different aspects of disasters (Annexure 2). Of these, two were on chemical and industrial disasters. The first included the *Bibliographical Guide to Disaster Planning, Management, Insurance Case of Bhopal, India* by the Coppa and Avery Consultants in 1985. This ten page paperback on the toxic gas leaks from the plant of the Union Carbide did not serve the purpose at hand. The other bibliography was exclusive to North America and the British Isles. There were nine that were annotated and carried a count of less than three hundred references. Among the seventeen, four were published in journals and all others were mainly internal publications of a research center or institute in the United States of America. A fact that emerged was that among them, not a single has been published in the format of a book. There was no bibliography on disasters at the global level and any exhaustive compilation of references on disasters was conspicuously absent. Since an exclusive bibliography on natural disasters in India has not been compiled or published, there was a temptation to make a random check of a few general bibliographies.

The results were equally disappointing. The three-volume set on *Social Sciences in Modern India* (1990) lists 13,316 references. It did carry an exclusive section titled 'natural calamities' which listed 44 references to famine and only 23 to natural disasters of India. The *Bibliography of Geography* also drew a blank – the one on the Himalayas contained 28 out of 4764 references on natural disasters.

In general, bibliographies compiled by institutions are richer in content compared to those compiled by individuals. This led to the selection of the *International Bibliography of the Social Sciences*, compiled and edited by the British Library of Political and Economic Science, the United Kingdom. Some 100,000 articles from over 2800 journals, and 20,000 books, are referred to annually for its compilation. It covers over seventy languages

Table 3.1 Meagre Presence of Research on Disasters in Bibliographies

Name of bibliography	**Total references**	**On natural disasters**	**On natural disasters in India**	**Per cent of total**
Library Research in Geography[1]	868	0	0	nil
Indian References Sources[2]	2076	0	0	nil
A Bibliography of Bibliographies on India[3]	1243	1	1	0.8
Bibliography of Geography[4]	585	0	0	nil
Theses on Indian Subcontinent, 1877–1971[5]	3769	0	0	nil
Bibliography on Himalayas[6]	4764	28	28	0.59
Social Science in Modern India[7]	13,316	23	23	0.17
Bibliography of Economics[8]	72,365	78	4	0.01
Bibliography of Sociology[9]	66,702	163	6	0.01
Total	**165,688**	**293**	**62**	**0.04**

[1]Martinson, T.J. (1972). *Library Research in Geography.* Metuchen, New Jersey: The Scarecrow Press Inc. [2]Sharma, H. D. *et al.* (1972). *Indian Reference Sources.* Varanasi and Jullundhar: Indian Bibliographic Centre. [3]Kalia, D.R. and M.K. Jain. (1975). *A Bibliography of Bibliographies on India.* Delhi: Concept Publishing. [4]Harris, C.D. (1976). *Bibliography of Geography.*Chicago: University of Chicago, Department of Geography. [5]Krishnan, G. and D. Rai. (1976). *Theses on Indian Subcontinent (1877–1971).* Delhi: Hindustan Publishing Corporation. [6]Gupta, R.K. (1981). *Bibliography of the Himalayas.* Gurgaon: Indian Documentation Service. [7]Sharma, S.K. and S.A. Abidi. (1990). *Social Science in Modern India (Volume 1, to 3).* Delhi: Deep & Deep Publishing. [8]International Bibliography of Social Science (1990–2003). *Bibliography of Economics.* Paris. [9]International Bibliography of Social Science (1990–2003). *Bibliography of Sociology.* Paris.

from more than sixty countries. This makes it a rich and commendable record of publications (Table 3.1). The bibliography has four divisions – anthropology, economics, political science and sociology. Published annually since 1952, this institution has produced fifty-five volumes of bibliography for each of its four divisions till 2006.

Since research on India's disasters is hard to come across, it was decided to bank on references from a ten-year sample of economics and sociology divisions of the bibliography. The result was, however, equally disappointing. When 72,365 references in eleven volumes of the *Bibliography of Economics* from 1990 to 2003 were searched (three were not accessible), what emerged were 78 references on natural disasters with only 4 that pertained to India. The situation with respect to the *Bibliography of Sociology* was no better. Of the 66,702 references in the eleven volumes, there were 163 references on natural disasters with only 6 that pertained on India. In the collective survey of almost 160,000 records in different bibliographies, there were only sixty-two items of research on natural disasters in India. An insignificant 0.04 per cent.

Nominal Presence in Handbooks and World Reports

A state of the art handbook or a world report on disasters by an internationally recognised institution seemed promising. This led to the search in the *Handbook of Disasters* (Rodriquez *et al.*, 2006). Rodriquez, the senior editor, along with Quarantelli and Dynes have rendered commendable service to the subject of disasters by editing the *Handbook*. In a way it is the first comprehensive work on disasters at the global level. It carries a list of 2067 references. Among these, barely seven were on natural disasters of India. Another laudable contribution is that of the International Federation of Red Cross and Red Crescent Societies, Geneva. It launched the first World Disasters Report in 1993. This annual report brings together the latest facts and figures of disasters. But when five of these reports, from 2001 to 2005, were examined for research on natural disasters of India, a small number of 13 references out of the total 457 was not of much help. Though a deep sea dive was taken into the publication records, handbooks and world reports, the catch that accrued was small. It was a signal to shift the search to the catalogue, another sizeable gateway to the literature on a subject.

Feeble and Fragmented in Catalogues

Library material is organised for the reader's use in two ways – by call numbers and by subject headings in the catalogue. The call numbers provide a classification of the subject and area that is used in shelving the books. The subject headings are cross-references in the catalogue that contains material pertinent to more than one subject. When a dozen leading libraries in Delhi were accessed, a consistent reality that emerged was the absence of the word disaster in the subject headings of the catalogues. Except for the recently inaugurated library of the National Institute of Disaster Management, New Delhi, which has introduced a category on natural disasters, none of the other libraries devote an exclusive rack, stack or catalogue to natural disasters. Absence from a call number catalogue, however, does not mean that the libraries do not contain books on the natural disasters of India.

One of the difficulties is that although books are available, they are scattered across many disciplines. Aggravating the problem is the fact that natural disaster as a field of study tends to meld with both the physical and social sciences. The geologist studies the plate tectonics and the psychologist investigates the emotional impact of an earthquake. Similarly, the architect looks at the theme from the point of view of safe buildings while the social worker investigates concerns of rehabilitation and livelihood. The boundaries of the research on natural disaster also overlap other larger set of collective crises like mass displacements and epidemics.

It needs to be recalled that while looking through the bibliographies of economics and sociology, references on natural disasters were strewn under different heads. In the case of economics, references on disasters were placed under demographic aspects, production and conservation, and insurance, while those of sociology found a site with social psychology, language, communication and media, ecology, geography, human settlement, public administration, international relations, social problems and even health care. In a way, the boundaries of disciplines are porous and it would be beneficial if knowledge is freed from these cubbyholes.This would give the study of disasters its due place in research. But while the critical thinking of reorganisation of disciplines may have to wait, it would be an asset to evaluate and learn from the publications of those working in disciplines akin to one's own. But before a benefit can be drawn, there is a need to know the material. This is specially so for research on disasters. There is a dearth in the number of books on this theme. This precludes the need for a separate category in the catalogue of a library, but there is also the more relevant fact that the system of classification in libraries was devised when subjects like disasters were nowhere on the horizon in India. As between a seed and a flower, it is difficult to discern what comes first and what follows, such is the case in a book and catalogue.

Nor on the Bookshelf

Till date, only a handful of libraries in India have an online service. The number of books on natural disasters of India on the online catalogue of the Delhi University Library System, when last checked on the second Wednesday of 2006, did not cross the mark of 100. A surf through some of the leading libraries across the world did not improve the strength beyond 130. India Habitat Centre frequently mails the current update on a disaster alert but disappointingly has a stock of only half a dozen books on natural disasters.

A title search through the catalogue of more than two dozen libraries led to specific books (Annexure 1). It was not easy to track relevant research tucked in contents of multi-subject books with titles not specific to disasters. Therefore, the contents of a book on ecology, environment, geography, cultural anthropology, psychology or even on regional development could carry some work on natural disasters of India. When these were rummaged the turn out again was small. Among the hundreds of such books that were spot-checked, forty-five carried one chapter, fifteen had two chapters, six had three chapters and four or more came from another twenty books. The choice of books was governed both by the possibility of their containing a chapter and the feasibility of a physical check. Since it was accessibility and availability rather than publication that guided the selection, this could be a source from which more information can be collected. The books showed a way to reach out for more. The references in their appendices gave a clue. Even the introduction for half a dozen books revealed that many had their beginning in the proceedings of a seminar or were products of a doctoral thesis. A clue is more than a link. This led to the search for proceedings.

Proceeding towards the Proceedings

India has hosted a large number of 'gatherings' where people from all walks of academia, government officials, policy-makers, consultants, and social workers came to discuss the theme of disasters. The interest is recent yet prolific. The busy timetable of a year for National Institute of Disaster Management, New Delhi, is a case in point. They have organised ninety conferences within a short span of 2005–2006. Generally, the outcome of most symposia in India is either a news item, an endnote or a proceeding. Few materialise into a book. Take *Coping with Natural Hazards: Indian Context*, edited by Valdiya (2004). It is an outcome of the seventy-first annual session organised by the Indian National Science Academy at the University of Pune, 2004. Similarly, the origin of the book *Disaster Management*, edited by Sharma (1994), is an outcome of the First Natural Disaster Country Workshop, organised by the Indian Institute of Public Administration, New Delhi. Since in this search, books were very small in number, proceedings have been included. But most proceedings have limited copies and circulation. There is a high probability that while sixty-three proceedings have been retrieved, these are but a tip of the large numbers lying scattered in offices of sponsors. A similar fate is that of theses and dissertations.

Not as a Thesis

A thesis is the staple of a doctoral work. The title of a thesis reflects the kind of research problem taken for investigation. It is the Association of Indian Universities that compiles the *Bibliography of Doctoral Dissertations.* India had 177,745 doctorates by the end of 2000. There are only 113 doctorates that focus on natural disasters. Apart from the doctorate, there are other academic courses that involve the writing of a dissertation. These are important, but unfortunately there is no organisation which compiles a list of these sources. This means that the number of dissertations accessible is only a miniscule percentage of a larger possibility. This is evident from the fact that while the search allowed a collection of 113 doctorates, only 40 dissertations of Master of Philosophy and 50 of post graduate level originating from a few select universities or institutions could be salvaged. In reality, the number of dissertations of a lower ranking degree would outnumber the doctorates (Annexure 3).

In consonance with this, when a search was made in the theses written by the holders of the degree of Advanced Professional Program in the Indian Institute of Public Administration, New Delhi, natural disasters emerged as an area of interest to only 13 of the total of 389 theses. Among the 1532 theses in the catalogue of the library of the School of Planning and Architecture, New Delhi, only 42 dealt with disasters as a theme. Established in 1958 the School offers an option to select one course from housing, transport, urban, environmental and regional planning. Even though the specialisation on housing carries contents like principles of designing houses in disaster prone areas, shelter design and management policies, out of 364 theses on housing, only 15 took up disasters for a thesis. Evidently, a thesis on a shopping mall or a holiday resort, took precedence over research on disasters. The Department of Geography, Delhi School of Economics, University of Delhi, was set up in 1959. Till date, it has awarded 212 theses under its Master of Philosophy programme. Of these, only 18 have taken natural disasters as a topic of thesis. Although the number of references from this arena of search has been small yet the addition of dissertations gave an indication of the academic institutions engaged in research in this field. A similar pragmatism led to reports.

Reports: Unreported

Disaster management in India, and the world over, has largely been viewed as a responsibility of the government. In India, disaster is a state subject but it is also supported by regular inputs from the central government. The result is that there are reports by both the state and the centre. As sponsors and donors, international organisations have displayed regular concern about disasters. The end product is reports. The temperament of the government machinery is adept at framing reports but these documents are rarely published or publicised. It is hard to even learn of their existence, let alone retrieve them for research. Most reports of the state government are archived in their respective headquarters, reducing the scope for common accessibility. Libraries shelve reports, but these are also arranged in accordance with the scheme of the catalogue rather than the title. Therefore, while reports have a mushrooming growth compared to books, they are more difficult to locate. Yet this was no reason to discard the hunt. Moving through the corridors of different government departments was the only way.

While the number of ministries in India has increased from eighteen in 1947 to fifty in 2006, India does not have an exclusive ministry to deal with disasters till date. No wonder that reports on natural disasters lie scattered in various departments of different ministries. The office to deal with disasters has been changed several times. Till the mid 1970s, the Department of Agriculture was responsible for matters relating to flood and drought. But in June 2002, the concern for disasters was transferred to the Ministry of Home Affairs. Yet these two are not the only ones. There are other ministries like that of Water Resources, Urban Development, and Science and Technology that house reports on disasters. Likewise, the Planning Commission of India carried a separate chapter on disaster management for the first time in its Tenth Plan (2002–07) document, but programmes and schemes like the Ganga Flood Control Commission (1973), Rashtriya Barh Ayog (1975), Drought Prone Area Programme (1975) and the National Water Policy (1988) have been on the cards for long. All these ministries, departments, plans and programmes generate a series of reports.

State governments too add to this share. The Maharashtra disaster management plan, the evaluation reports on the cyclones in Andhra and

Orissa, or devastating floods in Bihar, or droughts in Gujarat and Rajasthan pave the way for official reporting. There are reports also by scientific organisations. The Geological Survey of India, Kolkata, Indian Meteorological Department, Delhi, Indian Institute of Remote Sensing, Hyderabad, National Institute of Disaster Management, Delhi, Central Building Research Institute, Roorkee, Central Road Research Institute, Delhi, Central Arid Zone Research Institute, Jodhpur, have contributed to the reports. Project reports of the Indian Council of Social Science Research, New Delhi, National Institute of Mental Health and Neurosciences, Bangalore, and those of the National Institute of Oceanography, Goa, add to the stock of reports on disasters.

International donors and sponsors cannot be overlooked. Effort was made to search for reports of organisations like the International Federation of Red Cross and Red Crescent Societies, Oxford Committee for Famine Relief, World Bank, Asian Development Bank and numerous other cells of the United Nations Organisation.

The possibility that natural disasters of India could be of interest to international institutes extended the search into their work. But exclusive turn out was low. The Disaster Research Centre, the first social science research centre in the world devoted to the study of disasters was established at Ohio State University in 1963 and moved to the University of Delaware in 1985. The centre conducts field and survey research on different disasters including natural disaster. The Disaster Research Centre boasts of over 450 publications. When this list was read, there was not a single publication on India's natural disasters! Some others like the Earthquake Engineering Research Institute, Oakland, United States of America, Asian Disaster Preparedness Centre, Kobe, Disaster Research Centre, Delaware, Oxford Centre for Disaster Studies, United Kingdom, World Meteorological Organisation, Geneva, and International Centre for Integrated Mountain Development did furnish a report or two on natural disasters of India.

Given the scepticism attached to the search for reports, the final bounty of 430 findings made it worth the hours devoted to the search. Reports are three times the number of books and double that of doctorates. But many might have been missed.

Before moving further, a check of the sources and the sourced was due. Nearly a year had been spent toiling through bibliographies, handbooks, catalogue cards, doctoral records, books, chapter in books, proceedings and reports. Time had been invested tracking libraries, government departments, scientific organisations, international offices, and even browsing through the search engines of the Internet. The kitty had accrued a stock of nearly 1500 references. These were 134 books, 202 theses, 427 reports, 724 chapters in books, proceedings and reports. But were these enough? Should more be searched? From here to where?

A startling omission was that articles in journals were overlooked. A research without articles in journals is a sure failure. So obvious is the merit of including journals that there is just no need to ponder or debate this ground. All that is needed is to devise a method to trek through the dense jungle of journals.

In the Journal Jungle

To ask how many journals are there in the world is like asking how many species are there in a jungle. Many are born, many die, and many become extinct before they are even known. Some adapt, endure and evolve and thus continue. The tapestry is dense; nevertheless each occupies a habitat where its survival is best ensured. Yet the question remains of how to conduct a census of the journals in the world. To ensure identity in name and ownership, each journal has to be officially registered. This is done at the country level and there is no global body that has amassed information for all journals in the world. Modern information technology has made it easy to access a whole group of journals but there is no single stockist. Take the service of Science Direct. It offers more than a quarter of the world's scientific, medical and technical information online. It has over 2000 peer-reviewed journals. Springer Link offers 1848 journals, Taylor and Francis has 1650 journals, and Elsevier-Science Direct 2229 journals. But these sources are not static and they could add or delete journals as per their requirements.

The problem is that it is hard to quote an authentic figure for the total number of journals in the world. There are other issues as well. It is the bigger, economically profitable and popular species that attract world notice, but the jungle has many more of value. The factors of time, cost and skill hinder online presence of many journals from countries like India. The digital divide between journals that can afford to go online vis-à-vis those that are deprived prohibits a complete audit of the journals in the world. While the total number is hard to know and the destiny of journals may follow the way of the jungle, where the fittest survive, yet as of today, the jungle is dense and a rich repository for research. How does one narrow down to journals that carry the crop of research on natural disasters in India? (Annexure 4–7).

In Journals with Nomenclature and Synonyms of Disasters

A sure way is to identify journals that carry the words disaster, risk, hazard, natural calamity and emergency in their title. Their nameplate is proof of their dedication to this theme. If such a journal originates in India, then there is sure hope for a positive return. But India has rarely raised its fortunes. In spite of being a country with a record of disasters, its academia has failed to float a journal devoted to disasters. Newsletters, that too of recent origin, are all that India offers (Table 3.2).

But by and large these newsletters carry information about the works of their institutions rather than disasters. Gaining little, the search veered to foreign journals with a nomenclature of disasters. A fast track through the maze of world journals was to work through the search engines on the net. The problem was that journals with the word disaster in their title are mainly linked to works on medicine, trauma and business. The same situation applies to risk and emergency. Therefore when the search was specified to ones that could possibly contain works on natural hazard and natural disaster only eight fell into the group.

Initiated from different countries like the United Kingdom, United States of America and Japan, it was revealing that the birth of the first journal on this theme did not predate the 1970s. In fact journals like the *Disaster Prevention and Management, Disasters: The Journal of Disaster Studies,*

Table 3.2 Not a Single Journal on Disasters in India – Only Newsletters

Title of journal	Publisher	Year of inception	Periodicity of publication
Together, Towards a Safer India	Ministry of Home Affairs, Government of India, New Delhi	2003	Monthly
Possible-it is: An Initiative for Disaster Management in Northeast India	North-Eastern Council, Shillong	2003	Quarterly
Tidings	National Institute of Disaster Management, New Delhi	2004	Quarterly

Based on *4004*.

Policy and Management and the *International Journal of Mass Emergencies and Disasters* gained foothold only after the 1980s. In spite of the fact that the search covered different disciplines and academic societies, India had only a handful of references in their content (Table 3.3).

To bridge this lacuna, the decision was to work through the major components that go into constituting the identity of a natural disaster. Earthquake, cyclone, flood, drought, landslide, along with the waves of heat and cold, heavy rain and tsunami were listed as the first order constituents. The effort was then to look for journals that dealt with them exclusively. Since science has been segmented into a host of specialisations, there would be works concerned with the individual entities of the natural. This trajectory too did not yield dividends.

Table 3.3 Search in Journals that Focus on Disasters

Name	Focus	Year	Place of publication	Number of articles on India[1]
Disaster Prevention and Management	Disaster	1992	Bradford, U.K.	17
Journal of Natural Disaster Science	Disaster	1979	Kyoto, Japan	2
Disasters: The Journal of Disaster Studies, Policy and Management	Disaster	1977	London, U.K.	18
Natural Hazard Review	Hazard	2000	Reston, U.S.A.	1
Environmental Hazards	Hazard	1999	New Jersey, U.S.A.	1
Natural Hazard Journal	Hazard	1988	Netherlands	22
Science of Tsunami Hazard	Hazard	1982	Honolulu, U.S.A.	9
Natural Hazard and Earth System Sciences	Hazard	1970	Paris, France	2
Australian Journal of Emergency Management	Emergency	1986	Victoria, Australia	1
International Journal of Mass Emergencies and Disasters	Emergency	1983	Oklahoma, U.S.A.	2
Earthquake Spectra	Earthquake	1985	Oakland, U.S.A.	1
Bulletin of Indian Society of Earthquake Technology	Earthquake	1964	Roorkee, India	33

Based on *4004. The Natural Hazard Centre, Colorado publishes the Natural Hazard Observer and the Natural Hazard Review. Of the two, the latter was listed because observer did not have a single reference to India.[1] Depth of search influenced by accessibility of journal.*

There was no exclusive journal on flood, cyclone, drought or heavy rainfall. There was the Journal of the Japan Landslide Society but it is published in Japanese. Also available were the *Science of Tsunami Hazard*, published electronically since 1982 from Hawaii and the *Earthquake Spectra* from California. Both had only ten references on India with the former including nine of them. Dated 1964, the *Bulletin of Indian Society of Earthquake Technology* was a venture of the University (Indian Institute of Technology since 2002) of Roorkee. 33 references were sourced from its publications over the last nine years. A systematic search of twelve journals specific to the theme of disasters and its associated topics thus supplied only 109 references. The feeble number triggered the need to adopt another strategy.

It was evident that the search for research on disasters could not be a one-way street. Since the study of disasters is pursued by different disciplines, there was a possibility that a search along different roads could return a possible bonus. Guided by intuition, one moved across the thick journal jungle. At times, the author was known, at other times the journal seemed familiar; sometimes the reference came from a footnote stashed in an article and at other times a cross-reference gave a lead. Though apparently haphazard, there is a system and relationship in the way species come together in a jungle. This is true in the case of journals too. To reach out to peers, the channels of research have been organised according to disciplines and sub-themes. But the difficulty with the theme of disasters is that it combs through a multiplicity of disciplines. Many detours had to be taken before a reference could be reached. Journals were picked across the earth sciences, social sciences, medical sciences and technological sciences. While an attempt was made to keep a uniform distribution, it was not easy to achieve. After all, it is not the type of journal but the interest of authors that would determine their choice to work on India.

Journals on Different Disciplines

Publication of research on natural disasters has a strong association with journals belonging to the earth sciences. Nearly half the journals belonged to this group. Social sciences are a small share of the journals; a lead of three in every ten is taken by sociology, economics and geography in this case. A disaster calls for physical and emotional healing on emergency basis. Medical sciences ought to find a place in the research, but in the case of India only one in every eight journals has a stamp of its research input (Table 3.4). The limited returns forced a branching out in different directions.

Disasters are often categorised into those of land, water and air. Many scientists have the expertise to deal specially with these realms. It is the geologists, hydrologists, meteorologists and the environmentalists who have acquired the expertise to study the mechanism of an earthquake, landslide, flood, thunderstorm or cyclone. These specialisations are inclusive of subfields and journals like *Physics of the Earth and Planetary Interiors, Journal of Geodynamics, Indian Journal of Earth Science* and the *Earth and Planetary Science Letters*. When accessed, they did provide additional references on natural disasters in India.

There is a community of geologists and geophysicists for whom the *Journal of Geological Society of India*, *Bulletin of American Geological Society*, *Geology Today*, *Bulletin of Engineering Geology* and *Himalayan Geology* are venues for publications. Not to be left out are *Tectonophysics*, *Central Board of Geophysics, India*, *Geophysical Journal International*, *Journal of Indian Geophysical Union* and others in this family which generated references on India's disasters. The sum of references from the earth sciences pertaining only to land thus rose to 552.

Table 3.4 Search in Journals that Represent Different Interests and Disciplines

Themes	**Number of journals**	**Per cent of total**	**Number of articles**[1]	**Per cent of total**
Earth Sciences	*85*	*45*	*953*	*52*
Land	47	25	552	30
Meteorology	17	9	313	17
Interface	12	6	19	1
Water	9	5	69	4
Social Sciences	*57*	*30*	*597*	*32*
Sociology	28	15	140	8
Economics and Development	17	9	353	19
Geography	12	6	104	6
Medical Sciences	*23*	*12*	*69*	*4*
Technological Sciences	*15*	*8*	*110*	*6*
Risk/Hazard/Disaster	*11*	*5*	*109*	*6*
Total	**191**	**100**	**1838**	**100**

Based on *4004.*[1] *Relevant to natural disasters in India.*

Disasters are also related to the mechanism of the weather and climate. When the works of professionals in this realm were accessed, they contributed 313 references. The *International Journal of Climatology*, *Journal of Applied Meteorology*, *Monthly Weather Review* and *Mausam*, the latter published by the Indian Meteorological Department, New Delhi, were most rich in material.

Be it on the seacoast or rivers, hydrologists and water management experts are concerned with floods. This was the reason that journals like the *Hydrological Sciences Bulletin*, *Water Resources Management*, *Journal of Hydrogeology* and specially *Bhagirath* were scanned, and contributed another 69 references. The realms of land, air and water are all connected. The study of their interface led to journals like *Environmental Management*, *Tropical Ecology* and the *Proceedings of Indian National Science Academy*. When these along with a handful of others were searched, they provided 19 references.

Disasters are commonly treated as physical entities but their spatial dimensions apart from societal and developmental are more important. Drawing from the very definition of their discipline, geographers are popularly associated with the 'natural' as an area of investigation. When a dozen geographical journals, mostly those published in India, like *Transactions, Institute of Indian Geographers*, *Annals of National Association of Geographers, India*, *Geographical Review of India*, along with others like the *Singapore Journal of Tropical Geography* and the *Geographical Journal* were searched, the store of references accumulated a further 104 spread across a wider variety of journals like *Man in India*, *Indian Journal of Adult Education*, *Indian Journal of Social Work*, *Social Welfare*, *Social Action* and *Seminar* added another 140 to the pool. What augmented the references by a further 353 were works in journals like the *Indian Journal of Power and River Valley Development*, *Kurukshetra* and *Mountain Research and Development*, with the largest contribution coming from the *Economic and Political Weekly*.

A vital service needed during any disaster is that of medical professionals. The physical and emotional injury to people and the issues of health care are very critical for victims. This concern induced a search of journals of a special kind. The medical literature published in *The Lancet*, *British Medical Journal*, *Indian Journal of Medical Research*, *Indian Journal of Medical Ethics*, *Indian Journal for Practising Doctors* among a handful of others provided another 58 references. Revealing the state of research on psychiatric problems of victims of disasters was the fact that the *American Journal of Psychiatry*, *Archives of Indian Psychiatry*, *Journal of Traumatic Stress*, among a few others that were checked, yielded only 5 references. The arenas of *Indian Journal of*

Public Health, the *Nursing Journal of India* and a group of others on paediatrics and epidemiological issues were no better and had the measly 6 references. The medical journals provided an outflow of 69 references.

Technology is often seen as a solution to mitigate and manage disasters. The structure of houses in an earthquake prone zone or design for flood embankments have engaged the special interest of engineers. To forecast and predict disasters, the role of modern information technology like remote sensing and geographic information systems has also become popular. The group of journals clubbed under the umbrella of technology covered a wide latitude. They ranged from *Bulletin of Indian Society of Earthquake Technology*, *Indian Concrete Journal* and *International Journal of Remote Sensing* to the *Photonirvachak*, *Journal of the Indian Society of Remote Sensing*. Collectively these works added 110 references to the store of research on disasters in India.

Disasters make news. News follow disaster. Although the word journal and journalist have commonalities, yet the common prejudice among academicians is to sideline features in magazines. These are often considered to be thin on research and worthy only of instant consumption. However given the media response to disasters, the temptation to purview articles in magazine could just not be resisted. Displaying different points of view and coverage, seven magazines were selected. Originating in 1975, *India Today* can boast of being the magazine with the largest subscription in this country; the fortnightly *Frontline* (1984) also draws a huge readership. Nurtured by the Centre for Science and Environment, the weekly magazine *Down to Earth* has a special interest in subjects on the environment and does occasional features on disasters. The total print set of these three that is *India Today*, *Frontline* and *Down to Earth*, when checked, brought forth 530 references. The remaining four *Nature*, *New Age* and *Science*, *Science and Culture* were randomly searched yet they added close to 150 writings on natural disasters of India.

An ambitious foundation was thus laid. A search is a quest. A search is learning. But in the case of a search for research on disasters, it was an effort to smile through tears.

An Effort: A Smile through Tears

Any search is an uphill task. This was more so in the case of the search on research of disasters. India does not have an established centre for research on disasters. It lacks an exclusive journal on this theme, there is an absence of a professional body that takes interest in disasters. A bibliography, a place in the catalogue, is also missing. But some of the world's top ranking disasters have taken a toll on India. The fact that 4004 references on India's disasters could be put together is a reason to smile but this does not come without a wolley of tears once in a while! From the start to the end, problems plagued this search. References to disasters on India were fragmented. Without a respectable place in the curriculum or seats of learning, the study of disasters was scattered. The result is that the path to traverse had to be laid rather than followed. Enormous material had to be dug, shoved and sifted before a reference could be mined. The first tracks are the hardest to build.

It was only after 191 journals had been searched that a wealth of 1838 references could be mined. Not accounted here are all those journals that were checked but abandoned because they did not carry a single reference to disasters in India. There were 133 journals that gave less than 5 references each (Table 3.5, Annexure 8). These made the bulk of 69 per cent of the total journals. There were a few that were generous. More than 100 references each flowed from three journals namely, *Mausam* (243), *Economic and Political Weekly* (204) and *Current Science* (120). Books were equally deficient if not more so. Many books had to be kept aside as they did not contain a single reference to natural disasters of India. Even the eighty-six books that passed this inspection provided a turn out of only 348 references. This is just four references per book. But reality was harsher. In forty-five only a single reference came through while in another twenty-four the number was less than five per book. To find chapters on disasters of India in reports was more difficult. In 134 reports, the turn out was less than three chapters per

Table 3.5 Minimal Frequency of Articles on Disasters in Journals

Number of articles on disasters of India	Number of journals
Less than 5	133
5 to 10	26
11 to 20	16
21 to 40	6
41 to 80	7
More than 80	3
Total	**191**

Based on *4004*.

report. Universities too were unhelpful. Of the 269 Universities in India, only thirty-two produced doctorates on natural disasters. That seventeen of these had only one doctoral thesis is evidence of the sparse existence of research on natural disasters. Exceptions were Andhra University, Waltair, with fifteen and the Indian Institute of Technology, Roorkee, with sixteen doctorates. On a comparative scale, it is the Indian Institutes of Technology, that have been most forthcoming. India has seven Indian Institutes of Technology – Chennai, Kanpur, Guwahati, Mumbai, Kharagpur, Delhi and Roorkee. The latter four have twenty-seven doctorates on this theme to their credit. What is most disappointing is that in spite of the hard work, many a find had to be discarded. Though painful, there was reason to abandon.

Abandoned References

There is little reason to accommodate an incomplete, inaccurate or inconsistent reference. These were encountered frequently while searching through the literature appended in a book, cited in a thesis or acknowledged in reports. Many such cross-references had to be rechecked or crossed out. Take the case of the publications of the Geological Survey of India. Established in 1851 during colonial days, its prime objective was to investigate and document earthquakes in India. The three main outlets for its publications are special publications (including memoirs, records, and articles), bulletins and miscellaneous publications. However, these publications have been cited in a disorganised manner. Adding to the grouse is that publications of the Geological Survey of India are often jumbled with those of the Geological Society of India. The latter is a publication of the Association of Geologists that was established in 1958 in Bangalore. To make matters worse, these two organisations have not posted all their publications online. Even some of the leading departments of geology in the universities of India do not have a complete stock. It was only by a visit to Lucknow, a major Regional Centre of the Geological Survey of India, that corrections could be carried out. During such cross-checks, many works were abandoned while few could be retrieved. Some were salvaged but without all limbs intact. The latter are references where page numbers or the issue number of a journal could not be traced. A small minority of these was retained with the hope that they would serve some purpose. At times, a hint can be misleading, but it can also lead to valuable resources. While trying to play safe, tricksters too had to be set right.

Tricksters

A trickster is a journal that changes its name. Most journals do continue with a single name but there are some with names altered twice, thrice and even six times. The latter is not an exaggeration. From 1788 to 1953, the *Journal of the Asiatic Society* changed its name six times. There were cases of mergers and acquisitions. *The Journal of Himalayan Geology* was merged with *Himalayan Geology* in 1996 (Table 3.6).

Table 3.6 Tricksters – Journals that Changed Names

Journals	Name
Hydrological Sciences Journal	since 1981
Hydrological Science Bulletin	1972–1980
Bulletin of the International Association of Scientific Enquiry	1956–1971
Year of Inception	*1956*
Australian Journal of Emergency Management	since 1995
Macedon Digest	1986–1995
Year of Inception	*1986*
ISET Journal of Earthquake Technology	since 1998
Bulletin of the Indian Society of Earthquake Technology	1964–1997
Year of Inception	*1964*
Journal of the Earth System Sciences	since 2005
Proceedings of the Indian Academy of Sciences, Earth and Planetary Sciences	1934–2004
Year of Inception	*1934*
Journal of Asiatic Society	since 1953
Journal of Asiatic Society of Bengal	1935–1952
Journal and Proceedings of the Asiatic Society	1905–1934
Proceedings of the Asiatic Society	1865–1904
Journal of Asiatic Society of Bengal	1832–1864
Asiatic Researches	1788–1831
Year of Inception	*1788*
Mausam	since 1979
Indian Journal of Meteorology, Hydrology and Geophysics	1975–1979
Indian Journal of Meteorology and Geophysics	1950–1975
Year of Inception	*1950*
Journal Merger	
'Journal of Himalayan Geology' merged with 'Himalayan Geology'	1996

While a change is perhaps a welcome improvement regarding the ownership, status and coverage, it can create problems in identifying and locating a journal. This is especially so because libraries create a new entry in the catalogue while the maiden names may also be retained. It can be claimed that studying the life history of each journal should be part and parcel of the research but this realisation often dawns after encountering a problem. But then life often is like this, lived forward but learnt backwards. While there are adequate warnings that references should as far as possible be seen in the original, yet, like in most cases, wisdom is generally overlooked on grounds of shortage of resources.

Resource Crunch

India is a developing country. Libraries in India do not have adequate funds to bear the high cost of publications. The more established disciplines and subjects on the curricula consume most of the budget, leaving little for marginalised themes like disasters. A resource crunch is agony. Relevant books were difficult to reach. Journals too were hard to find. While librarians seem helpful, the libraries are archaic. There are missing volumes, torn pages and abrupt terminations of subscriptions. There are hardly any e-books on India's disasters and many journals are yet to go online. Shifting from one mode to another also has its own share of problems.

New Modes: Old Problems

Many advantages are cited for the use of the electronic mode of searching for literature. It is said to be quick, efficient and leads to a series of links and hyperlinks, but this is so only if one knows what one wants and how to work through the maze. While there are problems in accessing hard copies of books, the search for soft copies through the web carries its own share of hardships.

There was an enormous outpouring of irrelevant and duplicate material when the key words on natural disaster research in India were inserted in search engines. The reason for this is simple. There were works on the global scale, which simply cited India as an example. There were references on sites that provided data on disasters and so included India. Also there were references that pertained to disasters and India but were not specific to the word natural. This means that by typing in the key words a large number of search results promised a lot but delivered little.

Sifting the relevant from the irrelevant is indeed a task. There was need to be systematic and selective. Online search is a skill. But some problems are endemic. It is beyond the budget of individuals and even institutions in India may find it hard to subscribe to online documents. Access to publications on Elsevier comes with a price tag of US$ 4,134,405 per year. This is equivalent to approximately 169,510,605 rupees. Nearly the cost of purchasing sixteen middle-income housing flats in Delhi! The fluctuating rate of exchange adds to the problems. Many of the back volumes are either not available or access to them requires additional subscription. Such a situation is similar to that encountered with missing or misplaced journals in a hard copy format. This mars the quality of the search. Duplication and repetitions add to the frustration, but most daunting is this fear of omissions.

Regretfully it is accepted that this search is far from complete. It claims no universal coverage, is in no way exhaustive and it cannot assert to be up to date. But a question that could be asked is how many more are there than this 4004? The realm of the unknown is always expected to be more than that of the known, yet the best that can be hoped is that no significant work has been overlooked. Though the search does not reap a full harvest, it does assure that a lot has been covered. To gain a holistic grasp of research on disasters, this is an asset. This is a reason for more smiles than tears. However, a collection has to be sorted, shifted and organised before it can be made ready for use.

A Unit for Diversity

The 4004 references were dissimilar. There were ones like articles in a journal, chapters in a book, chapters in a report that stood like individuals, while there were reports, books and proceedings that came as collectives. Then there was the category of the unpublished in the shape of theses and dissertations, and the features in magazines.

Among the 4004 references, nearly two-thirds were in the form of articles in journals and chapters in edited books. Much more substantial works like books, theses and reports accounted for 19 per cent of the total. The share of the features in magazines formed 17 per cent of the references (Table 3.7).

More than 60 per cent of the references were works of single authors. This was specially so for theses, books and features in magazines. Reports were generally organisational efforts while articles in journals were equally

Table 3.7 Outlets of Publication of the 4004

Outlet of publication	Number of references	Per cent of total
Article in a journal	1838	46
Chapter[1]	724	18
Feature in a magazine	679	17
Report	427	11
Thesis/Dissertation	202	5
Book	134	3
Total	**4004**	**100**

[1] *In a book, proceedings or a report.*

divided between single and multiple authors. Of the 1838 articles, 24 per cent were in foreign journals. Among the 134 books, 10 per cent had foreign publishers. All the theses were produced within India (Table 3.8).

With these as the broad features of the 4004 references, what becomes more important is to trace the trend of research on disasters in India. When did research on disasters gain importance in India? What are the characteristics of this research? Which types of natural disasters have gained primacy in work and why? What are the approaches that have shaped this research? Which are the themes that have engaged researchers? What is the paradigm of disaster research in India? These and many such questions compel the need to standardise and organise the 4004 references.

Table 3.8 Number of Authors and Origin of Publication of the 4004

Category[1]	Author[2]					Origin[3]		
	Single	Two	> Two	Org.	Total	Indian	Foreign	Total
Article in a journal	52	25	23	0	100	49	51	100
Book	69	18	13	0	100	90	10	100
Chapter[4]	61	21	13	5	100	67	23	100
Feature in a magazine	86	9	4	0	100	71	29	100
Report	35	13	9	42	100	84	16	100
Thesis/Dissertation	100	0	0	0	100	99	1	100
Total	**61**	**19**	**15**	**5**	**100**	**80**	**20**	**100**

Based on 4004. [1]*Data in percentages* [2]*Authors of all 4004 considered, for background,* [3]*1181 references that had publishers were taken.* [4]*Chapters in books, proceedings and reports.*

Setting up a Norm

Research comes in a variety of baskets. There are articles, books, theses, reports, and chapters in books. Even among the theses, there are the doctorates, pre-doctorates and those pertaining to a postgraduate degree. To produce or publish in these dissimilar outlets involves a different set of time, investment and tenacity. It cannot be denied that a doctorate is a class apart when compared to a chapter in a book. The patience and dedication required by a book far exceeds that of a feature in a magazine. This variation creates the need for standardisation.

Standardisation imparts consistency and thus provides a base for comparison. To achieve this, equivalence has to be determined. But how can an article in a journal be compared to a feature in a magazine? How can a book be weighed along with a report? There is also the vexing problem of quality. An article in a journal may outshine a voluminous doctorate in brilliance. A book may be poorer in knowledge than a post-graduate thesis. Similarly, a Master of Philosophy dissertation from one university could be superior to a doctorate of another university. These are issues that need to be resolved. Even if a way out is found to complete the 4004 references, yet to grade various references into the scale of excellent to poor, is difficult if not impossible. What then could be an acceptable and sustainable mode to bring together these different outlets of research? The foremost task is to ascribe a relative weight to the different kind of research documents – books, reports and theses. As in the case of livestock units, passenger units, motor vehicle units, productivity units, and calorific units, an attempt can be to construct a research unit.

With a view to standardising the outlets of research, the following were the basic assumptions.

- Each kind of research product is of uniform quality. Hence, all books are to be deemed as of equal quality; all articles in journals are of equal quality and the like.
- It takes an equal effort to produce a similar type of research. That is all doctorates take a similar amount of intellectual input, time and effort.
- The diverse outlets of research demand a dissimilar degree of input, and thereby differ in their salience. Therefore, a doctorate is rated higher than a postgraduate thesis, a book is rated higher than a report and a magazine article is rated lower than an article in a journal.

With these suppositions, an article in a journal was adopted to define the standard norm of conversion. The article was selected because nearly 60 per cent of the 4004 references belonged to the category of articles. But more than dominance was the fact that it makes common sense to keep the categories of research as compact as possible so as to allow scope for analysis. In case a book would have been selected as the standard, then articles in journals would have to be added to make a book. A lot of fine grain analysis would be lost in this process.

Given these considerations, when an article was selected as one research unit, then the number of articles that could be generated out of a standard book, or for that matter a doctorate or a report, was possible to construct.

By virtue of being published in a journal, an article acquires certain distinguishing attributes. It is free of a lengthy introduction, conclusion, and other verbosities. The high premium on space and cost of production means that an article has to adopt a judicious selection of diagrams, tables or maps. Moreover, articles in journals are mostly peer reviewed. The implications of this are a vigilant choice of details and facts that are to be retained. Taking these properties into consideration and following a rule of the thumb, a standard book would give four articles of a journal, a doctorate would provide three, a Master of Philosophy thesis may furnish two, while it would be hard to even draw more than one from a post graduate thesis. A report is closer in effort and output to a doctorate, so it has been attributed an equivalence of three articles (Table 3.9). Features

Table 3.9 Setting a Norm – Standardising the Research

Outlet	**Weight per reference**	**Research references**	**Research units**
Article in journals	1	1838	1838
Chapter in book/proceedings/report	1	724	724
Postgraduate dissertation	1	49	49
Masters of philosophy dissertation	2	40	80
Doctorate thesis	3	113	339
Report	3	427	1281
Book	4	134	536
Feature in magazine	0.25	679	170
Total		**4004**	**5017**

Based on *4004*.

in magazines are often dismissed from the category of research but because of their special value in the context of disasters, they have been included. Considering that they carry more news than detailed research analysis, their value is scaled to a quarter or 0.25.

When the norm was applied to standardise the 4004 references, the latter number increased by almost 25 per cent. This is because the forty Master of Philosophy dissertations doubled while doctorates tripled, reports got multiplied three times. By attaching a weight of four, the books were the ones that quadrupled. But all types of research outlets did not swell. The features in magazines dwindled to a quarter of their original size. Ironing out the discrepancies, this differential shrinking and swelling converted the 4004 references into 5017 research units. Any effort of standardisation could of course be debated, improved and changed yet this innovation paves the way for a research on research.

IV

Research on Research

Research on a diversity of natural disasters encompasses 5017 research units. There is research on storm surges to landslips, cold wave to heat wave, earthquakes to cyclones, dust storms to rain storms. A subtle difference and the experts are quick to qualify them with another name. Thus, debris slide, landslip, rock-fall, slope-failure can all be categorised under a landslide. An alteration in velocity and a cyclone transforms to a super cyclone; a change in magnitude and a moderate earthquake is labelled as an intense type; a modification in the intensity and heavy rainfall is termed as a cloudburst.

Some researchers use the geographical setting as a characteristic of the natural disasters. Thus, coastal disasters are considered different from those in the mountains. In certain units, the element of nature dictates the choice of some of the research units. There are water related, weather associated or geomorphology associated disasters. There are still other ways to sort the natural disasters among the 5017 research units.

There are researchers who have worked on the slow type of disasters like a drought, cold wave or heat wave while many others have worked on sudden and quick types like earthquakes, cyclones and tsunami. Using the feel of a natural disaster they could be grouped into the shaky, windy and the wet kind. Earthquake, cyclone and heavy rain respectively would fall into these three different types.

Spatially, disasters could be scaled into the local, regional and macro-regional. In this, a landslide would belong to the micro and a cyclone to the meso hierarchy while a drought could acquire a macro rung at the national level. Researches have and could have used many ways to address disasters. Thus, different criteria like the speed and intensity of onset, the feel they import, the scale of their study could be used to classify disasters. But none of these appeared to be a meaningful fit to the 5017 research units.

Table 4.1 Variety of Titles in Research on Disasters

Natural disaster	Titles	Research units	Per cent of total
Specific		*4273*	*85*
Earthquake	Earthquake and seismicity	1269	29
Flood	Flood and flash flood	831	17
Drought	Drought, scarcity and rainfall deficit	723	14
Cyclone	Cyclone, super-cyclone, storm-surge, tropical storm, tidal-surge	547	11
Landslide	Landslide, avalanche, landslip, mass movement, subsidence, debris slide, rock-fall, slope-failure	434	9
Tsunami	Tsunami	182	4
Storm	Dust storm, thunder storm, storm, rainstorm, hail storm, tornado, norwester	52	1
Heavy rainfall	Heavy rainfall, cloudburst, rainfall	31	1
Heat and cold wave	Heat wave and cold wave	4	< 1
Assorted		*744*	*15*
	Disaster, natural disaster, coastal disaster, water-related disaster, hazard, natural hazard, multiple hazard, environmental hazard, geomorphic hazard, natural calamity, weather event, calamity, emergency and risk	744	15
Total	**48**	**5017**	**100**

Based on *4004*.

This is because researchers seem to have adopted two ways of addressing disasters. Either they were specific about the naturals they had opted to study or they used a general and assorted mode of expression, such as risk, hazard and disaster. The references with a specific form of address, be it to a cyclone, drought or earthquake, constituted 85 per cent of the research units, and the remaining were included into the assorted category. It could of course be questioned if any natural disaster arrives individually in such clear cut divisions; for example a cyclone is accompanied by storm, heavy rain, tidal-surge and a flood. Similarly, a heavy rain, landslide and flood are in unison; but the 4273 research units preferred to use them separately (Table 4.1). Besides the specific, the remaining

15 per cent of the research units favoured assorted terminology like hazard, risk, calamity, emergency or just the word disaster. Spread across the specific and the assorted, there emerged as many as forty-eight different types of natural disasters.

While scientists have distinguished natural disasters into a number of categories, many share a close affinity. Scrutinising on the basis of resemblance, it was possible to apportion forty-eight natural disasters into ten groups. Finding commonalities was not difficult. Thus, thunder storm, rain storm, dust storm, hail storm, tornado or norwesters all were brought under the label of a storm. Earthquakes and seismicity were another group. Under the hat of drought came rainfall deficit and scarcity. Cyclone became the overarching term for all research done on cyclones, super-cyclones, storm-surges, tropical storms and tidal-surges. Be it flood or flash flood, both were identified with the common surname. A landslide embraced all descriptions ranging from an avalanche, landslip, mass movement, slips and subsidence, debris slide, rock-fall, slope-failure, slope instability. The group of heavy rainfall absorbed those that addressed cloudburst or rain *per se*. Both the cold and heat waves were paired while the unique features of the tsunami made it stand by itself. This exercise gave a composition of nine specific types of natural disasters with a tenth under the shade of the assorted. Arranged alphabetically, these are cyclone, drought, earthquake, flood, heat and cold wave, heavy rainfall, landslide, storm and tsunami.

By virtue of their ability to contain a range of variety, the assorted group carried fourteen descriptions. Its assembly contained the disaster, natural disaster, coastal disaster, hazard, natural hazard, multiple hazard, environmental hazard, geomorphic hazard, natural calamities, weather events, calamity, emergency, risk, and water related types. The multiple group of hazards like the geomorphic hazards or the climatic hazard held sway among the assorted. In spite of the ability to diversify, it is the specifics and not the assorted that dominate the 5017 units. Even between the specifics, each of the natural disasters did not garner uniform attention. It ranged from a maximum of 29 per cent in earthquakes to a minimum of less than one per cent for the cold and hot waves.

Question of Adequacy

The breakdown of the composition of the research units brings forth the question of adequacy of 5017 research units. Are these enough? How can sufficiency be gauged? Is research ever enough? These are difficult questions to raise, let alone answer, but an inquiry is worth an attempt.

A comparison of the number of research units with the number of natural disasters in India can be done. While this would be interesting, the enormous problems of documentation and data on disasters make this a highly demanding if not impossible exercise. It would perhaps be relatively easier to trace the natural disaster with the least number of research units and then compare it with the number of disasters of that type in India. With four units to their credit, it is the cold waves and heat waves that have the lowest record among the research units. By comparison, a scrutiny based on the weather disaster reports brought forth the fact that from 1977–2002, nearly 8000 people had succumbed to death in a cold and heat wave in India. These deaths certainly deserve more attention.

Another indication of adequacy is to take a look at the parity with which the research on disasters has been conducted. The reason being that with consistency, there could be scope to achieve a balance between the destruction caused by a disaster and the number of research units it acquired. It has to be agreed that there is hardly ever a year when India does not experience a flood. The average annual death of nearly 1500 people and the impact on a population of close to 35 million from floods are a case for concern. These are the statistics provided by the Central Water Commission in 1953–99.

Against this, the number of research units in the kitty of floods was 831 during 1950s to 1990s or hardly 17 a year (Table 4.2). Disasters are commonly seen as episodes. The result is that research tends to arrive in

Table 4.2 Number of Research Units per Disaster Event

Number of research units	Number of events[1]
> 200	1
100–200	3
50–99	1
25–49	3
10–24	11
5–9	10
2–4	37
144	1
In fraction	38

Based on *4004*. [1]List of events in Table 4.3.

Table 4.3 Specific Disaster Events versus Number of Research Units

Year	Specific event	Persons dead	Research units	Year	Specific event	Persons dead	Research units
1819	Kutch earthquake	2000	10	1950	Assam earthquake	1526	34
1988	Bihar-Nepal earthquake	750	14	1999	Chamoli earthquake	100	38
1997	Jabalpur earthquake	38	14	1991	Uttarkashi earthquake	1600	46
2005	Mumbai floods	1000	16	1967	Koyna earthquake	100	58
1905	Kangra earthquake	20,000	17	1999	Orissa cyclone	9540	118
1934	Bihar-Nepal earthquake	10,500	17	1993	Killari earthquake	10,000	129
1897	Shillong earthquake	11,542	18	2004	Asian tsunami	10,749	159
1977	Andhra cyclone	10,000	32	2001	Bhuj earthquake	20,000	235

Based on *4004*.

the aftermath of an event. India has witnessed hundreds of natural disasters. But only 251 events find a place in the research. Of the 5017 research units, 1356 were works on individual events. Only 10 out of the 251 events received 25 or more research units each (Table 4.3). The fact that disasters as severe as the Kangra earthquake could hardly initiate 17 of the research units is a sure indicator of insufficiency. There could not be a better measure of inadequacy. Further, most of this research has accumulated in a short period of time.

Skewed Temporality

The 5017 research units are not an outcome of a day, decade or century. The increase in their number has neither been constant nor consistent. The publication year of the first unit that could be accessed dates to 1830 and the last that was searched belongs to 2006. With these as the cut off for the beginning and end, the research spans 177 years. This gives an average of 28 research units per year. Reflecting the inadequacy of research output is the fact that of the 177 years, 60 years did not make an addition of even a single research unit. The nineteenth century had the maximum number of gap years when, during 1830–99, there were no research units for 38 years. In the 100 years of the twentieth century, 22 years carried no research units. Even the output of units on the separate type of natural disasters displayed a variation. Research on floods came for 68 out of the 177 years. Its absence from 109 total years is no match to research on cyclone, landslide, drought and storm which did not surface for as many as 125 to 145 years. Research on heavy rain disasters was missing for 158 years (Table 4.4).

A superimposition of the major disaster events over the annual flow of research units provides some explanation to the critical breaks observed. There is a rise in 1935, which coincides with the Bihar-Nepal earthquake (1934). A second hill occurs in 1953. This is the result of the research on the Assam earthquake of 1950. The next prominent peak is in 1968, which follows the Koyna earthquake of 1967. Through the flat period of the 1970s and 1980s, a height is seen in 1979. This is the aftermath of the Andhra cyclone of 1977. Even the increased surge in research units in the post 2000 years follows the arrival of the Bhuj earthquake in 2001 and the tsunami in 2004. All these do explain the briskness that the arrival of a disaster event imposes on research. But the research interest is indeed short lived. The Koyna earthquake in 1967 attracts a total of 58 research

Table 4.4 Years without Research Units on Disasters of India, 1830–2006

Natural	Years with research units (no.)	Years without research units (no.)	Per cent years without research units
Heat and cold wave	5	172	97
Tsunami	7	170	96
Heavy rain	19	158	89
Assorted	31	146	82
Storm	32	145	82
Drought	45	132	75
Landslide	49	128	72
Cyclone	51	126	71
Flood	68	109	62
Earthquake	100	77	44
Total	**117**	**60**	**34**

Based on *4004*.

Figure 4.1 **Trend of Research on Natural Disasters in India, 1830–2006**

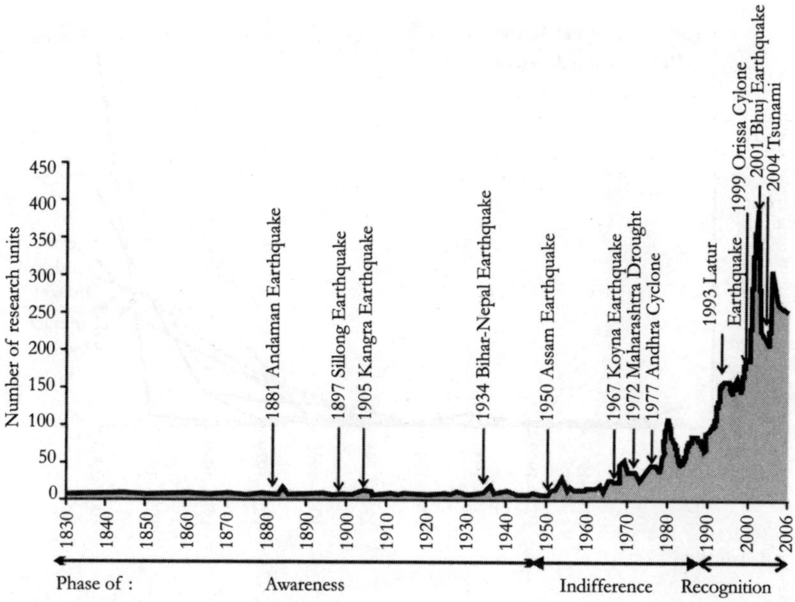

Based on *4004*.

units of which 71 per cent cluster in the five years succeeding the event. Similarly, the Andhra cyclone occurred in 1977 and drew a total of 32 research units, more than half of which were concentrated in the five years following the event.

When plotted on a longitudinal profile, the output of research units resembles a continuous flat plain from 1830 to 1953. There are a few sporadic bumps around 1884, 1889, 1905 and 1935. After a brief dip during 1940–49, the profile ascends through the 1950s and 1960s gaining some momentum in the 1970s. In 1970, the total research units were 460. By the end of 1980s, they still hovered around 1153. Then began the exponential climb from the 1990s to 2006 adding 3404 research units in sixteen years. The rate of accretion reached an all time high of 395 in 2001(Figure 4.1).

Research on the ten natural disasters follows a pattern similar to the one described earlier. For the first 160 years, a monotonous topography characterises the slow pace with respect to each of them. So laid back has been the research that a nick is rare in the profile. It is only from the 1990s to 2006 that the growth of research on the individual natural

Figure 4.2 **Trend of Research on Different Types of Natural Disasters, 1830–2006**

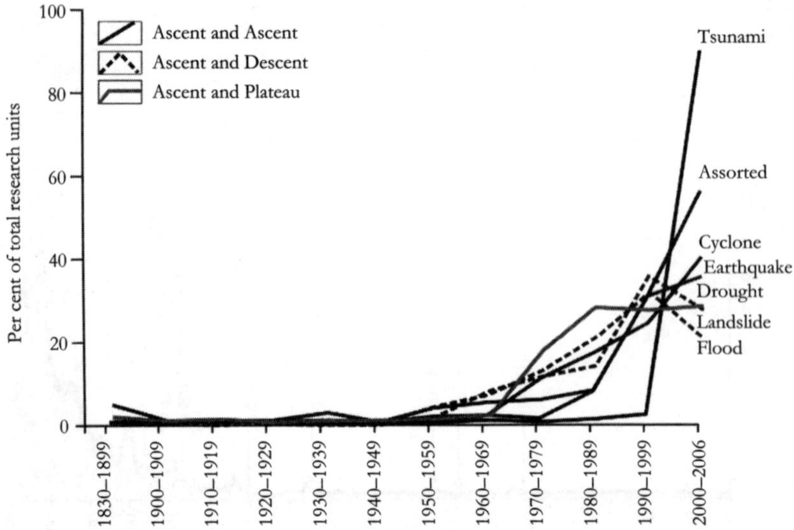

Based on *4004*.

disasters demonstrates a different pattern. With a heightened accumulation at the turn of the twentieth century, there is a variation in the disposition of all the ten disaster types from 1990–2006. A sustained zeal for research on earthquake, cyclone, tsunami and the collective group of the assorted put these into the group of the continuously ascending category. From 124 at the end of 1989, the research units on earthquakes take a steep climb to reach 467 in 1999. The numbers do not relent and surge headlong to add 553 in the first seven years of the 2000s (Figure 4.2).

The research units for cyclones stand at 99 at the end of 1989 but rise by adding 139 in the 1990s, and another 231 in the 2000s. The assorted category registers the greatest rise. From a low of 49 in 1989, the units increase nine times to reach 447 in 2006. Research on the tsunami achieves the sharpest ascent in the shortest time. From one in 1989, it soars high to reach 177 in 2006.

The graph displays an ascending and descending trend in the case of research units for landslides and floods. With an addition of 61 research units in the 1980s, the trend nearly doubles as another 155 are added in the 1990s. But during the initial seven years of the 2000s, the addition is only of 121 units and it causes a dip in the graph. As the research units fall from 262 in 1990–99 to 170 in 2001–06, even floods lose popularity among the research units in the first seven years of the twenty-first century.

Without losing or gaining, the research units for drought and storms stagnate through the last decade of the century. The number of research units on drought stays at nearly 200 in all three decades from 1980–2006. As the research units fluctuate from adding 12 at the end of 1989 to a decline of 10 in 1999, increasing again to add 15 by 2006, an unsteady trend characterises the research on storms.

Such aberrations, however, do not blur the dominance of production of research units from the 1990s to 2006. 68 per cent of the total research on disasters of India is concentrated in these seventeen years. Whereas the average annual addition through 1830 to 1947 was one per year, it increased to 35 per year during 1947–89. The production soared to 200 per year post 1990s, with the 2002s being most prolific, when there was an annual addition of 275 references. Each tributary of research on natural disasters bloats the graph of the post-1990s, but the maximum additions came from works on earthquake, which added 30 per cent research units. This was followed by 20 per cent coming from the assorted group and flood pouring another 13 per cent, collectively pushing up the number of research units by 2138. It cannot be denied that a majority of the 5017 emerged during 1990–2006. The temporal analysis reveals that research on disasters in India is not just inadequate but is also a late, recent and rushed entrant. As in the case of the temporality, do the research units display a place preference?

Spatial Bias

A disaster manifests itself at a place. It has a locale and spatially varying spread. India is a vast country. It is politically divided into twenty-eight states and seven union territories. Its population lives in 5161 urban centres and 638,635 villages. The Himalayas, Deccan Plateau, Ganga Plain, Thar Desert and coastal regions characterise its varied physiography. Which states of India have received maximum attention for research on disasters? Where among these mountains, plateaus and plains are researchers set to work? Has research on disasters in India taken scholars far and wide or have they confined themselves to a few spaces and places?

Many problems have to be resolved before attempting a spatial inquiry. A temporal analysis is made easy by the fact that mentioning the year is a compulsory requirement in any publication; in the case of place this is not mandatory. Leave alone a need in the title or subtitle, in certain cases even the text fails to refer to a specific place. There could be two reasons for this. Either the study is a definition, concept or classification which has no need to be bound to a geographical area or, more compelling, it is indifferent to place. The net result is that there are two kinds of research units, those that do and those that do not make a reference to the place. The former comprise nearly 39 per cent of the total research units. These are not place neutral. They are on India. After all, by the merit of being a research on India these gained a standing within 5017 research units. While 39 per cent carry an implicit suggestion, even the 61 per cent units that do make an explicit mention of a place identifying spatiality are not free from problems.

Identifying Spatiality

There are various types and interpretations of a place. A place could be rural or urban, it could be an administrative identity like a block, tehsil, district or it could be a political entity like a state. The place could also refer to a catchment, watershed and, in the case of natural disasters, a river, hill or the seacoast. The Himalayas or the Gangetic plain could well be the field for work. The study could also be covering areas with a suffix of 'prone' like the drought, flood and cyclone prone zones. The medley of scale, name and size is overwhelming. Such was the diversity encountered when the area of study covered by the research on disasters in India was viewed.

It was under the fourth Five Year Plan (1969–74) that the Government of India started programmes for the development of areas riddled with specific problems. It is then that the areas prone to drought or flood gained wide recognition. Nearly 178, or 4 per cent, of the research units have focused on the prone areas. Amidst these, it is the drought prone areas that have taken about 75 per cent of the attention, followed by floods at a lower 10 per cent.

There is research on urban areas like Delhi, Mumbai, Shillong, Srinagar, Aizwal, Guwahati, Kolkata, Chennai, Bangalore, Hyderabad, Coimbatore, Masulipatnam, and Kanniakumari. While some have focused on towns and cities, administrative units like the districts have occupied the attention of 180 research units. Exemplifying this is the research on districts such as Kutch (Gujarat), Kalahandi (Orissa), Uttarkashi and Chamoli (Uttarakhand), Kangra and Chamba (Himachal Pradesh). Many have been guided purely by state boundaries. As a case in point, there are 30 research units on the state of Uttar Pradesh and 40 on Maharashtra.

Rivers like the Brahmaputra have also provided opportunites and have drawn 43 research units. Hills and mountains like the Nilgiris, Western Ghats, Shivaliks or perhaps the entire chain of Himalayas have been an inspiration for work. The latter attracted as many as 794 research units (Table 4.5).

The waters that surround India have also been fascinating to study. The Indian Ocean figures in 96 research units while the Bay of Bengal and Arabian Sea are described in 79 research units. Therefore, akin to the diversity of India itself, the places of research have been varied. To comprehend the spatial frame of research, there was a need to identify a unit that could unfold the picture of regional diversity.

Mapping the Diversity

An appropriate, viable and basic spatial unit that can be pressed in service to arrange and prepare a regionally differentiating map of research units is a necessity. The task is not without structural constraints. For instance, if a research unit mentions the Himalayas as its field of study, then it would not be possible to divide the area into a scale of the mountain ranges or districts. On the obverse, if it is research on a district or block, then these can be clubbed to build a larger place unit like a state or the Deccan Plateau. Smaller bricks can be used to build larger units but vice-versa is not an option.

Table 4.5 Different Type of Areas Selected for Research

Type of areas	Research units[1]	Per cent of total (5017) research units
Himalayas	794	16
District	180	4
Prone area[2]	178	4
River	73	1
Road	46	1
Village	42	1
Block	18	n
Tehsil	3	n

[1]*These 1334 research units are indicative of the problems encountered in identifying spatial categories,*
[2]*like flood prone and drought prone area, n = negligible.*

Disasters in India are a state subject. A grouping of research units by states and union territories could help draw a comparative statement on the political unit that has received the maximum and the least attention by researchers. It was observed, however, that the states cut across diverse physiographic regions. In the case of research on disasters, it was also found that political units are less popular while works on the coast or a particular part of the plain area or a section of the mountains were more common. When accounting for earthquakes, as many as 794 units took the entire Himalayas into consideration. Similarly, when analysing cyclones, it was found that many research units adopted the coast of India as the frame of study.

It is an axiom that natural disasters do not respect political or administrative boundaries. In league with this was a need to plot 61 per cent of the research units that made a reference to any place within a meaningful frame. The intention was to create a spatial framework, which would allow a comparative assessment. The regionalisation scheme of the Census of India, 1981, was considered. While delineating regions, the census included the criteria of geological structure, physiographic features, drainage, forest and soil. These had been taken into consideration by many researchers on natural disasters. Further, the regions demarcated by the census, respected the boundaries of a district, thus giving an added feasibility of amalgamating administrative and political units.

The census had identified 4 macro, 28 meso and 101 micro-regions of India. The lack of specific details of a place precluded a micro level analysis and therefore the research had to be satisfied with the 28 meso-regions.

When the 5017 research units were sifted for their spatial attributes, a clear 1886 research units or 38 per cent were works at the national level and 30 research units or nearly one per cent shared concerns with neighbouring countries like Nepal, Bangladesh and Myanmar. These were not amenable to a regional analysis. Another 175 research units or 3 per cent referred to the oceans and were thus excluded. The spatial inquiry was made on the remaining 58 per cent of the total research units (Table 4.6).

The Census of India identifies 4 macro-regions namely, the Northern Mountains, Great Plains, Deccan Plateau and the Coastal Plains. Research on disasters in India has covered all these macro-regions. The Coastal Plains lead the way with 32 per cent of the total research while the

Table 4.6 Areas Included and Excluded for Spatial Analysis of Research

Category	Research units	Per cent of total
Included		
Administrative units, regions, prone areas, rivers, places, roads	2926	58
Excluded		
India level	1886	38
Ocean and seas	175	3
With neighbours[1]	30	1
Total	**5017**	**100**

[1]*Pakistan, Nepal, Myanmar and Bangladesh.*

Northern Mountains share another 27 per cent. Therefore, the macro-regions that frame the land and sea margins of India, soak up 59 per cent of the total research attention. These are followed by the Deccan Plateau which acquires 23 per cent, and the Great Plains that absorb 18 per cent. Coastal Plains cover nearly double the research units when compared with the Great Plains (Table 4.7).

While there is not a single macro-region that has been ignored, there are marked internal spatial variations. Extending over a total length of more than 6000 kilometres, the coastline of India begins from Sir Creek (Kutch), moves down to the tip of the Indian peninsula at Kanniakumari and then moves upwards to end where it encounters the Ganga-Brahmaputra delta of Bangladesh. Strung like a 'V' with a cleft at the tip near Kanniakumari, the Coastal Plains have been a favourite among researchers. The two cardinal divisions of the Coastal Plains, that is the western and eastern regions, are unequally matched. The western section with 8 per cent of the research units is limited when compared with the

Table 4.7 Spatial Variation of Disaster Research by Macro-regions

Macro-region	Research units	Per cent of total
Coastal Plains	931	32
Northern Mountains	794	27
Deccan Plateau	687	23
Great Plains	514	18
Total	**2926**	**100**

Based on *4004*.

eastern coast which has 38 per cent of the total research units. But what boosts the picture on the western half of India is the Gujarat region which has drawn 38 per cent of the total units of the meso-region. The islands of Andaman and Nicobar and Lakshadweep also have a share of 6 per cent in the Coastal Plains.

Along the northern boundary of India, the forbidding sweep of the Himalayas have engrossed 27 per cent of the total research units. There are various ways to sub-group this mountain chain. Built like a broad giant staircase, there are four massive ranges, the lower, middle, higher and the trans-Himalayan. Extending in length for almost 2500 kilometres, the Himalayas can also be divided into three sectional divisions, the western, central and eastern respectively (Table 4.8, Figure 4.3).

Following the 6 meso-regions sculpted by the Census of India, the Western Himalayas, Jammu and Kashmir, and Himachal Himalayas account for 25 per cent of the total research units on the Himalayas. The Central Himalayas have seized a sizeable 43 per cent of research units. Situated at the extremity, the Northeast Himalayas and Eastern Hill Zone have attracted 32 per cent of the total research units. In this west to east mountain chain, it is the Central Himalayas that hold top position among the research units.

Swept by the tributaries of the Ganga and Brahmaputra, the Great Plains are sliced into 7 meso-regions. Of these, the Lower Ganga Plain and Brahmaputra Valley that drew 75 per cent of the research units of the entire Great Plains. The remaining 25 per cent were scattered across the Haryana Plains, Middle Ganga Plain, Punjab Plain, Arid Rajasthan Plain and Upper Ganga Plain. On the Great Plains, the research interest on disasters declines from east to west.

Occupying the peninsula section, the triangular-shaped Deccan Plateau covers nearly 46 per cent of the land area of India. It is the only macro-region that displays extreme intra-variation of research attention. On the end of the spectrum is the Maharashtra Plateau, which draws 50 per cent of the total research units of the Deccan but on the other end are the northern Maharashtra and the north Madhya Pradesh Uplands, which do not have a single research unit. Defying any trend, a mosaic like pattern characterises the research on the Deccan Plateau (Table 4.9).

When the meso-regions are ranked according to attention received for research, it is the Eastern Coastal Region, Gujarat Region, Maharashtra Plateau and Uttar Pradesh (Uttarakhand) Himalayas, which come into

Table 4.8 Spatial Variation of Research in the Meso-regions

Macro/Meso-region	**Research units**	**Per cent of total in meso-region**
Northern Mountains	*613*	*100*
Uttar Pradesh Himalaya	262	43
Northeast Himalaya	145	24
Himachal Pradesh Himalaya	93	15
Jammu and Kashmir Himalaya	62	10
Eastern Hill Zone	48	8
Northwest Himalaya	3	0
Great Plains	*419*	*100*
Lower Ganga Plain	179	43
Brahmaputra Valley	133	32
Haryana Plain	42	10
Middle Ganga Plain	26	6
Punjab Plain	21	5
Arid Rajasthan Plain	12	3
Upper Ganga Plain	6	1
Coastal Plains	*860*	*100*
Eastern Coastal Region	411	48
Gujarat Region	324	38
Western Coastal Region	75	8
Islands	51	6
Deccan Plateau	*588*	*100*
Maharashtra Plateau	294	50
Andhra Plateau	81	14
Semi-Arid Rajasthan	75	13
Karnataka Plateau	38	6
Southern Madhya Pradesh Uplands	29	5
Orissa Highlands	23	4
Tamil Nadu Uplands	21	3
Bihar-West Bengal Uplands	16	3
Central Madhya Pradesh Plateau	10	2
Uttar Pradesh Uplands	1	n
Northern Maharashtra	0	0
Northern Madhya Pradesh Uplands	0	0

Based on *4004*. *n = per cent in negligible decimals.*

Figure 4.3 **Research on Natural Disasters, 1830–2006**

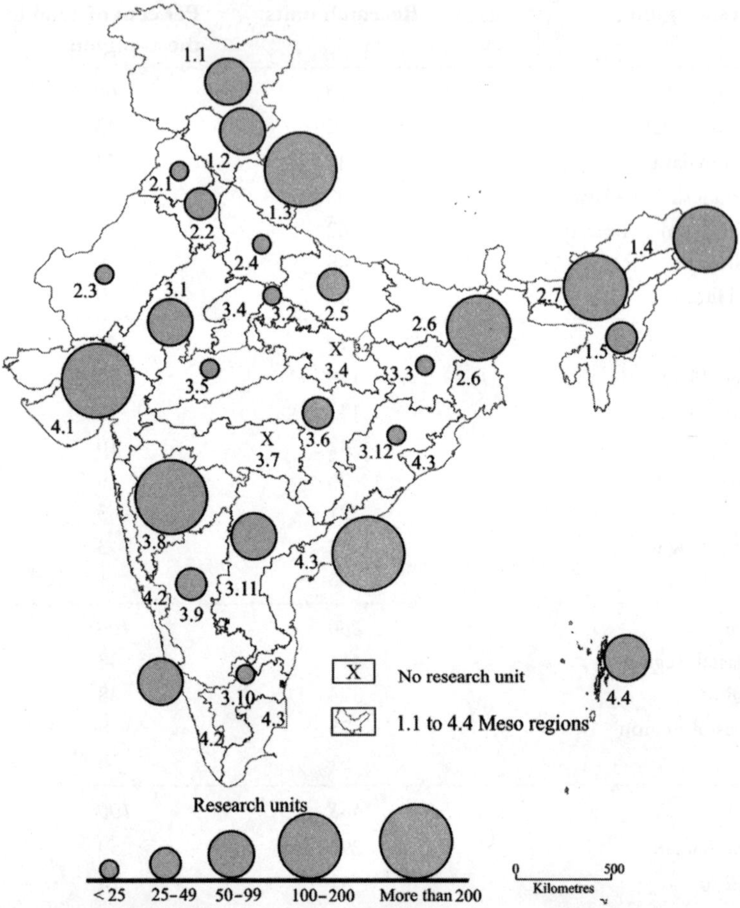

Meso Regions

Northern Mountains

1.1 Jammu and Kashmir Himalaya
1.2 Himachal Pradesh Himalaya
1.3 Uttar Pradesh Himalaya
1.4 North Eastern Himalaya
1.5 Eastern Hill Zone

Great Plains

2.1 Punjab Plain
2.2 Haryana Plain
2.3 Arid Rajasthan Plain
2.4 Upper Ganga Plain
2.5 Middle Ganga Plain
2.6 Lower Ganga Plain
2.7 Brahmaputra Valley

Deccan Plateau

3.1 Semi-Arid Rajasthan
3.2 Uttar Pradesh Uplands
3.3 Bihar-West Bengal Uplands
3.4 Northern Madhya Pradesh Uplands
3.5 Central Madhya Pradesh Plateau
3.6 Southern Madhya Pradesh Uplands
3.7 Northern Maharashtra

Deccan Plateau

3.8 Semi-Arid Rajasthan
3.9 Uttar Pradesh Uplands
3.10 Bihar-West Bengal Uplands
3.11 Northern Madhya Pradesh Uplands
3.12 Central Madhya Pradesh Plateau

Coastal Plain

4.1 Gujarat Region
4.2 Western Coastal Plain
4.3 Eastern Coastal Plain
4.4 The Islands

Table 4.9 Dominance of Research in Four Meso-regions

Meso-region	Research units	Per cent of total
Eastern Coastal Region	411	17
Gujarat Region	324	13
Maharashtra Plateau	294	12
Uttar Pradesh Himalaya	262	11
All Others	1186	47
Total	**1291**	**100**

Based on *4004*.

sight as the most important. These together capture 53 per cent of the total research units. The researchers display a marked bias for some geographical locales and have preferred to study selected natural disasters.

Disparity in the Study of the Type of Disasters

The number of disaster types subjected to research is not uniformly spread in the different macro-regions. Among the ten types, the Coastal and Deccan Plateau claim eight, the Himalayas seven and the Great Plains six. In the Coastal Plains, a large 36 per cent of the research units were devoted to earthquakes, followed by 31 per cent to cyclones. In the Deccan Plateau again, it is earthquake and drought that take an equal of 41 per cent each while floods take a lead of 61 per cent in the Great Plains. In the Northern Mountains, research on earthquakes and landslides acquire an almost equal share of 44 and 43 per cent of the research units respectively.

This arrangement summersaults at the meso scale. Within the Coastal Plain, while research units on cyclones take a lead of 64 per cent in the Eastern Coast, on the Gujarat Plains, earthquakes take a sweep of 84 per cent. Where islands are concerned, tsunami is the most preferred disaster whose study consists of 42 per cent of the research units. In contrast is the Deccan Plateau, where a dominance of drought in eight out of the ten meso-regions, brings greater evenness. The only exceptions are the Maharashtra Plateau and Southern Madhya Pradesh Uplands where more units on earthquakes are found.

Reflecting the overall characteristics of the meso-regions, drought runs first in the Arid Rajasthan Plain to be replaced by flood in the Lower and Middle Ganga Plains. The Brahmaputra Valley too is favoured for the study of floods, which gains a sizeable 48 per cent of the total, with earthquakes too attracting 44 per cent of the research focus.

Can the spatial distribution of research units provide a glimpse into the geographical distribution of disasters in India? While this is desirable, without the analysis of the latter, any comparative conclusion would be fraught with error. What can be reiterated is that media, local groups and politicians can more often than not influence the choice of areas. When funds are to be distributed, information and news is often manipulated to make certain areas appear worse off and hence more fund worthy than others. The selection of a field of study is also influenced by the accessibility and availability of data. The location of research centres and universities is still another factor to contend with. These practical considerations can also be a bottleneck for research in the most deserving areas. Therefore, while the temporal inquiry revealed the less and late arrival of research units, the spatial analysis brings to the forefront, a highly inequitable distribution. But partiality is confined not just in time and space. Even within the large variety of natural disasters, it was the study of earthquakes that captured maximum attention.

Quake Research

India has a *pot pourri* of disasters. Earthquakes, cyclones, floods, droughts, landslides are known but exceptionally intense cold and heat waves are also rampant. Hail storms and avalanches, lightning and cloudburst are frequent; tidal waves and tsunamis are also evident. Of the limited research that disasters attract in India, earthquakes grab maximum relative importance, followed by floods and droughts. Cyclones and landslides get low priority while other disasters are completely sidelined.

Of the 5017 research units, a commanding 29 per cent have focused only on earthquakes. With an average annual addition of 8 units per year, the research on earthquakes was double of that on flood, drought and the assorted group of disasters. Its top position in six of the ten decades from 1900 to 2000 gains a lead during 2000–06. In all, except for four out of the total of eleven decades from 1900–2006, the study of earthquake has held complete sway.

The emphasis given to the research on earthquakes is apparent across all types of research outlets. There are triple the number of atlases pertaining to earthquakes than to flood, cyclone and landslide. In the realm of maps also earthquakes are most systematically plotted. Nearly

Table 4.10 Dominance of Research on Earthquake in India

Indicator	Total	On earthquake	Per cent of total research units
Number of research units	5017	1469	29
Research units in different outlets			
Doctorates	*113*	*37*	*33*
Atlases	*7*	*3*	*43*
Catalogues	*13*	*5*	*38*
Research units in number of events	1356	757	56
Research units in research on macro-regions			
Deccan Plateau	*687*	*319*	*46*
Northern Mountains	*794*	*351*	*44*
Coastal Plains	*931*	*332*	*36*
Great Plains	*514*	*122*	*24*

Based on *4004*.

3098 earthquakes are plotted in the *Seismotectonic Atlas of India*. Earthquakes constitute 38 per cent of the total catalogues listed in the 5017 units (Table 4.10).

A large 33 per cent of the total individual doctorates have also dealt with earthquakes. All the Indian Institutes of Technology have a special orientation for research on earthquakes. Of their eighteen doctorates on disasters, thirteen were on earthquakes. Among all disasters, the professionals working on earthquakes are the only group who have formed a society called Indian Society of Earthquake Technology. Established in 1962 by a resolution at the second symposium of Earthquake Engineering held at Roorkee University, it started a six monthly (now quarterly) bulletin in 1965. The activities of the Society increased after the 1967 Koyna earthquake.

Earthquakes remain the core interest at the spatial level. In three-fourths of the macro-regions, earthquakes garner more attention. Except for the Great Plains where floods gain ground, in the rest of the 3 macro-regions, earthquakes draw 40 to 45 per cent of the total research units. In 8 among the 28 meso-regions, the study of earthquakes takes first position. Even within the small amount of research focused on border areas with Nepal,

Bangladesh and Burma, the study on earthquakes takes 63 per cent of the total research units.

There is no data or record to prove that earthquakes outstrip cyclones droughts and floods in frequency of arrival in India. Nor is it true that people killed, injured or displaced are highest for earthquakes. It is also a geographical fact that unlike Japan, Iran, and Turkey, India is not a country where quake prone regions form a sizeable share of the total area. On the obverse, the mainland of India is considered rather stable. Yet, works on earthquakes swallow a sizeable share of the research.

The research on India's disaster is thus skewed on three fronts. First, the post–1990 phase claims 68 per cent of the total research units. Secondly, 4 out of the 28 meso-regions take away 53 per cent of research units. And finally, among a distribution of ten natural disasters, nearly one-third of the research units fall in the kitty of earthquakes alone.

What explains this tilt in the pattern of research units? A quick way could be to qualify the importance of earthquake itself. An explanation in support of this could be that the Bhuj earthquake in 2001 and the tsunami in 2004 are recent enough to capture a large amount of research interest. That earthquakes are fascinating because the destruction associated with them appears more deadly could be another explanation. The early establishment of the Geological Survey of India could also be listed among the justifications. A clinching argument could be that researchers are people of free will and it is finally their whim and fancy that decides what will be pursued when and where.

But while all these reasons could be forwarded, a more pertinent and serious answer can only emerge if the genesis and evolution of the science of disasters in India is traced. What has been the origin of disaster studies in India? Who first championed its cause? Which among the physical and social sciences harboured an interest in natural disasters? Which disciplines were drawn to its call? What explains the waxing and waning of interest in the study of disasters?

It is difficult to sketch the roots of inquiry of disasters in India. Till date, the study of disasters in India has neither gained an effective recognition as a science nor has it been awarded an independent status as a discipline. There is no centre, university or department in India that has marshalled significant expertise on the study of disasters. Whereas these 5017 research units are the output of over 2000 authors, it is indeed difficult to pin a group of leading experts on the science of disasters in

India today. Its diluted professionalism is visible in the lack of an exclusive society or journal to its credit. Given these circumstances, it is through an inquiry of the 5017 research units that one can try to explain the trajectory of the study of natural disasters in India. Although an experiment, it is worth an attempt.

V

Evolution of the Study of Natural Disasters in India

The study of disasters runs parallel to the study of Indian society, economy, polity, music, paintings and sculpture. One need not go too far to seek the reason. As part and parcel of the planetary process, earthquakes, floods, droughts, and cyclones or for that matter, cold and heat waves are neither alien nor recent entrants to India. Arriving suddenly or building steadily, natural disasters have continued to take a toll on life in India since eons. Even though they are perceived as natural events, the fact that they are not viewed as 'normal' implies that they cannot be overlooked. The result is that references to natural disasters are interpolated into travel accounts, chronicles, and administrative reports and inscribed in edicts or weaved within legends. Strewn across this 'literature' is the source of the study of natural disasters in India. Be it an account, description or even a story, these involve research, imagination and thought. There is little doubt that if such sources were amassed from every language, class and creed of India, the wealth would enrich our understanding of the study of natural disasters.

While the retrieval, translation and analysis of these materials is a project worth a lifetime, yet its scope unfortunately falls out of the compass of this book. This is because in all such references, the description of disasters is embedded within larger texts and is incidental to the context. Secondly, most works are in the local dialects, regional languages or court language of the prevailing times. Evidence in support of this is a project sponsored by the Department of Science and Technology to catalogue the earthquake history of India in medieval times. Confined to the period between AD 1200 to AD 1800, the authors of this work confess, 'Primary sources of history for this period are available in the form of writings of court historians, travelogues, epigraphs and other documents. A sizeable number are in the Persian language. There are also writings in Sanskrit, Arabic, Turkish and regional languages of India. Many of these manuscripts are not published' (Iyengar, 1999). Only 27 per cent of the references sourced by this medieval history project were in English; there were 60 per cent in Persian, and the remaining 13 per cent in Arabic, Turkish, Urdu, French, Sanskrit, Hindi and Ahom (Table 5.1).

A perusal of these 220 references is proof to a double set of details: first that natural disasters have been an integral part of the study of numerous individual writings and secondly since Persian was the court language, works in English could not have been there.

In spite of the antecedence in the study of disasters in India, the first reference in the 5017 research units dates back to 1830. Set within the confines imposed by the type of language and mode of publication, the range of 5017 research units can be classified into three phases:

- Colonial (1830–1947): A Phase of Awareness
- Post-Independence (1947–1989): A Phase of Indifference
- Globalisation (1990 onwards): A Phase of Recognition

The personality of each phase is moulded not just by the difference in the time period but also by the total number of research units, their composition and the considerations that motivated or discouraged research. After all, the inclination to research an idea, theme or subject does not

Table 5.1 Linguistic Attribution of References Cited in the Project on Earthquake History of India in Medieval Times

Language of reference	No of references
Persian	133
English	59
Arabic	14
Sanskrit	8
Ahom	1
French	1
Hindi	1
Sindhi	1
Turkish	1
Urdu	1
Total	**220**

Based on R.N.*Iyengar, et al. Earthquake History of India in Medieval Times, 1999.*

evolve in a vacuum but is shaped by the state and society. Nowhere is this more applicable than in the case of research on natural disasters. There is synchrony between the number and type of research units and the ecology of the evolutionary phase of the science of disasters.

Colonial (1830–1947): A Phase of Awareness

The genesis of disaster studies in India can be traced to antiquity, but it was the British who laid a foundation for the study of natural disasters in India in the contemporary context. Evidence of their interest can be seen in the fact that all through the colonial phase they were the leading authors of the research units. From 1830–1900, all the 77 research units and even from 1900–47, 80 per cent of them flowed from their pens. Besides authorship, this phase has an exceptionally large presence of catalogues, records, notes and news. The reason for research units in the appearance of records and reports and their foreign authorship is simple to understand. India was a colony of the British from 1757 to 1947. They arrived in India under the banner of East India Company, but gained complete political supremacy by 1857. It was under British tutelage that the groundwork for the study of natural disasters was laid in India.

Famine in Research Units

To the colonial phase belong 159 research units. These form 3 per cent of the total units. If compared with the number of years of colonial rule, the research units crash to almost one per year. Certainly there should be more research credited to the British. It seems that if the works on famines are included, numbers would surely increase. During the British rule, India experienced repeated famines. In the 1770s, an estimated 10 million people died in the Bengal famine. Famines during 1875–1902 in different parts of India took a toll of 26 million people. In the second Bengal famine in 1943, over 3 million people perished of starvation. The mega scale mortality in the famines gripped the attention of the British administration. The Famine Commission appointed under the chairmanship of Sir George Campbell laid the basis for the first famine code in 1878. The code outlined the role of the administrative machinery to avoid extreme distress. The crux of the code for dealing with the conditions created by drought and famine was the initiation of famine relief works to provide employment to a large number of people.

The Famine Commission recommended the establishment of a separate Department of Agriculture in 1881. The continued administration for flood and drought by the Ministry of Agriculture even after independence in 1947 is also inherited from the colonialists. Following the famine of 1876–78, a special fund of 15 million rupees known as the Famine Relief and Insurance Fund was set apart every year with effect from 1882. Half of it was earmarked for development of railways and irrigation, if not spent on famine relief. In the post 1947 phase, the decision of the Finance Commission to consign funds for relief in natural disasters can also be traced to this practice introduced by the British.

The humongous scale of the famine captivated the wholesale attention of the British. A response to widespread starvation, disease and death became an administrative compulsion if not a compassionate concern. The descriptions fill their provincial and district gazetteers and the Imperial Census reports. The British launched the first regular Census in India on 10 February 1881, and writing of reports was a feature till 1931.

The reports from 1881 to 1931 incorporate an explanatory variable of the impact of disasters on people. In-depth observations by the census commissioners allowed them to write, 'influenza takes the oldest and youngest first and variations in age distribution are also due to epidemic, but the most potent factor is famine'. For the physically and mentally challenged, they wrote, 'the infirm are mostly ones of less physique and therefore are less able to resist the debilitating effects of famine'. Inter-community differences are also noted: 'the recent famine has proved the sturdy Maratha of Deccan to have far greater power of resistance than the softer people of Gujarat' (1901 Census Report). While an independent India borrowed many British practices, they failed to replicate their meticulous method of record keeping. The indifference of the Census Commissioners of 1961 to 2001 is evident from the fact that they did not write a single comprehensive report, let alone make a single mention of natural disasters.

The British officers scrupulously documented deaths and tragedies. It was their painstaking efforts put into the records that later made research on the reasons for the 1943 Bengal famine possible. On the recommendation of the commission appointed to inquire into the disastrous famine of 1876–77, the British set up the first climatic observatory at Alipore, near Calcutta (now Kolkata) in November 1879. Alipore, now a part of the municipality of the city, was the residence of the then Lieutenant

Governor of Bengal. Had one included famines as a type of disaster, the ledger of research units in the colonial period would have been filled with many more documents, reports, articles and books. But as explained earlier, famine is accepted as a man-made disaster and therefore does not fall in the jurisdiction of this research on 4004.

Trade, Ships, Cyclones and Forecasting

Famine was certainly a thorn in the crown of the British yet it did not dampen their enthusiasm to augment trade from this resource rich colony. In the nineteenth century, oceans were the only mode of quick and cheap movement of cargo. The ships had to cross the tropics in their 4308 nautical miles of voyage to and fro from London to Kolkata. The latitude of 23° 30´ N bisects India. Cyclones make their way on both sides of this landmass. Whereas the Arabian Sea is open, extensive and therefore calm, the Bay of Bengal is more turbulent. Flanked on all three sides by the coast of India, Bangladesh and Myanmar, the enclosure and smaller size of the Bay causes its waters to heat quickly in summer. The swirl of the wind caused by the eastern jet stream populates the Bay with frequent cyclones. Cyclones wrecked many trading ships along the coasts of India. A cyclone on 30 December 1760 took the lives of 11,000 people while three big ships were taken ashore near the coast of Pondicherry. When a severe cyclonic storm hit the Coringa coast, Andhra Pradesh, on December 1789, it took a toll of 20,000 people. Many boats anchored on the shore were driven inland and numerous sailors drowned. Several such cases of death and drowning were reported from the shores of India. The massive loss of 50,000 people and the damage to ships at the port of Kolkata in a cyclonic storm that lashed West Bengal between 2–5 October 1864 was devastating. As cyclones wrecked ships and treasures worth millions of pounds sank to the bottom of the ocean, it coaxed the British to monitor the velocity of winds.

The government appointed a committee in 1865 to formulate a scheme to develop a system of cyclonic warnings. On the recommendation of the committee, Kolkata became the first port where a storm warning system was set up in 1865. Thus storm warning messages were issued even before the establishment of the Meteorological Department in 1875. The warning system for west coast ports like Mumbai (earlier Bombay), Karachi, Vengurla, Karwar and Kumta came into operation in 1880. In 1882, besides Kolkata, the ports at Sagar Islands, Mud Port and Diamond

Harbour were also included in the list of those receiving storm warning messages. By 1886, the system of early warnings against cyclones was extended to cover all Indian ports. The Kolkata office was responsible for issuing storm warning to all the ports around the Bay of Bengal, while the west coast ports were served by the Mumbai headquarters.

To boost trade, it is imperative to increase production and facilitate the transport of goods. Climatic factors are important determinants of agricultural and forest produce. Hailing from a temperate land, the British were keen to gain knowledge of the climatic features of the tropics, specially the vagaries of the monsoon. The government perceived its revenue budget as a gamble with monsoons, flood and drought, and the illness and death of their officers in the scorching heat were also matters of concern. To observe, coordinate and compile the weather observations, the Imperial Meteorological Department was inaugurated in 1877 at Kolkata. Such was its importance that as the capital moved from Kolkata to Delhi in 1911, so did the headquarters of the meteorological department. The bid to siphon resources backed the urge for an administrative grip over the subcontinent. This motivated the British to invest in infrastructure. To govern and exploit their colony efficiently, the British built roads, railways, seaports, schools, hospitals and buildings. While laying the blue print of construction, the British became aware of the threat from earthquakes.

Construction, Earthquake and the Geological Survey of India

Governed by a scientific temper, the history of the science of earthquakes in India is also linked with the rise of British supremacy. It was during the colonial rule that close to two dozen major earthquakes rocked different parts of India. There was the 1819 Kutch earthquake, 1845 Assam earthquake, 1858 Cuttack earthquake, 1869 Cachar Hills earthquake, 1870 Assam earthquake, 1872 Kamrup earthquake, 1897 Shillong earthquake, 1905 Kangra earthquake and 1934 Bihar-Nepal earthquake. In most of these, the number of deaths and extent of damage was massive. The earthquake of 1885 commenced in Kashmir on 30 May and violent shocks were felt up to 16 August. Houses were destroyed and there was general panic, with people sleeping outside their houses for many days. It is said that 3500 people were killed, and the number of cattle, ponies and other domestic animals crushed by falling buildings and relief features was enormous. The Shillong earthquake in 1897 killed thousands and reduced buildings 'to rubble in a region the size of England' (Bilham R. and P. England, 2001).

Approximately 20,000 deaths took place in the 1905 Kangra earthquake. The epicenter of the Bihar-Nepal earthquake was some 10 kilometres south of Mt. Everest and to it are attributed 10,500 deaths.

The geologists were seen as best equipped to study earthquakes since they had an in-depth knowledge of the subject. This lodged the science of earthquake disasters into the folds of the Geological Survey of India. The birth of the Geological Survey of India can be traced back 1836, when the Coal Committee was constituted. However, it was in 1852 that the ambit of its functions was expanded to cover geological studies. Sir Thomas Oldham, the Surveyor General, initiated the study of earthquakes in India in the wake of the Cachar earthquake of 10 January 1869. The 1897 Shillong earthquake (M 8.7) was the first Indian earthquake which was instrumentally recorded at observatories outside India (Oldham, 1899). After this great earthquake, the first seismological observatory was set-up in the country under the aegis of the Indian Meteorological Department in 1898 at Alipore, Kolkata. Finding no accurate record on earthquakes existing in the country, Thomas Oldham prepared a catalogue with the title. *A Catalogue of Indian Earthquakes from the earliest time to the end of AD 1860*, which was also published in the same volume.

Other officers of the survey like Meddlicott, Blanford and Holland left an equally impressive legacy of studies on earthquakes. Most of these names figure among the 159 research units that belong to this phase. The fact that 30 per cent of the total research units in the colonial phase and 6 per cent of the total research units hail from the Geological Survey of India confirms the role of this institution.

Research, Societies and Documentation

When the British first established the 'modern' universities in India in 1857, they did not introduce disasters as a subject of study. Not a single title in the research units included the words disaster, hazard, risk or calamity. While a specialised subject on 'disasters' was unheard of, the introduction of the discipline of geology stepped into the research on earthquake. Geology was among the first group of disciplines introduced in the universities of British India. Forging close connections to it was the officer from the Geological Survey of India, Thomas Holland, who was appointed as the first Professor of Geology at the Presidency College, Kolkata. Incidentally, the introduction of geography into a university of India was paired up with geology. This head start of interest in the study of earthquakes, training of Indians as professional geologists, along with the Geological Survey of India, is largely responsible for the dominance of research units on earthquake when compared with other natural disasters. Had the British planted the concern for deaths and damage attributed to earthquake not in geology but with that of disciplines like sociology, political science and economics, the study of disasters would have acquired a different and perhaps more meaningful hue of analysis.

It was mainly the Department of Physics at the Presidency College that stimulated interest in the mechanism of atmosphere, in thunderstorms, heavy rains, cyclones and monsoons. Interestingly, it was again an officer of the Geological Survey of India, H.F. Blanford, who worked as the Professor of Physics in Presidency College from 1867–74. A geologist subsequently became the first meteorological reporter to the Geological Survey of India. This also explains how Blanford authored the *Catalogue of the recorded cyclones in Bay of Bengal up to the end of 1876,* published in the Proceedings of the Asiatic Society of Bengal (1877).

To disseminate research, the British were also instrumental in founding scientific societies. Had it not been for the Asiatic Society of Bengal, a clear 27 research units would not have found circulation. Sir William Jones founded the Asiatic Society of Bengal on 15 January 1784, in Kolkata. The Madras Literary Society was founded in 1812 at Chennai (then Madras). In 1830, it became an Auxiliary of the Royal Asiatic Society of Great Britain and Ireland, and has since been known as the Madras Literary Society and Auxiliary of the Royal Asiatic Society. The *Madras Journal of Literature and Science* carries 9 research units that belong to this phase. The British were authors of all of these.

The British wrote the first catalogue of earthquakes and compiled the first list of cyclones. They also set the tradition to document key disaster events. This is how the major natural disasters of the colonial period acquired a few research units. The 1869 earthquake of Cachar obtained 7 research units, 1819 Kutch earthquake had 7 research units, 1849 earthquake at Assam took 3 research units while the 1897 Shillong earthquake, 1905 Kangra earthquake and 1934 Bihar-Nepal earthquake have 10 units each to their credit.

While many events drew research there were periods of low output. A particularly lean chapter was 1914–18. During this period, the British were engaged in World War I and the research units fell to an abysmal zero. While the colonial interest was diverted to wartime needs, a parallel development was the orientation of the army and relief organisations in support of disasters. A Military Department was created in 1786; this was a forerunner of the Ministry of Defence. The use of army in emergency was frequently resorted to during the colonial phase. This practice continues up to the present. In eight years from 1994 to 2002, the army was called nearly 126 times for disaster management in India. Along with the inclusion of the army, the first international relief organisation, the Red Cross Society, also set up an office in India in 1920. Owing its birth to a humanitarian response to the horror following the Crimean War in 1856, it has been engaged in serving the victims of any tragedy worldwide. It is this society that authored the first world disasters report in 1993.

The colonial rulers were thus quite aware of the impact and interference of natural disasters on their trade, revenue and administration. Their institutes and societies are manifestations of this awareness. Laying the framework for monitoring and introducing the need for records, colonists introduced research on natural disasters. The cumulative count of their research units is small, yet a beginning was made. Pioneers endure the greatest hardship. When India gained independence, a legacy of institutions, organisations and societies with an avid interest in natural disasters accompanied her. India also inherited a meticulous method of bookkeeping and documentation. The records, memoirs, narratives, catalogues, gazetteers and imperial census are testimony to the colonial awareness about the need to study disasters.

Independence and the Opportunity Missed for Study of Disasters

India gained freedom on 15 August 1947. Hand in glove with India's independence was the partition. The Indian subcontinent got divided into India and Pakistan. Break-ups are rarely peaceful. Triggering the displacement of 10 million people, the aftermath of this decision led to death, riot and civil war. The tragedy was a man-made disaster with scars that have not healed till date. There is no doubt that the partition placed a serious burden on the newborn nation. But it also was an opportune time for India to get serious about establishing its apparatus for the management of disasters. Fresh in the memory of the generation of politicians and administrators of the 1950s were events like 1934 Bihar-Nepal earthquake, the cyclonic storm in West Bengal in 1942 and the second Bengal famine of 1943. In each of these disasters, over 10,000 people perished. Even India's first Prime Minister Jawaharlal Nehru in a letter of 18 August 1950 to chief ministers, writes 'Nature has been unkind to us. Just when we are expressing satisfaction at the progress made on the food front and the fine harvest, there came flood and havoc, or lack of rain or excess of it, and cyclone, and now a great earthquake in Assam' (Parthasarathi, 1986).

It is indeed ironical that while national leaders of India had personally witnessed some of the gravest disasters, yet when spelling out the principles to develop India they failed to lay the structures to pre-empt the recurrence of disasters. In spite of a legacy of institutions, the repeated experiences and the resounding siren of tragedies, India ignored the study of disasters. An opportunity was indeed missed.

Post-Independence (1947–1989): A Phase of Indifference

The low and vacillating number of research units from the mid to the penultimate decade of the twentieth century represent the indifference doled out to the study of disasters. In this stretch of forty-two years, the output of research units is only 1455. All through the four decades, the annual production of research units was 90 in 1947–57, 126 in 1958–67, 346 in 1968–77 and 893 in 1978–89. An addition of hardly 35 research units annually. Compared to the colonial phase, this growth was modest and the entire period was able to add only 29 per cent to the total stock. So deep is the comatose feeling towards disasters that research on all natural disasters becomes sluggish. The number of research units on earthquakes were 105 in 1949. From the 1950s to the 1980s, the additions inched along at less than 70 per decade. An average of barely 7 in a year. The pace of research on landslides, which barely collected 60 in any of the decades, was slow.

The numbers are starker when growth in other parameters relevant to the output of research is compared. Considering the number of universities, India had 37 universities around independence. By 2006, their number grew to 269. Even doctorates, an epitome of research, leaped from just 39 in 1920 to 4433 in 1960, with 63,708 added in 1990–2000. Calculating on the basis of the period from 1980–2000, India added 6371 doctorates to her stock each year.

The Geological Survey of India, one of the major progenies of the colonial era, made rapid expansion in the post 1960s period. It was restructured and a number of regional offices and specialised wings were set up. The number of geologists increased from 119 in 1951 to 1300 in

1980–81. A specialised division of landslides and seismotectonics was set up at the northern regional office of Geological Survey of India, Lucknow, in 1989, which was subsequently upgraded into a Centre for Geoseismology and Seismotectonic studies with sub-centres at regional levels.

Photo-geological interpretation came to the Geological Survey of India during the late 1960s, with the availability of aerial photographs and trained photo-interpreters. By the 1990s, the survey had started using Indian Remote Sensing Satellite data to aid specialised thematic mapping programmes in different domains of study.

The network of all kinds of monitoring and forecasting stations grew rapidly. Seismological stations grew from four in 1950 to forty in 1989. Special cyclone warning radars were set up along the east and west coasts of India. India inaugurated its first flood forecasting station in 1958 on the river Yamuna for the national capital of Delhi. On the recommendation of various committees, a flood forecasting and warning organisation was set up under the Central Water Commission in 1969. In this year, forty-one forecasting stations were added, taking their total number to forty-three. By 1990, 147 such stations had been installed across the major river systems of India. These and other such activities greatly expanded the potential for research on disasters. When the quantum jump in academic population, scientific organisations and technological advancement is pitched against a growth of 35 research units per year, the phase does seem antagonistic towards an inquiry on disasters.

All forms of science need support, sponsors and scientists. It is the politicians, administrators, planners, academicians and media that build the 'feel' and the concern for a subject or theme. These groups are the key in formulating policy and decisions on critical issues facing India. They are partners who chair all national and sub-national level decisions and enjoy international interaction and exposure. A handful of decision-makers are responsible for galvanising concern about critical issues that confront India. When a stock of their work for disasters was taken, what came forth was negligence and indifference. It is the apathy of all rungs of state and society that thwarted the development of the science of the study of disasters in India during 1947–89.

Administrative Laxity

The Constitution of India in its Article 74(1) states that a council of ministers with the Prime Minister as the head should aid and advise the President on all important decisions. The ministries conduct the business of the government. Changes in policy, volume of work and importance attached to certain issues, dictate the creation of a ministry. While the number of ministries in India has increased from eighteen in 1947 to fifty in 2006, India till date does not have a ministry to deal with disasters.

Up till the mid 1970s, the Department of Agriculture was responsible for matters relating to flood and drought. Since earthquakes and cyclones occur at random, these were to be dealt with as and when they strike. However, when the scale of death and damage is high, the Ministry of Defence, Health and Civil Supplies, among a dozen others, were pressed to service assistance. This circuitous route through departments within different ministries leads to efforts at odds with each other. The piecemeal approach reduced the value of disasters as a subject worthy of attention. This was evident in the corridors of planning too.

Failure to Plan

The Planning Commission, an apex body for planning in India, did not consider the issue of anticipating and managing disasters all through 1947–89. Established in 1950 to formulate a plan every five years for the most effective and balanced utilisation of the country's resources, disasters were not part of its seven point agenda. Had disasters been adopted as a subject worthy of attention there would certainly have been some influence on the output of research. A glimpse of this role of the Planning Commission on research can be seen through the case of the programmes of 'prone areas'. In a bid to provide special package of development, the Planning Commission launched a number of schemes under the Drought Prone Area Development Programme in 1975. The influence of this programme is represented by the fact that nearly 8 per cent of the research units that belong to 1947–1989, worked on the drought prone areas. Nearly a quarter of all books on natural disasters during this phase adopt the concept of the 'prone areas', a clear proof that research nourishes policy and the latter circuits back to research. But the number of such schemes was small and up to 1997, 'the plan documents did not include issues relating to the management and mitigation of natural disasters. The traditional perception was limited to the idea of calamity relief, which is seen essentially as a non-plan component of expenditure' (Tenth five-year plan). These were thus palmed to the Finance Commission. The responsibility of the latter was to provide funds from the Union to the States. It released the funds but on a levelled ground of calculations.

Levelled and Limited Funds

The Finance Commission gives additional grants and loans to cope with problems that do not fall within the ambit of the Planning Commission. The assignment of apportioning the non-plan resources to the states is also under the purview of the Finance Commission. From 1952–92, each of the ten finance commissions allocated some funds for disaster relief. But the modalities adopted to calculate the fund makes it more of an exercise in routine than an in-depth evaluation of the requirements of the states.

The commission calculates the amount that is to be allocated to the states by taking an average of the annual expenditure incurred on disaster over the previous few years. The second, fourth and fifth commissions took an average of ten, eight and nine years respectively. The seventh commission used a nine-year average, the eighth a five-year average, the ninth commission shifted to ten years and the tenth commission adopted the average aggregate ceiling on expenditure for six years. To compensate for the inflation and price rise, an addition of 10 to 20 per cent is added to this 'average'. The ninth finance commission marked 8.1 billion rupees each year to all the states for relief funds on account of natural calamities. If correction for inflation is calculated, the net purchasing power of the fund has remained almost the same.

In 1982–86, the states sought assistance of 250,950 million rupees but the actual amount given was 39,940 million rupees; only 15 per cent of the total. In Haryana, Himachal Pradesh, Kerala and Andhra Pradesh the sum given was only 12 per cent and for Uttar Pradesh it was the lowest at 10 per cent. This yawning gap between the requested and approved makes it difficult to estimate the actual fund required for disaster relief. Neither the Centre nor the State rolled out any method or machinery to monitor the amount, flow or destination of the fund.

Endorsing the view that disasters need to be responded to with the allocation of funds, the question of why the country should waste millions on relief and not find ways to mitigate disasters did not occur in the scheme of thinking. Within the lax administrative and planning machinery, the legal window also remained shut.

Absent: The Legal Window

The Government of India enacted nearly 1045 laws in this phase (*All India Reporter,* 1989). Reinforcing the indifference towards disasters was the absence of a single umbrella law for disasters and their management during this phase. Since a specific law was missing, there was a need to search how the legal system addresses the issue of disasters. The search takes one to the three sets of laws – laws initiated for environmental conservation, laws dealing with relief and compensation, and laws pertaining to insurance.

During the search, a mixed bag of laws incorporating forests, wildlife, minerals, water, air, and the occupational environment of workers, can be clubbed under the rubric of environmental laws in India. Laws that concern disasters can be grouped into four. The first group covers those that were designed to safeguard the specific 'element' of the environment, such as the Forest Act (1927). It was the British who formulated these laws to safeguard the abuse of a resource. The second group of laws are anthropocentric in nature and therefore concerned with safeguarding the adverse impact of environmental activity on human beings. The Atomic Energy Act (1962), the Specific Relief Act (1963) and the Mines Act (1957) were formulated to safeguard people, especially from occupational hazards. The third set of laws upholds the intrinsic value of the environment as a source of life and perpetuity. A fourth, more explicit legal assertion on disaster, is the group of laws on hazardousness, be it industry, its storage and transport or waste management. The Manufacture, Storage, and Import of Hazardous Chemical Rules and the Environmental Protection Act (1986), all belong to this category. Spawned in the shadow of the Bhopal gas tragedy, most of these laws deal with the theme of hazardous products but do not branch out to cover natural disasters.

The seed of these environmental laws was sown at the United Nations' Conference on Human Environment, held at Stockholm in June 1972, where India was a participant; the Water (Prevention and Control of Pollution) Act of 1974 gave the statute book its first real foundation for environmental protection. Other major enactments that followed were the Forest (Conservation) Act, 1981, and the Air (Prevention and Control of Pollution) Act, 1981. On a call from the Stockholm Conference, the Government of India incorporated the right to a wholesome environment into the fundamental right to life under Article 21 of the Constitution. Peppered within these two groups of laws, there are some references to the mitigation of disasters. After all, if a law is created to preserve pristine forests, the care of flood, drought, landslide and avalanche automatically follows. Similarly, clean air and water are the best antidotes for epidemics.

In general, insurance acts are a group of laws that could provide buffers against the intensity of a disaster. An insurance policy can be taken for fire, lightning, explosion of gas in domestic appliances, aircraft, riot, damage by any vehicle or animal, earthquake, subsidence and landslide, flood, inundation, storm, cyclone, terrorist activity and other perils. Life and property can thus be insured against a host of possible disasters.

The first insurance company on Indian soil, the Oriental Insurance Corporation, was started in Kolkata in 1818 by Europeans to help the widows of their community. Up to the end of the nineteenth century, foreign insurance companies enjoyed a virtual monopoly and discriminated against Indians, whose life was insured with 10 per cent extra premium. Higher premiums provoked a group of nationalists who floated the Bharat Insurance Company in 1896 at Lahore in Punjab (now Pakistan). From this nascent beginning, the number has grown and today India has over 33 insurance companies, each doing lucrative business. Motivated to mount profits, they levy high rates of premiums and interest on policies, which are beyond the reach of millions in India. This makes one question the role of insurance in safeguarding the common man, especially the farmers or fishermen who face recurring floods, droughts or cyclones.

The need for crop insurance has been debated for a long time and was introduced only in 1973 on an experimental basis. Based on the individual approach, the programme was discontinued in 1976 as it proved to be nonviable. In 1979, a scheme based on the insurance of a homogeneous yield of an area was started. Under this, if the area yield is less than the guaranteed yield, all insured farmers irrespective of the yield are compensated equally. If the area yield is more than the guaranteed yield, no indemnity is paid, even if some of the farmers have lower yield. However all crop insurance policies in India are taken privately and only a few commercial farmers with large size land holdings can reap this benefit (Rusagi, 1988).

The question of crop insurance being provided to all farmers by the government has often been raised. A comprehensive crop insurance scheme was proposed to the ninth finance commission. As per this proposal, every farmer will have to take up a scheme for insuring the crop against natural calamities in which the Centre and the State will disburse the premium in the ratio of 90:10. The commission objected to this proposal on the grounds that the stipulated premium was heavy and only the insurance companies would garner benefits. Insurance therefore remains a prerogative of the wealthy.

Some scope on this front was created by the introduction of the Public Interest Litigation. Prior to 1980s, only the aggrieved party could personally knock on the doors of justice. With the Public Interest Litigation Act, any citizen of India or any consumer group or social action group can approach the apex court of the country seeking legal remedies in cases where the interests of the public are at stake. While the law does create the space to negotiate for a less risky India, neither lawyers nor citizens have filed any case for disasters. The failure of people to use the forum of public interest for responding to disasters devalued the importance of disasters.

Till the end of this phase, there was no talk or effort, leave alone a bill or law, on disasters. The lack of response by the administration and the judiciary reduced the importance of disasters. Disasters serve as templates through which society shows its concern. They expose conditions that need alteration. While bureaucrats and lawyers could be insensitive, it was expected that at least professionals, like news reporters and media personnel, would steer the conscience of society towards the tragedy of disasters. Unfortunately, this did not happen.

Sell more than Tell

The high rate of illiteracy in India does not prevent it from being the second largest producer of newspapers and periodicals in the world. With nearly 8000 newspapers in English and a larger number in regional languages, the press is a powerful medium in India. The total claimed circulation of newspapers is close to 118 million. While it is difficult to calculate the space that newspapers give to disasters, a survey was carried out taking India's most popular news magazine. The feeble attention devoted to disasters was evident from the covers of *India Today*. This magazine flagged its first edition in 1978. About 4 million people subscribe to its English edition. The magazine switched from fortnightly to weekly periodicity in May 1997. In the twenty-five years of publication, 787 editions of the magazine have been rolled out. The theme of the cover page of this magazine was selected as the frame of reference for an empirical assessment of the attention given by the media to disasters.

Of 787, only 750 issues of *India Today* were accessible. Just 22 considered disasters as important enough to be on cover page. While it is expected that politics, elections, terrorism, violence and even international issues would bite a sizeable portion of the premium space of a cover, yet

Table 5.2 Meagre Attention to Disasters on the Cover Page of *India Today*, 1978–2003

Theme of cover page	Number	Per cent of total
Politics	288	38
Violence	83	12
International affairs	72	10
Society	72	10
Commerce	50	7
Sports	40	6
Disasters	*22*	*3*
Others	123	14
Total	**750**	**100**

Based on *India Today 1978–2003, Missing numbers 1981 Jan–Dec, 1982 Jan–May, 30 June, 15 July, 30 Dec, 1982. Of 22 cover pages on Disasters, 7 were on the Kargil War.*

compared with disasters, there were three times more cover pages on society, fashion and films while sports, specially cricket, captured double the number. Almost every year, floods grip 2.3 per cent of the total area of India and affect 34 million or 3.3 per cent of its population. Yet till 2003, *India Today* had not devoted a single cover to this theme (Table 5.2).

There is no doubt that print media can play a crucial role in shaping public opinion and even policy but this depends more upon the political culture and democratic temperament that prevails in a country. Perhaps what Amartya Sen said for the role of media towards hunger also rings true for disasters. He laments, 'the subject of persistent hunger of a third of the rural population of India does not attract much attention in newspapers…these standard events in India seem to be not newsworthy'. The journalist rationale for the same runs as follows 'poverty and hunger have been around for a long time and exist on a forbiddingly vast scale – in media parlance you need some kind of news on topical peg…' (Sen, 1981). Shackled within a political-economic control for the print media, what sells is obviously more important than what tells. This is true also for Indian cinema, which has remained an unused mirror for disasters.

Cinema: An Indifferent Mirror

With 14,975 films in its spool till 2004, India is among the largest producer of films in the world. The estimated daily viewership of Indian cinema covers nearly 23 million people (Rajadhyaksha, 1999). Nearly 2 per cent of India's population goes to the cinema every day. Cinema commands a strong hold over the minds of the people of India. The song, dance, figure of speech, the hero, heroine and villain are virtual icons for the Indian people. But has Indian cinema sensitised the viewers towards disasters?

Disasters have been bypassed as a theme for cinema in India. This is evident from a reading of the *Encyclopaedia of Indian Cinema.* India has produced 14,975 films, the encyclopaedia provides abstracts of 1758 films. Among these abstracts of feature films, only 37 use disasters as a frame of context. In about half of these, a disaster is in the backdrop and the remaining attribute some direct importance. Nearly half of these 37 films are on war, with another one-fourth on famine and drought. Earthquake, flood or a cyclone could be a stimulating canvas for the setting of a film but these occur in only 15 per cent of the sample. Even within this miniscule 0.2 per cent, the portrayal of a disaster leaves much to be desired. The details of why the disaster happens or how it impacts are mostly brushed aside. Disasters are used as extraneous occurrences that bring tragedy and are at times also portrayed as events of divine intervention (Table 5.3).

Parallel cinema or low budget films make an effort to project a more complete truth but while these films draw accolades at film festivals, they never receive national fame. A large viewership of Indian cinema thus wallows within an unreal world of myth and superfluous entertainment.

However, films on disasters find no sponsor and perhaps no audience. This may be the reason for the decline in their numbers. The decades of the 1940–60 were most prolific in production, but thereafter, their number

Table 5.3 Cinema with Disasters as a Theme – Few to Count

Theme	**Number**	**Per cent of total**
War	18	49
Drought and famine	10	27
Earthquake	2	5
Flood	2	5
Hurricane and storms	2	5
Mining	1	3
Bridge collapse	1	3
Fire	1	3
Total	**37**	**100**

Based on *Rajadhyaksha, A. and P. Willemen, Encyclopedia of Indian Cinema, 1999.*

Table 5.4 Films on Disasters – Declining Frequency over Decades

Decade	Number	Per cent of total
1920–30	2	5
1931–40	9	24
1941–50	7	19
1951–60	7	19
1961–70	4	11
1971–80	3	8
1981–90	4	11
1991–95	1	2
Total	**37**	**100**

Based on *Rajadhyaksha, A. and P. Willemen, Encyclopaedia of Indian Cinema, 1999.*

trickled down to three or four per decade. Cinema thus remains an unused vehicle to evoke concern about disasters in India (Table 5.4).

While the glitzy glamour and the neon hoarding of commercial films steal the limelight from documentaries and audio-visuals, it is the latter that usually capture events of national significance. How many short films, documentaries and audiovisuals does India have on disasters? (Table 5.5).

Table 5.5 Disaster Themes in Documentaries

Disaster	Number
Drought/famine	9
Earthquake	7
Flood	6
Bhopal gas leak	4
Floods and drought	3
Cyclone	3
Fire	2
Railway safety	2
Avalanche	1
Landslide	1
Others	12
Total	**50**

Based on *Audiovisuals on the Environment, 1997 Indira Gandhi Conservation Monitoring Centre, WWF For Nature, India, video/films/cd's on disasters, National Institute of Disaster Management, Centre for Science and Environment and Catalogue of Selected Films, Films Division, 1995.*

The following analysis is confined to documentaries produced by four major producers. These are the National Institute for Disaster Management, Centre for Science and Environment, Indira Gandhi Conservation Monitoring Centre at World Wide Fund, and the Catalogue of Selected Films by the Films Division. All four centres together produced only 50 audio-visuals on disasters in India. 38 out of this group of 50 have depicted a specific disaster while the remaining 12 demonstrate the response that needs to be adopted in the face of disaster. When compared with the 37 feature films, the 50 audio-visuals may seem a large number, but within their own group, the relative number is small. The World Wide Fund on its reels has 1637 audio-visuals but those on disasters form only 1.4 per cent of the total. Similarly, the catalogue of Films Division lists 574 films, but ones on disasters are only 0.6 per cent of the total. Every day, every region, every agent, carries an instance of some or the other disaster. All these disasters could be a subject for films and documentaries but this has not been the case.

Despite the overload of disasters, India has produced only 37 films and 50 audio-visuals on this significant theme. This is less than 126 hours of film time.

Evidently a gross neglect of disasters is a voice that echoes through the five year plans, the census reports, news magazines and films. Soaked with indifference, it could well be argued that although research depends on equipment, funds and finances of the sponsors yet it is the task and responsibility of the researcher to identify critical problems like disasters that confront society.

Honed by sensitivity, led by intuition, research is a passion independent of signals from programmes, commissions and coverage by the press. Slow but sure are the centres of learning and research. But nowhere is the indifference more alarming than in the portals of education. Disasters have been peripheral to inquiry within teaching and research. The academic community could not see, let alone nourish the study of disasters in India.

Teaching and learning about disasters was completely neglected. All through 1947–90, right through the schools, colleges and even the most important professional and competitive examinations did not contain a course or syllabus that mentioned the study of disasters.

Peripheral to Teaching

In India, disasters were not taught in the main field of pedagogical endeavour from school to university. A perusal of the syllabus of the

Central Board of Secondary School Education prior to 1990 was undertaken to begin the exercise. The influence of the board can be seen from the growth in the number of schools affiliated to it. In 1962, the year it came into existence, there were 309 affiliated schools. The numbers stood at 6847 in 2003.

Each year, over 0.35 million students appear for the pre-board higher secondary class X examination while 0.9 million students sit for the class XII senior secondary examination. When the course content of board examination was checked, geography offered a section on natural hazards, psychology listed human behaviour in the face of natural disaster, while history contained a section on war. Inching close to adulthood, as students leave school, their exposure to disasters was limited to news reports, general knowledge quizzes or a clipping pasted on the notice board (Table 5.6). Where would a student pursue a course on disasters in India after school?

A syllabus mirrors the prevailing concerns of society. A symptom of unresponsiveness is the absence of disasters as a stream of study across academia in India. Among the 269 recognised central and state universities of India and their 11 thousand colleges, such was the indifference that the need to conceptualise a regular course on disasters and their management was not felt. The only way one could pursue the study of natural disasters was to follow an indirect route. Thus, an

Table 5.6 Missed Opportunity to Incorporate Disasters in the Curriculum from School to University to National Examinations

	Students taking the exam (in million) 2003–2004	**Papers offered**	**Papers mentioning disasters**
Class X	0.35	5	0
Class XII	0.9	30[1]	3 (Psychology, History, Geography)
Universities	8.3	n.a	1[2]
UGC NET	0.1	309	9
UPSC	0.02	396	28

Based on *CBSE School Curriculum, 2004, 2005, Class X, XI, XII, Syllabus and Guideline, UGC Junior Research Fellowship and Lecturership Exam, 2002–03, Rao's IAS Syllabus, 2003.* [1] *In three streams, science, commerce, humanities.* [2] *Anna University – a. Remote Sensing and GIS for Disaster Mitigation and Management in M.Tech (Remote Sensing) b. Disaster Mitigation Measures in M.Sc. (Environmental Science) c. Three electives viz., Disaster Resistant Structures, Wind and Cyclone Effects on Structures and A seismic Design of Structure in M.E (Structural Engineering).*

interest in earthquakes would involve the study of geology, geography or civil and structural engineering; drought and famine would mean a route through the atmospheric or the hydrological sciences or agricultural economics, sociology and geography.

The curriculum of India's two leading competitive examinations, Civil Services and National Entrance Test, do not contain disasters as a field of knowledge.

It augurs well to frame courses of study in accordance with the need of the country. It would be relatively easy to drive home the concern and concept of disasters in a country where disasters are day-to-day occurrences. This too was not done. In school, college and university, the study of disasters was fragmented and peripheral.

Each year, thousands of research projects are sanctioned across India. The chief patron that shoulders 75 per cent of financial support to scientific and industrial research is the Government of India. From the Department of Science and Technology, a database for sixteen years from 1985 to 2001 was picked to assess the status of research on disasters. In this list of 20,031 projects, only 160 carried a disaster in its title.

The punitive research on disasters is evident from the total funds released for this purpose. During 1985–2001, 284 million rupees were granted for projects on disasters. This is a mere 1.11 per cent of the total funds released. Fifty-five projects on disasters had a grant support of less than half a million rupees each, while 77 fetched between 0.5 to 1.5 million. Only 28 took 1.5 million or more. The small amount per project on disasters perhaps reflects more on the scale of work intended by the scientist than the stinginess of a funding agency.

The Indian Council of Social Science Research is a body that primarily supports projects for social scientists. Since its inception in 1969, the council has sanctioned 1982 research projects. Ones on the theme of disasters were only 14, or not even one per cent of the total (Table 5.7).

Of the many natural disasters that occurred in these forty-two years, 107events were taken for research. Among these, only three, the Koyna earthquake 1967, Assam earthquake 1950, and 1977 Andhra cyclone fetched more than 30 research units each. The larger group of 74 per cent of the events drew a shrivelled response of hardly one research unit. There were disasters like the Kutch earthquake of 1958, the floods of North Bengal of 1968, Darjeeling landslide of 1968 and the Andhra cyclone of 1969, which found no mention in the research units of this phase.

Table 5.7 Low Involvement of Social Scientists in Disaster Research, 1969–2003

Year	Total projects	On disaster themes
1969–74	382	2
1975–81	527	4
1982–91[1]	450	4
1991–96	299	1
1996–01	253	3
2001–03	71	0
Total	**1982**	**14**

Based on *Records of the Indian Council of Social Science Research, Delhi, 1969–2003.*
[1]*Information for 1987–88 is not available.*

Administrators, planners and media have the power to lever change within society but they have done precious little for the cause of disasters. Since the edifice to teach the subject was not built, the quantum of research was bound to be low. This explains the continuous low output of research units from 1947 to 1989.

The concern from all quarters tampered the graph of growth of research units all through these decades. Except for apathy, there is just no cause for taking a recess from research during this phase. Although India had gained freedom, she had not become free of disasters. From 1947–89, disasters roamed and reigned over India.

It is clear that India has witnessed many disasters, each worthy of study and concern. A plea for nonchalance cannot be accepted on grounds of lack of awareness. The people at the helm of affairs in India did have ample opportunity to travel and learn from the world outside. While they did travel, they failed to learn about the practices to manage disasters. Disasters are not unique to India. Countries outside India have been finding ways to mitigate disasters as disasters are widespread.

Failure to Learn from Other Countries

By international standards India is late on most fronts. The Japanese formed a seismological society in 1887, this was an initiative taken after the Yokohama earthquake of 1880. India set up a formal society for earthquake research in 1962 and does not have a single journal on disasters. Some of the leading international journals on disasters have celebrated their silver jubilee. *Disasters: The Journal of Disaster Studies, Policy and Management* published from the United Kingdom was launched in 1977. The *Journal of Natural Disaster Science* was inaugurated in Japan in 1979. The *Science of Tsunami Hazards* was started in 1982 in the United States of America. The birth of a dozen other journals confirm that publishing on disasters has attained professional maturity in the international arena.

The National Institute of Disaster Management, India, was born in 2003. This is forty-one years after the United States of America established a Disaster Research Centre at Ohio State University in 1963. As early as 1951, a Disaster Prevention Research Institute was established at Kyoto University. The National Research Institute of Earth Sciences and Disaster Prevention, Osaka, was also set up in 1963. The Japanese set up a formal link for the social sciences with the Disaster Research Centre in Ohio.

Japan's first disaster related legislation was enshrined in 1880. Its first attempt to prepare a comprehensive national disaster management system can be traced to the 1961 Disaster Countermeasures Basic Act. The latter was prompted by the destructive typhoon Vera that stormed the Japanese coast of Honshu in 1959 and left nearly 5000 dead.

India could have profited from an earlier adoption of some features of the Federal Emergency Management Agency. The United States of America formalised this act, generally considered the first piece of disaster legislation, in 1979. Its inception can be traced to the Congressional Act of 1803. In the century that followed, the United States government passed over 100 *ad hoc* legislations in response to hurricanes, earthquakes, floods and other natural disasters. It was in 1979 that the executive order merged many of the separate disaster-related responsibilities into a new Federal Emergency Management Agency (FEMA). It has taken India nearly 200 years to adopt the FEMA model and install a National Disaster Management Authority in 2005.

A number of countries especially in Asia, Africa and Latin America, remained indifferent to natural disasters in this phase. But how long could this situation go unnoticed? The United Nations Organisation, an apex body with a membership of 192 countries of the world, declared the 1990s as the International Decade of Disaster Reduction. The subject of disasters entered a phase of recognition.

Globalisation (1990 onwards): A Phase of Recognition

India's new economic policy in 1992 put on air its decision to join the race of globalisation. The philosophy of globalisation is free trade, open markets, fair competition, and barrier less movement of money, people and ideas. The promise of globalisation is not dependent on the public sector. While the government is a facilitator, the key player is the private sector.

How is globalisation related with the recognition of disasters? At the outset, there may seem no apparent bond, but globalisation has ushered an undercurrent of a risk society, 'disasters are a deep institutional crises of the phase of industrial modernity' (Beck, 1998). The risk of collapse of financial markets, the danger of trans-national trading, the breach of trust between investors, and the threat arising from large-scale economies of production, transport and transference are inherent to globalisation. But while financial, technological and ethical risks emerge on the horizon, the most serious and perhaps most unexpected was the risk of terrorism. The latter did not lurk in shadows but actually struck at people and places in different parts of the world. Involuntarily or unconsciously, the phase of globalisation converted the risk of terrorism into a serious man-made disaster. World powers were unable to and perhaps unwilling to dissolve the root problems that breed terrorism. While this is easier said than done, what has been geared instead is a global decision to pre-empt, prepare and confront this disaster. Terrorism dictated an emergency about the need to revamp decisions and mechanism for disaster management. Natural disasters rode on the tide of this concern. Therefore, while the principal aim of the International Decade of Natural Disaster Reduction was to 'capitalise on existing knowledge of ways to mitigate natural events, and to foster the systematic transfer of this knowledge to those countries and communities recognisably most at risk', globalisation with its excess baggage of risks endorsed the need for urgent action. The deep cry of humanity from a triple set of natural disasters was a catalyst for recognition. There was the Orissa cyclone in 1999, the Bhuj earthquake in 2001 and the tsunami in 2004. What made their presence felt was the representation from three different naturals – the cyclone, earthquake and a tsunami. Arriving at intervals of two years, the short space of occurrence meant that they could not be easily forgotten. The media mounted pressure. Exposing the agonies of the victims in disasters, the media created an undeniable public alert. This was especially so in the case of the tsunami.

The tsunami took place on 26 December at 6:30 am on a Sunday morning. Since the epicentre was near Sumatra, the up welling waves of the Indian Ocean affected the coast of not just India but as many as eight countries in Southeast Asia and six in Africa as well. It could not have been a more unfortunate day, date and time. Thousands of local and foreign tourists flocked the beaches of South Asia to celebrate their Christmas holidays. With the death of 275,000 people, it turned out to be a Black Sunday. The international and regional catchment of tsunami put India in serious spotlight. On the coast of India, nearly 10,273 people died.

Disasters could not be concealed in this phase. A synergy of factors coaxed India to give due importance to and recognise the need to do something for disasters. Independent India's first ever Cabinet Committee (on economics) was constituted in 1949 and it was only in 1991–92, that a committee of such status was formed on natural calamities with the aim to examine and direct programmes for reducing the adverse impact of natural calamities. It was also in 1991–92 that a National Advisory Council headed by the Union Agricultural Minister to identify and recommend programme for natural disaster mitigation was created. All this culminated into the need for a High Power Committee to review existing arrangements for preparedness and mitigation of Natural Disaster Management at the national, state and district levels. Constituted in August 1999, it was the first time that a High Powered Committee on disaster management had been formed. This was the first organised effort at the highest level to evolve a systematic, comprehensive and holistic approach towards all disasters. By including experts from different ministries, it ensured a greater level of representation. The original mandate of the High Powered Committee was confined to the preparation of management plans for natural disasters. But with the approval of the Prime Minister on April 17, 2000, its terms of reference were enlarged to include non-natural or man-made disasters. The power of this highly placed committee set the ball rolling to accommodate the cause of disasters at different rungs of administrative decisions.

In the Plans

Disaster management was incorporated as an integral part of planning at the national, regional and state levels. The first five-year plan was launched in 1951, it took nine five-year plans and seven annual plans before the

Planning Commission focused directly on disasters. The tenth five-year plan document of 2002–07 carried an exclusive chapter on disaster management. It was for the first time that the Perspective Plan for the National Capital Region, Delhi 2021, decided to include a chapter on disaster management of earthquakes, floods and *andhis* (storms).

The state governments need to prepare their respective plans as well. Differing in the depth of details, a feature common to all the plans is to first assess the type and scale of disasters within the respective administrative boundaries. This is followed by an outline of the resources that could be harnessed in the face of a disaster (Table 5.8).

The state governments also got busy preparing plans. A proposal of action for administrators, along with a list of dos and don'ts, is also included in these plans. In the lead were the district authorities of Andaman and Nicobar Islands, followed by the Maharashtra Disaster Management Plan and the Multi-pronged Disaster Management Plan of Orissa. Other states followed suit. The years of inception of all these plans were in post 1990s, with the majority belonging to the 2001–06.

In the Finance Commission

The Finance Commission introduced some radical changes. From its inception on 22 November 1951 to 2005, eleven finance commissions

Table 5.8 A Handful of Plans Initiated for Disaster Management, 1996–2001

Year	Title of the Disaster Management Document
1996	District Management Plan for Andaman and Nicobar Islands, District Administration, Nicobar
1998	Maharashtra Disaster Management Plan, Risk Assessment and Vulnerability Analysis, Government of Maharashtra, Mumbai
1998	Multi-disaster State Response Strategy, Maharashtra Emergency Earthquake Rehabilitation Programme, Government of Maharashtra, Mumbai
1998	Latur District Disaster Management Information Database
	Latur District Disaster Response Plan, Maharashtra Emergency Earthquake Rehabilitation Programme, Government of Maharashtra, Mumbai
1999	Mumbai Disaster Management Plan, Brihan Mumbai Municipal Corporation
2001	District Disaster Management Plan, Aizwal
2001	Disaster Management Plan, Chamba
2001	Municipal Corporation of Delhi Plan
2001	Multi-pronged Anti-disaster Plan, Orissa

allocated funds for relief. But it was only in 1990–95 that a specific fund called Calamity Relief Fund was laid out. The tenth finance commission, for the first time, formulated a distinct funding mechanism for taking care of calamities of rare severity by creating the National Fund for Calamity Relief. The problems of defining calamities of rare severity led the eleventh finance commission (1997–2002) to revamp the fund to create the National Calamity Contingency Fund, with an initial corpus of 5 billion rupees. The National Centre for Calamity Management was initiated in 2000. These were not cosmetic alterations. In 2005, the twelfth finance commission (2002–07) enlarged its mandate to include mitigation and prevention of calamities and not just relief and rehabilitation. In line with this, in 2004, many state governments like those of Bihar and Sikkim changed the name of their departments from Relief and Rehabilitation to the Department of Disaster Management. By including the necessary process of drawing up disaster management and mitigation plans, the eons old relief codes were rechristened as disaster management codes.

In the More Prestigious Ministry

The concern for natural disasters was taken from the Ministry of Agriculture and transferred to the Ministry of Home Affairs in June 2002. This was not a small feat, after all matters relating to crops and loss of livestock due to natural calamities and coordination of relief measures necessitated by drought had existed since 1969 onwards within the Ministry of Agriculture. Although all ministries are important, yet the Ministry of Home Affairs enjoys a higher status, as it includes portfolios like that of Internal Security, Central Reserved Police Force, Border Security Force, Centre-State Relations, among others. Sharing space with the more prestigious gives immediate visibility.

In International Partnership

The search for additional support to build its base for disaster management led India to take on international projects. The Government of India and the United Nations Development Programme launched the Disaster Risk Management Programme (DRMP) in August 2002. With 7 million dollars, this multi-funded project was to cover the 169 most multi-hazard prone districts of seventeen selected states. Its twelve-point programme included setting up disaster management authorities, outlining plans, installing hazard cells, establishing emergency operations centres, along with holding mock drills and training exercises. The Urban Earthquake Vulnerability

Reduction Programme was included within the agenda. This programme envisaged strengthening communities, urban local bodies and the administration in mitigation, preparedness and response across thirty-eight cities in the country. The cities selected were those that had a population above half a million and fell under seismic zones III, IV and V. Learning from this initiative could enhance the national capacity building programme. A Disaster Management Support Project, a collaborative effort between the Disaster Management Division of the Home Ministry and the United States Agency for International Development, was also signed on 19 January 2004. The main thrust of this 16-million-dollar project was to prepare India for disasters.

To undertake a mid-term review of the International Decade of Natural Disaster Reduction the World Conference on Natural Disaster Reduction was held in May 1994 at Yokohama, Japan. The Yokohama Strategy and Plan of Action for a Safer World was adopted here. As the international decade of natural disaster reduction drew to an end in 2000, it was replaced by the International Strategy for Disaster Reduction. However, this was not the end. Knowing well that much more efforts were needed to accomplish any change, a world conference on disaster reduction was held in Kobe Hyogo, Japan from 18 to 22 January 2005. The selection of the date and place were not incidental. It was on 17 January 1995 that the Great Hanshin-Awaji earthquake struck the Hyogo prefecture, including the City of Kobe, resulting in thousands of fatalities. The people of the city of Kobe commemorated the tenth anniversary of the tragedy on 17 January 2005. The Hyogo was selected to reiterate the role of international community in disaster reduction. The Hyogo Framework of Action and Hyogo Declaration set out a result-oriented plan of action for the next decade. Seeking support and reiterating commitment were delegates from India, present at all such venues.

Ways to reach out to international agencies were also charted. Although India became a member of the Asia Disaster Preparedness Centre, Bangkok, as soon as the latter became an independent entity in 1999, it took her a decade to join the International Search and Rescue Advisory Group. Established in 1991 as a sequel to the Armenia earthquake of 1988, this group is a global network of more than eighty countries and disaster response organisations under the United Nations. India joined this organisation in its meeting at Asia-Pacific Regional Group, Seoul (Republic of Korea) in early November 2000. India was elected as its chairperson in September 2006. When the South Asian Association for Regional Cooperation was created in 1988, India was a founding member. In 2006, when the SAARC Disaster Management Centre was looking for a place to anchor, New Delhi became a preferred location.

In Alert

While international comrades were sought, the structure of how to manage disasters drew from the notion of perceiving these as akin to war.

The number of dead and scale of damage are common features. The need for search and rescue, relief and compensation by both cannot be denied. The fact that both are times of emergency can also not be ruled out. What reinforces the similarity is the need and presence of the army. It is a fact that disasters in India have till date been managed by the defence forces. This perception of resemblance of a natural disaster with a war has seeped into the very design being laid for management of disasters in India. In war, there are no runners-up. The art of warfare demands a perfection of all strategies. This is achieved through telephone battle, blackboard exercise, sand models, and exercise with troops, without troops, simulation methods and other mock drills. A similar plan is laid for natural disasters. Take the case of mock drills. The nomenclature resounds the philosophy of the need for practice and rehearsal as a prelude to disaster preparedness. This was launched on a experimental basis on 6 October 2006 at Delhi. Since then, two dozen such 'mock' exercises simulating a disaster have been conducted in different parts of India. Plans are afoot to conduct these with larger number of stakeholders and covering a wider territorial expanse. Along with these experiments, what is being charted at the country level is the identification of resources, centres and manpower, which could be roped in at the time of a disaster.

The Disaster Response Network was launched on 1 September 2003. This is a website which carries a list of resources available at the district and state level that can be mobilised in times of an emergency. Today it carries an inventory running into thousands, with items grouped into eight categories, namely, fire, flood rescue, health services, nuclear biological and chemical, search and rescue, shelter, telecommunication and transportation. Fourteen regional response centres will be operational which will facilitate the quick movement of such equipments and resources to the site of a disaster.

Along with cataloguing and networking through information technology, a National Disaster Response Force was also created in 2005. In tune with the military experience, this comprises eight battalions of central paramilitary forces. Each battalion comprises about 1,000 troopers and support personnel. Two battalions each from the Border Security Force, Indo-Tibetan Border Police, Central Industrial Security Force and Central Reserve Police Force have been deputed to mount a professional response during disasters. Each of these battalions would have a specialised response team consisting of 45 personnel, including doctors, paramedics and engineers, among others. In all, there will be 144 specialists' search and rescue teams in the eight battalions stationed at Arrakonam (Tamil Nadu), Mundali (Orissa), Greater Noida (Uttar Pradesh), Chandigarh, Barasat (West Bengal), Guwahati (Assam), Pune (Maharashtra) and Gandhinagar (Gujarat).

To help the state governments, a go ahead signal was given in 2005 to create ten national disaster mitigation resource centres. These regional centres are to provide assistance to neighbouring states. Located at Bhanu (near Chandigarh), Noida, Gandhinagar, Pune, Nagpur, Hyderabad, Chennai, Bhubaneswar, Kolkata and Guwahati, each centre will cater relief for 25,000 people, except Bhanu and Kolkata which will cater for 50,000 personnel, as these include the adjacent high altitude regions. The location pattern is such that instead of each of the twenty-eight states having a separate centre, these ten would serve a group of neighbouring states. To coordinate the entire disaster emergency operations, the existing control room at the national level was also upgraded in 2003 as the national emergency operations centre. Equipped with satellite-based mobile voice, data and video, it would communicate with state emergency operations centres. Since in a calamity communication is the first casualty, effort is being made to transfer the National Emergency Communication Network to a satellite-based communication channel.

Along with the infrastructure, the proposal to revamp the 1.2-million Civil Defense and Home Guards voluntary organisation and associate them with the activities of disaster mitigation and response is on the cards.

Forging international links, and restructuring the administrative, financial and media apparatus are laudable efforts but the curtain raiser on the seriousness of deliberation on disasters is the announcement of the Disaster Management Act.

The Act

India acquired its first comprehensive Disaster Management Act in 2005. The legal decision was sent to the Rajya Sabha on 11 May 2005 and was approved on 28 November 2005. The bill moved through the Parliamentary Standing Committee and a phase of amendments to reach the Lok Sabha on 12 December 2005. The President gave his assent on 23 December 2005. In less than eight months, it became the Act 53 of 2005.

The act specifies a series of plans including those for preparedness, mitigation, response, capacity building and emergency that need to be built and implemented. Organised in eleven chapters, the Disaster Management Act outlines the legal and institutional framework for management of disasters in the country. Spelling out the channels of administration that need to be created, the act specifies that the role of centre and national level authority and executive committee is to lay down guidelines for managing disasters while the State is to implement these and define the modalities of the plan. The role of the district is to execute the plan and spell out the way to carry forward the management. Specifying a variety of plans at different geographical scales, the act is the first of its kind of legal document on disasters in India.

Keeping in view the federal polity, it was enacted under the entry 'Social Security and Social Insurance' in the Concurrent List of the Constitution of India. This gives the state governments the added scope of launching their own legislation. The disaster management policy is close on heels. While the draft policy was discussed at a national level workshop held at Hyderabad on 27 October 2006, it will soon be placed before the cabinet for ratification. Among the salient features of the act, the decision to set up a National Disaster Management Authority was also present.

An Authority

The National Disaster Management Authority was created with the Prime Minister of India as chairperson in 2005. The responsibility of the authority was to lay down the plans, policies and guidelines for disaster management. Serviced by the Ministry of Home Affairs, the authority would approve the National Disaster Management Plan. It laid down guidelines to be followed by ministries and departments of the Central Government and any state authority in drawing up the state plan. It would also coordinate, enforce and implement policies and plans for disaster management and arrange to oversee the provision of funds for mitigation measures, preparedness and response. To meet any threatening disaster situation, the Central Government National Disaster Response Fund was created. This fund shall also be applied by the National Authority in relation to projects which are undertaken exclusively for the purpose of mitigation of disasters.

The process of these activities brewed the idea of introducing study and research on disasters within the curriculums of India.

An Institute

To augment the capacity for natural disaster management, India set up its first ever National Centre for Disaster Management in March 1995. The centre was upgraded to the status of National Institute of Disaster Management in October 2003, and formally inaugurated on separate premises within the campus of the Indian Institute of Public Administration on 11 August 2004. Disaster management cells were established in all the twenty-eight state training institutes. Since its inception in 1995 till 2003, the national institute along with its state counterparts carried out 107 workshops, enrolling nearly 3324 participants (Table 5.9).

In 2006, the Lal Bahadur Shastri National Academy of Administration, Mussoorie, was designated as the nodal training institute for different modules of disaster management. The pace of training has not slowed. The two annual reports of 2004–05 and 2005–06 of the Ministry of

Table 5.9 Workshops Conducted by the National Institute of Disaster Management, New Delhi, 1995–2003

Year	Number of workshops/ training programmes	Number of participants	Average per workshop	Total days
1995	7	59	8	24
1996	5	144	29	22
1997	13	279	21	53
1998	10	127	13	36
1999	15	682	45	59
2000	16	658	41	53
2001	20	709	35	72
2002	16	532	33	54
2003	5	134	27	22
Total	**107**	**3324**	**31**	**395**

Based on *Annual Reports of Indian Institute of Public Administration, 1995–2003.*

Home Affairs list nearly 100 workshops and training programmes conducted to propagate the cause of disaster management in India. To support this activity, each state level administrative institute is provided 1.5 million rupees annually to conduct 100 programmes in a year. While an earthquake shook Latur on 30 September 1993, it acquired a disaster management training centre only on 30 September 2006. Better late than never. The same could be said for courses of study in schools and colleges.

In a Curriculum

The message of disaster management was folded into the curriculum of formal education. The Central Board of Secondary Education has accorded affiliation to 8278 schools and adds nearly 200 more to its rolls each year. Although established in 1952, the board introduced disaster management in their curriculum for eighth standard from the academic year 2003–05, for the ninth from 2004–05, and for class ten in the following academic session. The National Council of Educational Research and Training was not to be left behind. It included disaster management in its syllabus in 2005. State education boards across India followed suit.

There were also universities that responded by introducing a course on disaster management (Annexure 9). The Indira Gandhi National Open University, New Delhi, started India's first distance learning programme on Disaster Management in 1999. The Indian Institute of Ecology and Environment, New Delhi, too offers a Master of Science in Disaster Mitigation in affiliation with the Sikkim Manipal University of Health, Medical and Technological Sciences. The Guru Gobind Singh Indraprashtra University began a Master of Business Administration in Disaster Management in 2005. The Sastra Deemed University, Thanjavur, began a Master in Technology in Disaster Mitigation in 2003. Established in the 1960s and 1970s, these universities commenced courses on disasters only after 2001.

The Red Cross Society came to India in 1920. With over 700 centres spread across the country, its work for relief over eighty-six years is no doubt commendable. In 2006, the Indian Red Cross Society started a one-year diploma in disaster preparedness and rehabilitation at its headquarters in Delhi. Other efforts are in the pipeline. The Medical Council of India is introducing courses on emergency health management and building the scope to provide in-service training for hospital managers and professionals.

Other departments pitched in to promote education. In 1997, India's Postal Department printed eight million postcards and two million letter cards in different languages to take the slogan 'natural disasters preparedness brings safety' to the households of India.

Keen to encourage path-breaking strategies, even the Department of Science and Technology, for the first time in 2003, announced the 'development of innovative technologies for mitigation and management of natural disasters' as an integral part of its policy.

Disasters swiftly gained fame as a subject across India (Table 5.10). The acknowledgement for disasters came from all quarters. The parliament,

Table 5.10 Significance Conferred to Disasters, 1830–2006

Indicator	**Phase I**	**Phase II**	**Phase III**	**Total**
	Awareness	**Indifference**	**Recognition**	
	1830–1947	**1947–1989**	**1990–2006**	
Number of research units	159 (3)	1455 (29)	3403 (68)	5017 (100)
Books	3 (2)	27 (20)	104 (78)	134 (100)
Doctorates	0 (0)	45 (40)	68 (60)	113 (100)
Atlases	0	3 (43)	4 (57)	7 (100)
References in international journals	10 (2)	124 (29)	297 (69)	431 (100)
Foreign collaborations	1 (1)	7 (11)	59 (88)	67 (100)
Usage of the word Disaster	0 (0)	38 (6)	547 (94)	585 (100)
Usage of the words Disaster Management	0 (0)	16 (4)	345 (96)	361 (100)
Training at State Administrative Institutes	0	0	27	n.a
National Disaster Mitigation Resource Centres	0	0	9	n.a
Disaster Management Acts	0	0	1	n.a
National Institutes for Disaster Management	0	0	1	n.a
National Disaster Management Authorities	0	0	1	n.a
SAARC Centres for Disaster Management	0	0	1	n.a
Place in syllabus	No	No	Yes	n.a
Courses for study	No	No	Yes	n.a

Based on *4004. Figures in parenthesis denote per cent of total, n.a denotes not applicable.*

administration, planners, educationists, doctors and the triforces have all begun to discuss and chart some guidelines for management. India traversed this route of giving importance to the subject of disasters in a short, span of time. The time period for decisions gained rapid ground after 2000.

What is of significance is that India traversed this route of giving importance to the subject of disasters in a very short span of time. When this widespread recognition is superimposed on the growth of research units, what emerges is a marked similarity. The growth of research units is directly proportional to the trend of recognition.

A Sharp Growth

A precipitous increase in research units characterises the 1990s to 2006 phase. From a base of 1613 in 1989, the units grew to 3090 in 1999. Moving to a new height, 1927 units were added by 2006. Parallel to the rapid decisions India took for disasters is the addition of 3403 research units in a short span of seventeen years.

The units are twenty-one times that of the colonial phase, and two and half times more than those in the forty-two years preceding 1990. The annual averages are more revealing. There is an all time high of 200 per year during 1990–2006 compared to the tottered pace of 35 per year during 1947–89.

A thinner slice of the data and three high points can be spotted in the chart of growth. One is positioned in mid-1995, another at 2002, while the third is sited around 2004. The point of growth of research units intersects sharply with the arrival of the triple disasters. Nearly 15 per cent of the research units of the post-1990s focus on the three natural disasters of 1999 Orissa cyclone, 2001 Bhuj earthquake and 2004 tsunami. But these events are not the sole reason for the boost in numbers. India has seen this variety and scale of natural disasters several times in the recent past. The death of 10,000 people in the Latur earthquake (1993), the loss of another 10,000 in the Andhra Pradesh cyclone in 1977, and the death of 3000 in Uttarkashi earthquake are not small in comparison. But it was the disasters post-1990 that got special notice. If the research units per event are compared, then Orissa cyclone has drawn 118, Bhuj 235 and tsunami 159 units. When the research units for Bhuj are compared with Latur, the 129 research units of the latter are half the number; Andhra Pradesh cyclone of 1977 with 32

Figure 5.1 **Phase of Recognition to Disasters, 1990–2006**

research units drew one-seventh of Bhuj, while Uttarkashi earthquake with its 46 units is one-fifth that of the research units on Bhuj.

Disasters have punctured and punctuated virtually every decade of India. A relevant question would be why did Killari, Uttarkashi and Andhra not secure more research units? Why did the Orissa cyclone, Bhuj earthquake and tsunami draw heightened attention? While all events are important, they do not receive similar amount of attention. The number of research units per event was 3 in the colonial phase, 3 in the phase of indifference but it rose to 9 in the phase of recognition. It just cannot be denied that it was the widespread recognition given to disasters in general which increased the number of research units per event.

The sudden rise in the number of research units is common to all types of natural disasters. Cyclones added 370 research units during 1990 to 2006. This is twice that in the phase of indifference. Research units on earthquakes registered an increase of 1020 research units during 1990–2006; this again is double that of the phase of indifference. The comparison becomes sharper when it is noted that the multiplications were achieved in less than one-third of the time. Tsunami is the fastest growing entrant for research interest. The leaders were tsunami, cyclones, and earthquakes.

The phase of recognition was not confined only to growth. The research units acquired a diverse language, a range of disasters were researched, people of varied academic fields showed interest in its study, different professionals authored their research and India's disaster research found international acceptance.

A Diverse Language

A pronounced change has been the implant of 'a new-language' into the research units post the 1990s. This is neither a tall claim nor a hastily built opinion. To work this out, the titles of the 5017 research units were scrutinised with a one-point question, how do the research units address a natural disaster?

The word disaster is used in varied implication which could be explored in 50 to 55 synonyms. When the 5017 research units were examined, a five-fold classification emerged. First, there were units that were 'conventional' and called a cyclone, earthquake or landslide by its nomenclature. The second group comprised those that used a general descriptor like physical, natural, environmental or a group of weather disasters. The vocabulary of risk, hazard, calamity, emergency, crisis, event and catastrophe defined the third category. The fourth simply used the term disaster. The fifth added the word 'management' as a suffix to disaster.

While used in a segregated form, in practice there were compound cases. Therefore, a research unit could use cyclone and disaster, and also add the word emergency to its title. When multiple words were taken into the tally and results analysed, the conventional titles comprised 86 per cent of the total. Leaving this standard group, the remaining 14 per cent, were interesting. They were the trend breakers. Allowing the scope for multiple counting, in such research units 84 per cent used the word disaster, 52 per cent referred to management, descriptors like environmental or physical were important for 31 per cent, whereas 19 per cent used the phrase hazard, risk, calamity, crisis or catastrophe.

In the lexicon, hazards and disasters are the most popular words. Both have a different purpose and origin. Hazard refers to the 'possible threats that can accrue from any agent'. In the 1950s, geographers working at the University of Chicago popularised a behavioural approach to natural hazard research. In 1974, the Commission on Man and Environment, of the International Geographical Union, designated natural hazard investigation as one of its principal concerns. To investigate the concept of perception of natural hazards, the commission selected forty principal investigators in twenty countries of the world. India was chosen for work on flood hazard. This explains why one-third of the research units by Indian geographers is focused on floods. Many of them adopt perception as the channel for understanding how humans adapt and cope with natural hazards. Whereas the word hazard has affinity with geographers, it was the group of sociologists at the University of Ohio who pioneered research on disasters in 1963.

While geographical ideas were imported from the United States of America to India, in contrast the sociological school was unable to impact disaster studies in the mainstream research in India. While hazard had a more direct entry, only 66 research units used this word but the expression 'disaster' is used by 585 units, nearly nine times more. But as the word disaster has gained much wider circulation, it is used more in the parlance of a common and generalised expression rather than rooted in the societal or development context.

The etymology of the word disaster can be traced to the French word *desastre* which is the combination of two terms 'des' meaning bad or evil and 'astre' meaning 'star'. So when combined, the expression is 'bad or evil star'. In earlier days, a disaster was considered to be the loss due to some unfavourable star. In India, researchers did use the word disaster but it was fused with the group of naturals like flood, earthquake and references were in the form of 'flood disaster' or 'earthquake disaster'.

The word disaster is neither unheard, unfamiliar or difficult to recall, yet when a chronological audit of the relevant research units was taken, the research unit that first used the word arrives only in 1972 in a title which read 'Disastrous Weather Events of 1971', published by *Indian Journal of Meteorology and Geophysics*. The second came in 1978, 'Planning for Disaster Prone Area', in a postgraduate thesis of the School of Planning and Architecture, New Delhi. The first article in a journal that employs the word disaster in reference to India makes an entry in 1981. The article entitled 'From Disaster to Development: Notes from India' was published in the journal *Disasters: The Journal of Disaster Studies, Policy and Management*.

The belated arrival of the word disaster in the research units is also evident from the fact that it was completely absent from the research units of the colonial phase and had a small presence of 3 per cent in the phase of indifference. It is in the 1990s that its usage rose to 12 per cent and thereafter, post 1999, to 19 per cent of the total research units. With 93 per cent of the usage of the word 'disasters' clustered in the post 1990s, its recent walk into the research units cannot be challenged. The inclusion of the word 'management' is more novel. Not a single research unit of the colonial phase utilises this expression. Even in 1947–89, it is able to enter with only one per cent of the total research units. The expression 'disaster management' thrives in the post 1990s and more so in the post 2000 phase. While the word 'management' originally meaning a governing body (of a theatre) has an origin way back in 1598, yet it seems that its popular application to as wide arenas as business, hotel, funds, investments, et cetera is what seems to have bought it with ease as a suffix to disasters.

The above information, leads to the conclusion that the J-curve, along with a distinct change in the language of the research units from 1990 onwards, is an indicator of the recognition that disasters gained. But this is not all. A marked increase in the number of articles published in foreign journals is an index of this larger recognition.

To the Foreign Realm

The British initiated the research on natural disasters in India. While it had a foreign beginning, for years the research remained secluded and sheltered within the confines of publications and outlets within India. Even today, out of the 5017 research units, only 431, that is 9 per cent, are derived from a foreign journal. The share is not only small but also scattered. The crop from no less than 100 foreign journals has provided these 431 units. The winds of recognition to disasters that blew in the 1990s carried the research of disasters to the foreign realm. The largest chunk of 47 per cent of the research units that were published in foreign journals belong to the seven years since 2000. It was during these years since 1990 that 53 foreign journals were added to the list of publications on India (Table 5.11).

Most of these journals had been in the publishing business for many years. The *American Journal of Psychiatry* dates back to 1844, but when the modus of search on the Internet through key words was used, the earliest title on India emerged only for 1996. Though a limit to accessibility of a number of back issues could be a factor responsible for this, yet the picture remained unaltered when entire content pages were scanned in some journals. The *Australian Journal of Emergency Management*, which circulated its first volume in 1986, included a research on India's natural disasters for the first time in 2003. The *Journal of Natural Disaster Science*, Kyoto, Japan was started in 1979. Till the end of 2006, it had carried a total of 249 articles; of these only two featured India's disasters. The first belongs to 1989 and the second to 2002. The *International Journal of Mass Emergencies and Disasters*, which came into production in 1983 and has carried a total of 398 articles till date, contains just two articles on India's natural disasters. The first article came in 1998 and the second in 2004. Within the 177 years of the temporal span of the research units, the sliver of 2004–06 added 40 articles to the foreign list each year.

A similar situation applies to reports with international donors and sponsors. Out of the 67 joint reports of foreign organisations archived in these research units, 59 belonged to 1990–2006. Forging international

Table 5.11 Articles on Disasters of India in Foreign Journals, 1990–2006

Phase	Years	Number of articles	Per cent of total
Awareness	1830–1947	10	2
Indifference	1947–1989	24	29
Recognition	1990–2006	297	69
Total	**171**	**431**	**100**

Based on *4004*.

interests was not the only way of diversifying. Within its own domicile, a number of disciplines, professional and specialists pooled in to write and read on disasters.

Variety in the Domicile

When the resume of 245 authors of chapters in twenty-four edited books that were published post 2000 were checked, the results revealed a spectrum. The medley cut across thirty disciplines and professions. The list consisted of engineers, doctors, geologists, geographers, meteorologists, planners, architects, social workers, journalists, environmentalists, economists, army officials and even a theologist! With the change of guard, once skimpy chapters in edited books and an occasional article in proceedings gave way to exclusive titles on disaster management. The encyclopaedias and the specific titles of atlases on disasters also made their appearance in this phase.

The body of students that rushed to enrol for a course on disaster management came from an equally assorted academic background. The Indira Gandhi National Open University, New Delhi, offers a postgraduate diploma in disaster management. Because of the flexibility of being a distance learning programme, the course attracted 516 candidates in the year 2006–07. When the subject information provided in the application forms of 195 students was put together, it covered the physical sciences, humanities, social sciences, commerce, engineering, medicine, commerce, law, mass communication, military science and public administration (Annexure 10).

The weekend programme of Master of Business Administration in Disaster Management offered by the Guru Gobind Singh Indraprastha University had an uptake of 54 candidates in 2006–07. Since priority was given to people employed with the government, its profile drew members from the Municipal Corporation of Delhi, Labour Commission, Central Vigilance Commission, Indian Council of Agricultural Research, Indian Army, Delhi Jal Board, Indian Railways and the Fire Department. There were also a small number of teachers and those employed with the private sector. In a grade different from the course of two years is the introduction of an adjunct offer to study disaster management along with an undergraduate degree to students at Miranda House, University of Delhi. It was in vogue with students of Geography, Hindi, Political Science and History, among others (Table 5.12).

Table 5.12 Research by Authors from Different Disciplines, 1990–2006

Discipline/Profession	No. of authors	Per cent of total
Engineers	37	16
Geologist	27	12
Technologist[1]	25	11
Government employee	23	10
Social scientist [2]	19	8
Geographer	16	7
Ecologist [3]	12	5
Meteorologist	12	5
Non government organisation	12	5
Journalist	11	5
Doctor	11	5
Others [4]	20	9
Total	**225**[5]	

Based on *chapters from twenty-four edited books published in 1990–2006* [1]*remote sensing,* [2]*sociologist, political scientist, anthropologist, economist, planner* [3]*biologist, agricultural scientist and biotechnologist, marine scientist* [4]*lawyer, architect, business administrator, theologist among others.* [5]*Details for fifteen authors not furnished.*

The recognition given to the subject of disasters is a nation wide phenomenon. The catchment of the students enroled at the Indira Gandhi National Open University carried a representation from all the states of India. While the northern region was the leader drawing nearly half the enrolments, the southern, eastern and western regions sent more than 60 students each. Though less in number, the residential address of students from as far as the Andaman Islands and northeastern states added to the flag of a nationwide representation (Table 5.13).The subject of disasters, its awareness, education and need for management, occupies front rank in India today. The pressing problem, shift in the mood of the administration and change in the importance given to disasters has created an opportune time for the subject. Disasters as a theme have achieved the goal of recognition. Administrators and planners agreed to give it a due place. Even within foreign journals and collaborations, the works on India have created a habitat. India has given more attention to disasters during the first decade of the twenty-first century than it did for the entire twentieth century.

Table 5.13 Catchment of Student Enrolment in the Course on Disaster Management at Indira Gandhi Open University, New Delhi

Region	State/Union Territory	Enrolment	Per cent of total
Northern region	Haryana, Punjab, Chandigarh, Jammu and Kashmir, Uttarakhand, Uttar Pradesh, Delhi	246	48
Western region	Gujarat, Maharashtra, Rajasthan	92	18
Southern region	Pondicherry, Goa, Karnataka, Andhra Pradesh, Tamil Nadu, Kerala	69	13
Eastern region	Orissa, West Bengal, Bihar	60	12
Central region	Jharkhand, Madhya Pradesh	26	5
Northeastern region	Mizoram, Sikkim, Arunachal Pradesh, Manipur, Nagaland, Meghalaya, Assam	22	4
Islands	Andaman and Nicobar	1	n
Total		**516**	**100**

Based on *Admission List of the Post Graduate Diploma in Disaster Management, IGNOU (2005–2006), n = negligible*

Are these activities a foundation or a banner for propaganda? Are these mere words or is there action? Is it a wave that will subside or sustain? Time tests truth. In this case the wait may not have to be long. Another natural event and India would know how prepared, geared and management savvy she has become on the front of disasters. But while the future will tell us if we allowed natural events to turn into disasters, it would be worth our while to take stock of the 4004.

VI

The 4004

What are the 4004 references about? What themes about the disasters do they address? Is there a common paradigm that dominates the thinking in these 4004? What is the appropriate way to comprehend a large and disparate variety? A human construct to simplify and organise complex reality is to devise a classification. But there is the problem of correct identification of both the individual and the group to which they belong. There also are situations of incompletion, misinformation and duplication. It is difficult to decipher details from titles that read like 'solid foundation', 'without a warning' and 'caught in a trap'. Since these are penned as features in magazines, the design of these captions no doubt catch attention, but they also pose a problem in sorting. While these titles are bold yet mysterious, the ones authored by administrators and government officials are cautious and tight-lipped.

Labels like 'Flood Picture in States' or 'Notes and News' or 'The Koyna Earthquake', '1959 October floods of Damodar river', 'Assam earthquake of 1950', 'The Bihar flood story', 'The Indian Earthquake', are mind teasers. Articles in journals that promised too much were also problematic. Take for example, research units that specified that their work was on 'cause, characteristics, impact, response and management', ironically covered the article in six to ten pages of a journal. These, either addressed issues with extreme generality or did not sufficiently meet their claims.

The troubleshooters were also those that occupied ecotones. 'Impact and response to Bhuj earthquake to rural and urban areas of the district of Kutch', 'Genesis of floods in West Bengal and their impact on society' and 'Landslides in the Himalaya: causes, types and management'. Thus, a research unit on cause and management or impact and response precludes perfect sorting. Over ambitious titles failing miserably to meet their purpose, or studies too narrow in their focus added to the woes of classification. Another hurdle was the excessive use of abbreviations, technical terms and terminology. Acronym.com and scientific glossaries helped decode the technicalities and abbreviations but for quite a few problems, either the research or the researcher had to be tracked. Though significant effort was made, not all instances of research fell in neat compartments.

Any classification, therefore, is neither perfect nor neat. All are guided by subjective discretion. A classification begins with drawing boundaries between the features that get along well and ones that stand apart. A multi-pronged approach was adopted to identify the similar and dissimilar within the listed 4004. While the titles were read, interpreted and analysed, a keen eye was kept on the sources of references and the academic profile of the researcher. To double check and weed out doubts, wherever possible, the reference itself was read. In spite of the ploughing and furrowing, it was not easy to pin the theme of the research unit. Only by picking the nuances, sifting the differences, sorting out the intention and shuffling the units in a variety of ways did a thematic classification evolve. With this as the lens, the 4004 research references on disasters came to fall within eight different themes, which are the Basics, Occurrence, Characteristics, Cause, Impact, Response, Forecasting and Management (Table 6.1).

The methodology of attentive sorting generates other schemes of classification. At the outset, the 4004 references could be grouped into ten different naturals. Arranged alphabetically, these are cyclone, drought, earthquake, flood, heat and cold wave, heavy rain, landslide, storm, tsunami and the assorted, the latter contain works on hazard, risk or disaster and other selected synonyms.

Table 6.1 References for each Category

Category	References	Number
Basics	0001–0089	89
Occurrence	0090–0771	682
Characteristics	0772–1060	289
Cause	1061–1653	593
Impact	1654–2450	797
Response	2451–2992	542
Forecasting	2993–3236	244
Management	3237–4004	768

Another classification that emerged was that of the outlets of research. The 4004 come in six different moulds: book, report, thesis, articles in journals, chapters in edited books, proceedings and reports, and features in a magazine (Table 6.2). While these two classifications are important in themselves, there was a need to group the references according to the theme of the research. With this as the lens, the 4004 research on disasters came to fall within eight different themes. But there was a need to go beyond. After all it does become important to capture the spirit of the research. What in the disaster is being studied? What was being addressed in the occurrence? What was being forecasted and predicted? When the 4004 research were scrutinised for their philosophical underpinnings, they fell into three categories – those that dealt with only the 'natural' aspect, those exclusive to the 'human' dimension and ecological ones that were inclusive of both natural and human components in a disaster (Table 6.3). Therefore, when viewed as pieces in a kaleidoscope, a slight tilt and these 4004 references present a different picture. The pages to follow unfold these nuances.

The 4004 references have been classified in three principal ways. First by theme, second by type of natural disaster and third the spirit of the research. The classification of theme carries eight divisions – basics, occurrence, characteristics, cause, impact, response, forecasting and disaster management. Each of these is further divided into the ten types

Table 6.2 Research References for Type of Outlet

Theme	Book	Report	Thesis	Article[1]	Chapter[2]	Feature[3]	Total
Basics	3	25	0	38	14	9	89
Occurrence	9	59	24	436	123	31	682
Characteristics	6	27	19	196	33	8	289
Cause	10	59	40	313	100	71	593
Impact	29	38	33	306	108	283	797
Response	20	53	30	167	69	203	542
Forecasting	4	18	7	125	54	36	244
Management	53	148	49	257	223	38	768
Total	**134**	**427**	**202**	**1838**	**724**	**679**	**4004**

[1] *in a journal;* [2] *in a book/report/proceedings;* [3] *in a magazine.*

Table 6.3 Dominance of Natural over Human in Research References

Phase	Natural	Human	Ecological	Total
1830–1947	121	14	0	135
1948–1989	704	294	57	1055
1990–2006	1244	1394	176	2814
Total	**2069**	**170**	**233**	**4004**

Figure 6.1 **Structure of the 4004**

Structure of 4004

THEME

Classification

First level

Basics
Occurrence
Characteristics
Cause
Impact
Response
Forecasting
Management

Theme: Type

Second level

Cyclone
Drought
Earthquake
Flood
Heat and cold waves
Heavy rain
Landslide
Storm
Tsunami
Assorted

Arrangement of each 4004

Final level

Serial No.	Year	Paradigm	Author	Title	Type	Place: publisher
0001 ↓ *4004*	1830 ↓ 2006	■ Natural ● Human ▲ Ecological	surname arranged alphabetically	*in italics*	Book Report Thesis Journal Magazine	City: Government Institutional Private

of natural disasters, that is cyclone, drought, earthquake, flood, heavy rain, heat and cold wave, landslide, storm, tsunami and the group of assorted. A three-fold arrangement, namely that of works that were natural, human and ecological, defined the spirit of the work. While the first two have been arranged as heading and sub-heading, the selection of the spirit of the research runs as a parallel indent to each of the 4004. The form of citation is largely self-explanatory. For all three types of classification, the year is given first, followed by the name of the author, the title of the work, along with the edition and number of volumes, when such information is called for. This structure is not imposed but has evolved from a research on 4004.

Basics

Basics are the fundamentals and foundations of any intellectual pursuit. References that explore and debate the concepts and principles of research are included within the 'basics'. Reviews of the history of science, critiques on the status of research and commentaries on the methodology of research are also part and parcel of the basics. Since dictionaries, glossaries, bibliographies and encyclopaedias carry a universal trademark they too have been bracketed within this group. Data plays a supportive yet essential role to research. Documents, catalogues, statistical handbooks and information stored in atlases and maps too are part of the spine of the basics. Among the 4004 research references, the basics are miniscule 88 and constitute only 3 per cent of the total. (Tables 6.4, 6.5 and 6.6)

Table 6.4 References for each Category

Category	References	Number
Conceptual	0001–0037	37
Bibliography and Encyclopedia	0038–0043	6
Catalogues and Data	0044–0073	30
Atlas and Maps	0074–0089	16

Table 6.5 Research References on Basics

Category	Book	Report	Thesis	Article[1]	Chapter[2]	Feature[3]	Total
Concepts	0	5	0	18	8	6	37
Bibliography and Encyclopedia	3	1	0	2	0	0	6
Catalogues and Data	0	11	0	12	4	3	30
Atlas and Maps	0	8	0	6	2	0	16
Total	**3**	**25**	**0**	**38**	**14**	**9**	**89**

[1] *in a journal;* [2] *in a book/report/proceedings;* [3] *in a magazine.*

Table 6.6 Dominance of Natural over Human in Basic's Research References

Phase	Natural	Human	Ecological	Total
1830–1947	6	0	0	6
1948–1989	25	0	0	25
1990–2006	43	15	0	58
Total	**74**	**15**	**0**	**89**

Basics: Conceptual

0001 ■ 1959 Tandon, A.N. *Development of Seismological Organisation in India.* Roorkee: Proceedings of Earthquake Engineering Seminar.

0002 ■ 1968 Tandon, A.N. and S.N. Chatterjee. Seismicity studies in India. *Indian Journal of Meteorology and Geophysics*, 19: 273–280.

0003 ■ 1979 Krishnan, A. Definition of drought and factors relevant to specification. In *Proceedings of the International Symposium on Hydrological Aspects of Droughts.* Delhi: Indian Institute of Technology.

0004 ■ 1984 Schware, R. Flood information systems: Needs and improvements in Eastern India. *Environmental Management*, 8 (1): 55–65.

0005 ■ 1991 Singh, S. Environmental hazards and disasters: Classification and management. *National Geographer,* 26 (1): 1–16.

0006 ● 1992 Sinha, D.K. Natural disasters: Scientific and cultural moorings. *Science and Culture*, 58 (5): 81–82.

0007 ■ 1992 Tandon, A.N. Seismology in India: An overview upto 1970. *Current Science,* 62 (1–2): 8–16.

0008 ■ 1993 Lakshmipathi, T. Signals ignored: Quake divides earth scientists. *Frontline,* (November 5): 122.

0009 ■ 1993 Martin, M. and K. Cherail. Quake causes rift among seismologists. *Down to Earth,* (October 31).

0010 ● 1994 D' Souza, R. Inundated by colonial misconceptions. *Down to Earth,* 2 (19): 51–52.

0011 ■ 1994 International Federation of Red Cross and Red Crescent Society. India: Earthquake myths and realities. In *World Disasters Report.* Geneva. 119–124.

0012 ■ 1994 Krishna, J. General misconceptions about earthquakes. *Indian Concrete Journal,* 68 (11).

0013 ■ 1997 Kale, V.S. Flood studies in India: A brief review. *Journal of Geological Society of India,* 49(4): 359–370.

0014 ● 1997 Nair, R.B. Acts of agency and acts of God: Discourses of disaster in a post-colonial society. *Economic and Political Weekly,* 32 (11): 549–50.

0015 ● 1998 Special Correspondent. Twisted tale: Natural disaster. *Economic and Political Weekly,* 33 (13): 681.

0016 ■ 1999 Department of Science and Technology. *Earthquake Research in India.* New Delhi: Government of India.

0017 ■ 1999 Sukhija, B.S. Earthquake recurrence: State-of-the-art and directions for future. *Current Science,* 77 (8): 1013–1014.

0018 ● 1999 Suresh, V. and K. Singh. The International Decade for Natural Disaster Reduction 1990–2000: A path for nations to tread on. *Shelter,* special issue, October, 16–17.

0019 ■ 2000 Bilham, R. and V.K. Gaur. The geodetic contributions to the study of Indian seismotectonics in India. *Current Science,* 79 (9): 1259–1269.

0020 ● 2000 Bose, A. Are natural disasters man-made? *Economic and Political Weekly,* 35 (43 and 44): 3793–3794.

0021 ● 2000 Gill, Y.S. The science of disasters. In S. Parasuraman and P.V. Unnikrishnan (eds). *India Disasters Report: Towards a Policy Initiative.* New Delhi: Oxford University Press. 85–87.

0022 ● 2000 Sharma, V.K. *et al. India: International Decade of Natural Disaster Reduction and Beyond.* New Delhi: Indian Institute of Public Administration.

0023 ■ 2001 Balaram, P. Earth science, earthquakes and aftershocks. *Current Science,* 80 (3): 317–318.

0024 ■ 2001 Jayaraman, K.S. India promises more earthquake research. *Nature*, 409 (February 22): 968.

0025 • 2001 Morrison, K.D. Naturalizing disaster: From drought to famine in South India. In G. Bawden and R.M. Reycraft (eds). *Environmental Disaster and the Archaeology of Human Response.* Albuqerque: Maxwell Museum of Anthropology. 21–33.

0026 ■ 2001 Rajendran, C.P. and K. Rajendran. Earthquake research in India. In Ganesh *et al.* (eds). *Proceedings of the Workshop on Popular Lectures On Facts and Myths of Earthquake, Indian Context.* Tiruchirapalli: Bharathidasan University. 35–39.

0027 ■ 2001 Sharan, R.B. Need for researches on hazard zonation of landslide and earthquake. In *Proceedings of the Role of Earth Sciences in Integrated Development and Related Societal Issues.* Special Publication No. 65 (2). Kolkata: Geological Survey of India. 39–45.

0028 ■ 2002 Shewale, M.P. Drought studies in the Indian Meteorological Department. In R. Nagarajan (ed). *Proceedings of the National Workshop on Recent Trends in Drought Assessment Monitoring and Management.* Mumbai: Centre of Studies in Resources Engineering, Indian Institute of Technology. 16–23.

0029 ■ 2003 Gupta, H.K. and G.D. Gupta (eds). *Earthquake Studies in Peninsular India Since 1993.* Bangalore: Memoirs of the Geological Society of India, No. 54.

0030 ■ 2003 Mahadevan, T.N. Earthquake studies in Peninsular shield since 1993. *Current Science,* 85 (10): 1494–1495.

0031 • 2003 Samal, K.C. and M. Shiblal. Vulnerability reduction for sustainable development: The context of natural disaster. In *Orissa Human Development Report 2003.* Bhubaneswar: National Centre for Development Studies.

0032 • 2005 Ahluwalia, A.D. The need for tsunami museum in South Asia. *Journal of Geological Society of India*, 65 (4): 505.

0033 • 2005 Kapur, A. Insensitive India: Attitudes towards disaster prevention and management. *Economic and Political Weekly*, 40 (42): 4551–4560.

0034 ■ 2005 Narain, S. Science after the tsunami. *Down to Earth,* 13 (17): 5.

0035 ■ 2005 Warr, C.E. *et al.* The Asian tsunami, academics and academic research. *Singapore Journal of Tropical Geography*, 26 (1): 244–248.

0036 ■ 2006 Gupta, G.D. and H.N. Srivastava. Improved understanding about Indian earthquake hazard. In H.N. Srivastava and G.D. Gupta (eds). *Management of Natural Disasters in Developing Countries.* New Delhi: Center for Science and Technology of the Non-Alignment Movement.

0037 • 2006 Rautela, P. Redefining disaster: Need for managing accidents as disasters. *Disaster Prevention and Management*, 15 (5): 799–809.

Basics: Bibliographies and Encyclopaedias

0038 ■ 1973 Chakraborty, A.R. Bibliography of earthquakes in India. *Indian Minerals*, 27 (1): 127–144.

0039 ■ 1978 Special Correspondent. Bibliography on landslides in India. *Indian Minerals*, 32 (2).

0040 ■ 1988 Gangopadhyay, K.K. *Bibliography on Earthquakes in India.* Miscellaneous Publication, No. 60. Kolkata: Geological Survey of India.

0041 • 1998 Sinha, P.C. (ed). *Encyclopaedia of Disaster Management* (Volumes I–X). New Delhi: Anmol Publications.

0042 ■ 2006 Bose, C.P.B. (ed) *Modern Encyclopaedia of Disaster and Hazard Management* (Volumes I–V). New Delhi: Rajat Publications.

0043 • 2006 Goel, S.L. (ed). *Encyclopaedia of Disaster Management* (Volumes I–III) New Delhi: Deep and Deep Publications Pvt. Ltd.

Basics: Catalogues and Data

0044 ■ 1845 Smith, R.B. Register of Indian and Asiatic earthquakes for the year 1843. *Journal of the Asiatic Society of Bengal,* 14: 604–622.

0045 ■ 1877 Blanford, H.F. Catalogue of the recorded cyclones in Bay of Bengal up to the end of 1876. *Proceedings of the Asiatic Society of Bengal*, 46 (2): 332–338.

0046 ■ 1881 Special Correspondent. List of earthquakes recorded in Assam during years 1879 and 1880. *Proceedings of the Asiatic Society of Bengal*, 50 (2): 61–67.

0047 ■ 1883 Oldham, T. *A Catalogue of Indian Earthquakes from the Earliest Times to the End of 1869 AD.* Memoirs of the Geological Survey of India, No. 19 (3). Kolkata.

0048 ■ 1934 Ferolli, D. Seismological records of the recent earthquake. *Current Science*, 2 (8): 296.

0049 ■ 1936 Davison, C. On the earthquakes recorded by the pendulum seismograph at Shillong (Assam) from 1903 to 1931. In *Records of the Geological Survey of India*, 69 (2): 184–202.

0050 ■ 1964 Johnson, V.P. and S.R. Raghavan. Earthquake records at Hyderabad. *Journal of Indian Geophysical Union,* 1 (1): 57–60.

0051 ■ 1971 Blanford, H.F. *Catalogue of Large Floods in India, 1971.* Memoirs of the Indian Meteorological Department. New Delhi: Central Water Commission.

0052 ■ 1972 Flierl, G.R. *et al.* Deadly surge in the Bay of Bengal: Dynamics and storm tide tables. *Nature*, 239 (September 22): 213–215.

0053 ■ 1975 Dhar, O.N. and B.K. Bhattacharya. A catalogue of the highest ever recorded floods in Indian rivers: A preliminary appraisal. *Indian Journal of Power and River Valley Development,* 25 (12): 402–409.

0054 ■ 1976 United Nations Educational, Scientific and Cultural Organization. *World Catalogue of Very Large Floods.* Paris.

0055 ■ 1979 Henderson, F. *Disastrous Storms in the Bay of Bengal: A Listing of Major Cyclones by Month through 1979.* Virginia: Evaluation Technologies Inc.

0056 ■ 1981 Dhar, O.N. *et al.* A catalogue of major and devastating floods of India and their space-time distribution. In *Proceedings of the International Conference on Flood Disaster.* New Delhi: Indian National Science Academy.

0057 ■ 1983 Guha, S.K. and P.C. Basu. *Catalogue of Earthquakes (M 3.0 and above) in Peninsular India.* Mumbai: Atomic Energy Regulatory Board.

0058 ■ 1983 Bapat, A. *et al. Catalogue of Earthquakes in India and Neighbourhood from Historical Period upto 1979.* Roorkee: India Society of Earthquake Technology, Indian Institute of Technology.

0059 ■ 1983 Srivastava, H.N. and K. Ramachandran. New catalogue of the earthquakes in Peninsular India during 1839–1900. *Mausam*, 34 (3): 351–356.

0060 ■ 1985 Central Water Commission. *Statewise Data on Flood Affected Areas and Damages (1953–84).* New Delhi: Government of India.

0061 ■ 1987 Department of Relief and Rehabilitation. *Flood-Drought Statistical Handbook.* Patna: Government of Bihar.

0062 ■ 1989 Central Water Commission. *Water and Related Statistics.* New Delhi: Government of India.

0063 ■ 1989 National Remote Sensing Agency. *Drought Bulletins: 1989 Onwards for Various States.* Hyderabad.

0064 ■ 1991 Murthy, T.S. and M. Rafiq. A tentative list of tsunamis in the marginal seas of the Northern Indian Ocean. *Natural Hazard Journal,* 4 (1): 81–83.

0065 ■ 1991 Ramachandran, K. *et al.* New catalogue of the felt Indian earthquakes during 1901–1971. *Mausam,* 42 (2): 171–182.

0066 ■ 1992 Ramamoorthy, A.S. *Data Utilization in Flood Studies: Natural Resource Management, New Perspectives.* Bangalore: National Natural Resources Management System.

0067 ■ 1996 Lakhera, R.C. and P.K. Champati Ray. Satellite census of landslides prone areas in Himalaya. *Himalayan Geology,* 17: 209–216.

0068 ■ 1998 Prasad, K.A scientific approach to collection of data on flood damages. In B.N. Verma (ed). *Flood Devastation and Agricultural Development in Eastern India.* Delhi: B.R. Publishing Corporation. 3–12.

0069 ■ 2000 Rajendran, C. P. Using geological data for earthquake studies: A perspective from Peninsular India. *Current Science*, 79 (9): 1251–1258.

0070 ■ 2001 Special Correspondent. Flood plains of India. *Down to Earth,* 10 (11): 56.

0071 ■ 2002 Das, B. Untapped data. *Down to Earth,* 11 (8): 47–48.

0072 ■ 2006 Murty, T. S. *et al.* Historical tsunami data for the Indian Ocean, with special reference to India. In *Proceedings of the Brain Storming Session on the Great Tsunami Event of 26 December 2004*. New Delhi: Department of Science and Technology, Government of India. 33–40.

0073 ■ 2006 Rastogi, B. K. and R. K. Jaiswal. A catalogue of tsunamis in the Indian Ocean. *Science of Tsunami Hazard*, 25 (3).

Basics: Atlas and Maps

0074 ■ 1972 Kalia, K. L. *et al.* Quantitative seismic map of India. *Bulletin of the Seismological Society of America,* 62 (5): 1119–1132.

0075 ■ 1978 Kalia, K. L. *et al.* Atlas of isoseismal maps of major earthquakes in India. *Geophysics Research Bulletin,* 16: 235–269.

0076 ■ 1979 Indian Meteorological Department. *Tracks of Storms and Depressions in the Bay of Bengal and Arabian Sea.* Delhi: Indian Meteorological Department.

0077 ■ 1984 Khattri, K. N. *et al.* A seismic hazard map of India and adjacent areas. *Tectonophysics*, 108 (1–2): 93–134.

0078 ■ 1985 Central Water Commission. *Flood Atlas of India.* Delhi: Government of India.

0079 ■ 1991 National Atlas and Thematic Mapping Organisation. *India: Natural Hazards (Map).* Kolkata.

0080 ■ 1992 Krishna, J. Seismic zoning maps of India. *Current Science*, 62 (1–2): 17–23.

0081 ■ 1997 Arya, A. S. *et al. Vulnerability Atlas of India.* New Delhi: Ministry of Urban Affairs and Employment, Government of India.

0082 • 1998 Arya, A. S. *et al. Guidelines for Improving Hazard Resistant Construction of Buildings and Land use Zoning for Earthquake, Windstorm and Flood Hazards.* New Delhi: Ministry of Urban Affairs and Employment, Government of India. National Government Publication.

0083 ■ 1999 Bhatia, S. C. *et al.* A probabilistic seismic hazard map of India and adjoining regions. *Annali di Geofisica,* 42 (6): 1153–1164.

0084 ■ 1999 Kumar, M. R. and S. C. Bhatia. A new seismic hazard map for the Indian plate region under the global seismic hazard assessment programme. *Current Science,* 77 (3): 447–454.

0085 ■ 2000 Dasgupta, S. *et al. Seismotectonic Atlas of India and its Environs.* Special Publication No. 59. Kolkata: Geological Survey of India.

0086 ■ 2001 National Remote Sensing Agency. *Atlas of Landslide Hazard Zonation Mapping in the Himalayas of Uttaranchal and Uttar Pradesh States using Remote Sensing and Geographic Information Systems Techniques.* Hyderabad.

0087 ■ 2003 Kumar, R. and S.C. Bhatia. A new seismic hazard map for the Indian plate region under the global seismic hazard assessment programme. In H.K. Gupta (ed). *Disaster Management.* Hyderabad: University Press (India) Private Limited. 85–99.

0088 ■ 2003 Pandey, Y. *et al.* Preliminary seismic microzonation map of Delhi. In *Proceedings of the Map India 2003*, GIS Development, Survey of India and National Consortium on Remote Sensing in Transportation.

0089 ■ 2006 Nath, S.K. *Seismic Hazard and Microzonation Atlas of the Sikkim Himalaya.* New Delhi: Seismology Division, Department of Science and Technology, Government of India.

Occurrence

Tracing the roots to the Latin word *occurrere* an occurrence could be an episode, event or a happening. The instance of occurring has both a temporal and spatial dimension. In temporality, an occurrence could be annual, seasonal, cyclic or sporadic and random. The spatial patterns could be dispersed or clustered. If an occurrence is frequent at a place, it acquires a geographic indication and provides an additional expressive name to the place under reference.

The references that belong to the group classified as 'occurrence' orbit around the question of when and where. In the context of disasters, a research that explores the long term trend of drought, or the sequence of pre-and after-shocks of an earthquake or those that build the changing decadal profile of floods belong to occurrences. References seeped into the identification of the regions are also typified within this group. After all, research that demarcates the areas prone to floods or drought does carry a spatial expression. Similarly, research that describes the picture of disaster represents a strong element of the occurrence. Amidst the 4004 research references, 681 can be segregated into the class of 'occurrences'. This is 17 per cent of the total references (Tables 6.7, 6.8 and 6.9).

Table 6.7 References in each Category

'Natural' Disasters	**References**	**Number**
Cyclone	0090–0177	88
Drought	0178–0232	55
Earthquake	0233–0478	246
Floods	0479–0599	121
Heat and Cold Wave	0600–0601	2
Heavy Rain	0602–0613	12
Landslide	0614–0727	114
Storm	0728–0739	12
Tsunami	0740–0750	11
Assorted	0751–0771	21

Table 6.8 Research References of Occurrence

Natural disaster	**Book**	**Report**	**Thesis**	**Article**[1]	**Chapter**[2]	**Feature**[3]	**Total**
Cyclone	1	7	1	76	1	2	88
Drought	2	5	0	27	17	4	55
Earthquake	2	17	13	170	27	17	246
Floods	1	9	5	87	17	2	121
Heat and Cold Wave	0	0	0	1	1	0	2
Heavy Rains	0	0	0	9	1	2	12
Landslide	0	18	2	42	51	1	114
Storm	0	0	1	10	1	0	12
Tsunami	0	0	0	8	1	2	11
General Hazard	3	3	2	6	6	1	21
Total	**9**	**59**	**24**	**436**	**123**	**31**	**682**

[1] *in a journal;* [2] *in a book/report/proceedings;* [3] *in a magazine.*

Table 6.9 Dominance of Natural over Human in Research References on Occurrence

Phase	**Natural**	**Human**	**Ecological**	**Total**
1830–1947	62	0	0	62
1948–1989	252	0	0	252
1990–2006	367	1	0	368
Total	**681**	**1**	**0**	**682**

Occurrence: Cyclone

0090 ■ 1934 Banerji, S.K. The South Indian cyclone of 13–16 December 1933. *Current Science*, 2 (9): 335–338.

0091 ■ 1943 Sen, S.N. Monsoon cyclones. *Science and Culture*, 9 (2): 90.

0092 ■ 1950 Special Correspondent. Mumbai cyclone of November 1948. *Indian Journal of Meteorology and Geophysics*, 1 (1): 59–72.

0093 ■ 1952 Special Correspondent. Masulipatnam cyclone of October 1949. *Indian Journal of Meteorology and Geophysics*, 1 (2): 145–159.

0094 ■ 1953 Special Correspondent. Severe Nagapattinam cyclone of 30 November 1952. *Indian Journal of Meteorology and Geophysics*, 4 (4): 87–97.

0095 ■ 1954 Desai, B.N. and Y.P. Rao. The severe Nagapattinam cyclone of 30 November 1952. *Indian Journal of Meteorology and Geophysics*, 5 (2): 200–202.

0096 ■ 1959 De, A.C. and S.N. Sen. Cyclonic storm of 13–14 September 1958 in the Bay of Bengal: A radar study. *Indian Journal of Meteorology and Geophysics*, 10 (4): 393–398.

0097 ■ 1961 Special Correspondent. Severe cyclones in the Bay of Bengal in May 1961. *Indian Journal of Meteorology and Geophysics*, 12 (3): 494–510.

0098 ■ 1962 Sarma, V.V. and V.C. Bedkar. Midjet cyclone over Chennai, 20 November 1960. *Indian Journal of Meteorology and Geophysics*, 13 (4): 472–480.

0099 ■ 1963 Special Correspondent. Severe cyclone in the Bay of Bengal, May 1963. *Indian Journal of Meteorology and Geophysics*, 14 (4): 415–521.

0100 ■ 1967 Rama Rao, M. *et al.* Satellite study of the cyclonic storm activity in the Indian seas between 24 September and 5 October 1966. *Indian Journal of Meteorology and Geophysics*, 18 (4): 485–490.

0101 ■ 1971 Special Correspondent. Severe cyclonic storms in the Bay of Bengal during October and November 1970. *Indian Journal of Meteorology and Geophysics*, 22 (1): 123–127.

0102 ■ 1972 Special Correspondent. Cyclonic storm. *Indian Journal of Meteorology and Geophysics*, 23 (1): 103–109.

0103 ■ 1973 Das, P.K. *et al.* Cyclones and depressions of 1972 Bay of Bengal and Arabian Sea. *Indian Journal of Meteorology and Geophysics*, 24 (4): 327–344.

0104 ■ 1973 Raghavendran, V.K. A statistical analysis of number of tropical storms and depressions in the Bay of Bengal during 1890–1969. *Indian Journal of Meteorology, Hydrology, and Geophysics*, 24 (1): 125–130.

0105 ■ 1973 Special Correspondent. Severe cyclonic storm in the Bay of Bengal. *Indian Journal of Meteorology and Geophysics*, 24 (2): 189–194.

0106 ■ 1974 Alexander, G. *et al.* Cyclones and depressions of 1973: Bay of Bengal and Arabian sea. *Indian Journal of Meteorology and Geophysics*, 25 (3–4): 347–362.

0107 ■ 1974 Das, P.K. Bay of Bengal. *Quarterly Journal of Royal Meteorological Society*, 100: 437–449.

0108 ■ 1975 Datta, R.K and R.N. Gupta. Tracking tropical storms in the Bay of Bengal in calculation of probabilities of its striking east coast by storm analogue technique. *Indian Journal of Meteorology, Hydrology, and Geophysics*, 26 (4): 345–348.

0109 ■ 1975 Mishra, D.K. *et al.* A satellite study of intensities of cyclonic storms in the Bay of Bengal. *Indian Journal of Meteorology, Hydrology, and Geophysics*, 26 (4): 455–464.

0110 ■ 1976 Alexander, G. *et al.* Cyclones and depressions of 1974: Bay of Bengal and Arabian Sea. *Indian Journal of Meteorology, Hydrology, and Geophysics*, 27 (2): 113–126.

0111 ■ 1976 Bhattacharji, P. *et al.* Cyclonic storm of 14–15 August 1974 in the Bay of Bengal: A radar study. *Indian Journal of Meteorology, Hydrology, and Geophysics*, 27 (1): 9–14.

0112 ■ 1976 Special Correspondent. Arabian Sea cyclone of October 1975. *Indian Journal of Meteorology, Hydrology, and Geophysics*, 27 (1): 220–224.

0113 ■ 1977 Alexander, G. *et al.* Cyclones and depressions of 1975 Bay of Bengal and Arabian Sea. *Indian Journal of Meteorology, Hydrology, and Geophysics*, 28 (1): 3–20.

0114 ■ 1977 Government of Andhra Pradesh. *Statement on Cyclone and Tidal Wave on 19 November*. Hyderabad.

0115 ■ 1978 Pant, P.S. *et al.* Cyclones and depressions in the Indian seas in 1976. *Indian Journal of Meteorology, Hydrology, and Geophysics*, 28: 613–628.

0116 ■ 1979 Raghavan, S. *et al.* Radar synoptic study of the Nagapattinam cyclone of 12 November 1977. *Mausam,* 30 (1): 21–30.

0117 ■ 1980 Krishna Rao, P. Life history of the 12 May 1979 tropical cyclone as seen by advanced television infrared observation satellite. *Mausam*, 31 (1): 31–44.

0118 ■ 1980 Mooley, D.A. Severe cyclone storms in the Bay of Bengal 1877–1977. *Monthly Weather Review*, 108 (10): 1647–1655.

0119 ■ 1980 Pant, P.S. *et al.* Cyclones and depressions over the Indian seas in 1977. *Mausam,* 31 (3): 337–356.

0120 ■ 1980 Srinivasan, V. *et al.* Cyclones and depressions in the Indian seas in 1978. *Mausam,* 31 (4): 495–506.

0121 ■ 1981 Mukherjee, A.K. *et al.* Cyclones and depressions over Indian seas in 1979. *Mausam,* 32 (2): 115–126.

0122 ■ 1982 Ghosh, A. *et al.* A tornado over Orissa, April 1978. *Mausam,* 33 (2): 235–240.

0123 ■ 1982 Mukharjee, A.K. *et al.* Cyclones and depressions over Indian seas in 1980. *Mausam,* 33 (1): 3–12.

0124 ■ 1982 Singh, R. and C.P. Thorat. Depressions and cyclonic storms of the Indian seas: A climatological analysis. *National Geographer,* 17 (2): 169–180.

0125 ■ 1983 Fernandes, A.A. *et al. Storm Surges along the East Coast of India.* Dona Paula, Goa: National Institute of Oceanography.

0126 ■ 1983 Mooley, D.A. *et al.* A study of cyclonic storm incidence on the different sections on the coast around Bay of Bengal. *Mausam,* 34 (2): 139–152.

0127 ■ 1983 Mukherjee, A.K. *et al.* Cyclones and depressions over Indian seas in 1981. *Mausam,* 34 (1): 1–8.

0128 ■ 1984 Mooley, D.A. and C.M. Mohile. Cyclonic storms of the Arabian Sea, 1877–1980. *Mausam,* 35 (2): 127–134.

0129 ■ 1984 Subbaramayya, I. and S.R.M. Rao. Frequency of Bay of Bengal cyclones in the post-monsoon season. *Monthly Weather Review,* 112 (8): 1640–1642.

0130 ■ 1984 Veeraraghavan, K. *et al.* Tracking of a cyclone over Arabian sea by enhanced satellite pictures. *Mausam,* 35 (1): 67–70.

0131 ■ 1985 Ramasastry, A.A. *et al.* Cyclones and depressions over the Indian seas in 1983. *Mausam,* 36 (1): 1–6.

0132 ■ 1986 Murty, T.S. *et al.* The storm surge problem in the Bay of Bengal. *Progress in Oceanography,* 16 (4): 195–233.

0133 ■ 1986 Ramasastry, A.A. *et al.* Cyclones and depressions over the Indian seas in 1984. *Mausam,* 37 (1): 1–8.

0134 ■ 1988 Das, N. *et al.* Cyclones and depressions over Indian seas and the Indian sub-continent in 1986. *Mausam,* 39 (1): 1–8.

0135 ■ 1988 Sharma, G.S. and A.S. Murthy. Storm surges along the East Coast of India. In M.I. El–Sabh and T.S. Murty (eds). *Natural and Man Made Hazards, Proceedings of the International Symposium.* Holland: D. Riedel Publisher. 257–278.

0136 ■ 1989 Das, N. *et al.* Cyclones and depressions over the Indian seas and the Indian sub-continent in 1987. *Mausam,* 40 (1): 1–12.

0137 ■ 1989 Saha, A.K. Cyclones over India. *Science and Culture,* 55 (8): 265–299.

0138 ■ 1989 Subbra Reddy, E.V. Synoptic study of the October 1982 Bay cyclone. *Mausam,* 40 (3): 263–266.

0139 ■ 1990 Indian Meteorological Department. Important events: Tropical cyclone 4–9 May 1990. *Mausam,* 41: 93.

0140 ■ 1991 Gupta, G.R. *et al.* Cyclones and depressions over the Indian seas during 1989. *Mausam,* 42 (1): 1–16.

0141 ■ 1991 Gupta, G.R. *et al.* Cyclones and depressions over the Indian seas and neighbourhood during 1990. *Mausam,* 42 (3): 227–240.

0142 ■ 1991 Indian Meteorological Department. Important events: Tropical cyclone 14–18 December 1990. *Mausam,* 42 (1).

0143 ■ 1991 Mc Pherson, K. (ed). *Which Way Does the Wind Blow in a Cyclone? Long Term Trends in India.* Perth: The Indian Ocean Centre for Peace Studies.

0144 ■ 1992 De, U.S. and D.S. Desai. Cyclones and depressions over Indian seas and neighbourhood during 1991. *Mausam,* 43 (3): 231–238.

0145 ■ 1993 De, U.S. and D.S. Desai. Cyclones and depression over the Indian seas and neighbourhood during 1992. *Mausam,* 44 (3): 225–230.

0146 ■ 1993 Pisheroty, P.R. *Tropical Cyclone.* Mumbai: Bhartiya Vidhya Bhawan.

0147 ■ 1993 Regional Specialised Meteorological Centre on Tropical Cyclones. *Report on Cyclonic Disturbances (Depressions and Tropical Cyclones) over North Indian Ocean in 1992.* New Delhi: Indian Meteorological Department.

0148 ■ 1994 De, U.S. and D.S. Desai. Cyclone and depressions over Indian seas and neighbourhood during 1993. *Mausam*, 45 (3): 199–204.

0149 ■ 1994 Mohanty, U.C. Tropical cyclone in the Bay of Bengal and deterministic methods of projections of their trajectories. *Sadhana: Academy Proceedings in Engineering Sciences,* 19 (4): 567–582.

0150 ■ 1994 Raj, Y. E. A. and S. M. Jamadar. A severe cyclonic storm risk index for coastal stations of Indian subcontinent. *Vayumandal,* 24 (1–2): 20–25.

0151 ■ 1995 De, U.S. *et al.* Cyclones and depressions over the Indian seas and neighbourhood during 1994. *Mausam*, 46 (3): 227–234.

0152 ■ 1995 Gray, W.M. *Tropical Cyclones.* Geneva: World Meteorological Organisation.

0153 ■ 1995 Suresh, R. *et al.* Bay of Bengal cyclone 14 December 1993. *Vayumandal*, 25 (3–4): 68–72.

0154 ■ 1996 De, U.S. *et al.* Cyclones and depressions over the Indian seas and neighbourhood during 1995. *Mausam*, 47 (3): 221–228.

0155 ■ 1997 Dikshit, S.K. *et al.* Cyclone and depressions over North India during 1996. *Mausam*, 48 (3): 343–350.

0156 ■ 1998 Dikshit, S.K. *et al.* Cyclone and depressions over Indian seas and neighbourhood during 1997. *Mausam,* 49 (3): 279–284.

0157 ■ 1998 Singh, O.P. *et al.* Bay of Bengal cyclone cluster of November 1992. *Mausam,* 49 (2): 173–176.

0158 ■ 1999 Gupta, M.C. *et al. Orissa Super Cyclone 1999.* New Delhi: Indian Institute of Public Administration.

0159 ■ 1999 Thapliyal, V. *et al.* Cyclones and depressions over the North Indian Ocean during 1998. *Mausam,* 50 (3): 233–242.

0160 ■ 2000 Editorial. Cyclone in Tamil Nadu. *Down to Earth,* 9 (15): 11.

0161 ■ 2000 Gupta, A. *Tropical Cyclones in the Indian Seas: Observations and Predictions.* Delhi: Indian Institute of Technology (Unpublished Ph.D thesis).

0162 ■ 2001 Agarwal, A. Cyclones in India. *Down to Earth,* 10 (20 and 23): 56.

0163 ■ 2001 Patel, I.P. and G. Krishnakumar. Cyclone storms and depressions affecting Orissa state. *Mausam,* 52 (2): 424–426.

0164 ■ 2001 Patwardhan, S.K. and H.N. Bhalme. A study of cyclonic disturbances over India and the adjacent ocean. *International Journal of Climatology*, 21 (4): 527–534.

0165 ■ 2001 Rao, D.V.B. *et al.* Trends and fluctuations of the cyclonic systems over North Indian Ocean. *Mausam,* 52 (1): 37–46.

0166 ■ 2001 Singh, O.P. Long term trends and frequency of monsoonal cyclonic disturbances over North Indian Ocean. *Mausam,* 52 (4): 655–656.

0167 ■ 2001 Thapliyal, V. Cyclone and depression over North Indian Ocean during 2000. *Mausam*, 52 (3): 455–462.

0168 ■ 2002 Thapliyal, V. *et al.* Cyclone and depressions over North Indian Ocean during 2001. *Mausam,* 53 (2): 256–270.

0169 ■ 2003 Alam, M.M. *et al.* Frequency of Bay of Bengal cyclonic storms and depressions crossing different coastal zones. *International Journal of Climatology*, 23 (9): 1119–1125.

0170 ■ 2003 Kane, R.P. Periodicities in the frequency of Indian monsoonal cyclonic disturbances and storms. *Mausam,* 54 (2): 443–446.

0171 ■ 2003 Thapliyal, V. and A.B. Majumdar. Cyclones and depressions over North Indian Ocean during 2002. *Mausam,* 54 (3): 579–584.

0172 ■ 2004 Banerjee, S.K. and S.K. Kundu. A case study of 12 November 2002 cyclonic storm in Bay of Bengal using doppler weather radar. *Mausam,* 55 (3): 671–80.

0173 ■ 2004 Jayaanthi, N. *et al.* Cyclones and depressions over North Indian Ocean during 2003. *Mausam,* 55 (3): 389–396.

0174 ■ 2004 Narkhedkar, S.G. *et al.* Orissa Super Cyclone of October 1999 as revealed by Indian remote sensing P4 satellite data. *Indian Journal of Radio and Space Physics*, 33 (1): 35–42.

0175 ■ 2004 Sadhuram, Y. *et al.* Seasonal variability of cyclone heat potential in the Bay of Bengal. *Natural Hazard Journal*, 32 (2): 191–209.

0176 ■ 2005 Jayanthi, N. *et al.* Cyclone and depression over North Indian Ocean during 2004. *Mausam,* 56 (3): 517–526.

0177 ■ 2005 Malleswararao, P. and B.S. Jasmin. Cyclone prone areas in the east.

Occurrence: Drought

0178 ■ 1968 Bagchi, K. and R. Bhattacharya. Drought incidence in Rarh plateau (West Bengal). *Geographical Review of India,* 30 (4): 66–70.

0179 ■ 1973 Rao, K.N. Spectral analysis of Palmer drought index for India. *Indian Journal of Meteorology and Geophysics,* 24: 257–270.

0180 ■ 1975 Rao, K.L. Rainfall and drought areas in India. In K.L. Rao (ed). *India's Water Wealth: Its Assessment, Uses and Projection.* Delhi: Orient Longman, 4–16.

0181 ■ 1975 Subrahmanyam, V.P. and A.A.L.N. Sarma. Incidence of aridity and drought in the climatic spectrum of the South Indian region. *Tropical Ecology,* 16: 120–27.

0182 ■ 1977 Ratnam, B.P. and S.N. Joshi. Drought frequency in Hagari and Bijapur. *Annals of Arid Zone,* 16 (2): 171–175.

0183 ■ 1977 Sharma, A.A. Variability of droughtness in the climatic spectrum of South Indian region. *Annals of Arid Zone,* 16 (4): 395–400.

0184 ■ 1978 Singh, T. *Drought Prone Areas in India: Aspects of Identification and Development Strategy.* New Delhi: People's Publishing House.

0185 ■ 1979 Arunachalam, B. Identification of drought prone areas in Maharashtra: A search for methodology. In N.B.K. Reddy (ed). *Proceedings of the All India Symposium on Drought Prone Areas of India.* Tirupati: Rayalseema Geographical Society. 157–166.

0186 ■ 1979 Choudhary, K.R. Concepts, approaches and methodology for delimitation of sub-areas in drought prone areas. In N.B.K. Reddy (ed). *Proceedings of the All India Symposium on Drought Prone Areas of India.* Tirupati: Rayalseema Geographical Society. 47–50.

0187 ■ 1979 Gupta, H.S. Trend analysis of rainfall and probability of drought in the Chhatisgarh region (Madhya Pradesh). In N.B.K. Reddy (ed). *Proceedings of the All India Symposium on Drought Prone Areas of India.* Tirupati: Rayalseema Geographical Society. 157–180.

0188 ■ 1979 Naganna, C. Delimiting drought prone areas in Karnataka and a mitigation strategy. In *Proceedings of the International Symposium on Hydrological Aspects of Drought*. New Delhi: Indian Institute of Technology. 486–493.

0189 ■ 1979 Nigam, R.K. Drought prone area of India: Problems and perspectives. In N.B.K. Reddy (ed). *Proceedings of the All India Symposium on Drought Prone Areas of India.* Tirupati: Rayalseema Geographical Society. 69–78.

0190 ■ 1979 Reddy, N. B. K. *et al.* Delimitation of drought prone areas of India: A geographical approach. In N. B. K. Reddy (ed). *Proceedings of the All India Symposium on Drought Prone Areas of India.* Tirupati: Rayalseema Geographical Society. 37–42.

0191 ■ 1979 Satapathy, P. P. Identification of drought prone areas of Bihar. In N. B. K. Reddy (ed). *Proceedings of the All India Symposium on Drought Prone Areas of India.* Tirupati: Rayalseema Geographical Society. 215–224.

0192 ■ 1979 Singh, R. B. *et al.* The delimitation of drought affected area. In N. B. K. Reddy (ed). *Proceedings of the All India Symposium on Drought Prone Areas of India.* Tirupati: Rayalseema Geographical Society. 191–200.

0193 ■ 1979 Soundarvalli, G. Delimitation of drought prone areas in Tamil Nadu. In N. B. K. Reddy (ed). *Proceedings of the All India Symposium on Drought Prone Areas of India.* Tirupati: Rayalseema Geographical Society. 181–90.

0194 ■ 1979 Subrahmanyam, V. P. and B. H. Malini. Delimitation of drought prone areas: Techniques and methodology. In N. B. K. Reddy (ed). *Proceedings of the All India Symposium on Drought Prone Areas of India.* Tirupati: Rayalseema Geographical Society. 31–36.

0195 ■ 1980 Rama Krishna, Y. S. *et al.* Studies on the incidence of droughts over Western Rajasthan. *National Geographical Journal of India,* 26 (1–2): 44–49.

0196 ■ 1980 Sastri, A. S. R. S. *et al.* A modified scheme of drought classification applicable to arid zone of Western Rajasthan. *Annals of Arid Zone,* 19 (1–2): 65–72.

0197 ■ 1980 Subrahmanyam, V. P. and B. H. Malini. Drought prone areas of Andhra Pradesh. In N. B. K. Reddy (ed). *Proceedings of the All India Symposium on Drought Prone Areas of India.* Tirupati: Rayalseema Geographical Society. 117–120.

0198 ■ 1981 Das, P. K. A historical perspective: Drought and famine in India. In R. Pearse (ed). *Tropical Drought.* Geneva: World Meteorological Organization. 3–12.

0199 ■ 1981 Malini, H. B. Patterns of aridity spread and severity of drought in Andhra Pradesh. *Indian Geographical Journal,* 56 (1–2): 30–36.

0200 ■ 1983 Dikshit, J. Identification of drought prone areas and prediction of crop expectation in Maharashtra. *Geographical Review of India,* 45 (4): 48–81.

0201 ■ 1983 Mooley, D. A. and B. Parhasarathy. Droughts and floods over India in summer monsoon season, 1871–1980. In A. S. Parrot *et al.* (eds). *Variations in the Global Water.* Netherlands: D. Reidel Dordrecht. 239–252.

0202 ■ 1983 Patnaik, M. and J.K. Routray. Intensity of drought and spatial variation in agricultural development: A case study of Kalahandi district in Orissa. *Transactions, Institute of Indian Geographers*, 5 (2): 165–170.

0203 ■ 1983 Ratnam, B.P. and S.N. Joshi. Drought frequency for *rabi* seasons in Dharwad and Bijapur. *Annals of Arid Zone*, 22 (1): 103–106.

0204 ■ 1984 Guha, M.K. *et al.* Deterministic time variation of drought. *Mausam*, 35 (3): 350–353.

0205 ■ 1984 Mooley, D.A. *et al.* Drought over Peninsular India during 1861–1980 and associated circulation features. *Mausam,* 35 (3): 337–342.

0206 ■ 1986 Jayaseelan, A.T. and S. Thiruvengadachari. *Recent Drought History of Andhra Pradesh State.* Hyderabad: National Remote Sensing Agency.

0207 ■ 1987 Gregory, S.A. Further note on extreme rainfall deficits in India. *Theoretical and Applied Climatology*, 38 (3): 174–179.

0208 ■ 1987 Parthasarathy, B. *et al.* Droughts and floods in the summer monsoon season over different meteorological sub-divisions of India for the period 1871–1984. *Journal of Climatology*, 7: 57–70.

0209 ■ 1987 Ramachandra, C.M. Another spell of drought. *Frontline*, (May 2): 27–28.

0210 ■ 1989 Chowdhury, A. *et al.* Variability of drought incidence over India: A statistical approach. *Mausam,* 40 (2): 207–214.

0211 ■ 1989 Gregory, S. Changing frequency of droughts in India, 1871–1985. *Geographical Journal*, 155 (3): 322–334.

0212 ■ 1991 Gregory, S. Patterns and intensification of marked rainfall deficit condition in Maharashtra, during the summer (June-September). *Transactions of the Institute of Indian Geographers*, 13 (1): 1–27.

0213 ■ 1992 Katyal, J.C. *et al. Occurrence of Droughts and Floods in India Between 1870 and 1990*. Hyderabad: Central Research Institute for Dryland Agriculture.

0214 ■ 1992 Penchalairam, T. and Y.N. Ramanaih. The spatial analysis of rainfall in the drought prone area of Cuddapah district (Andhra Pradesh). *Transaction of the Institute of Indian Geographers*, 14 (1): 65–77.

0215 ■ 1993 Chadha, S.K. Recurrence of droughts in Uttar Pradesh. In S.K. Chadha (ed). *Understanding Environment.* Delhi: Amar Prakashan. 99–103.

0216 ■ 1993 Reddy, A.V.S. *A Report on the Recurring Droughts of Anantpur: Impending Danger of Desertification.* Hyderabad: Government of Andhra Pradesh.

0217 ■ 1993 Sharma, S.N. Meteorological droughts in Rajasthan. In I. Pal *et al* (eds). *Changing Contours of Arid Ecology: Physical and Cultural Heritage*, Volume I. Jaipur: Asian Environmental Council.

0218 ■ 1994 Mooley, D.A. Origin, incidence and impact of droughts over India and remedial measures for their mitigation. *Sadhana: Academy Proceedings in Engineering Sciences*, 19 (4): 597–608.

0219 ■ 1995 Chaudhary, J.L. Drought classification studies in Bastar district of Chhattisgarh region. *Geographical Review of India*, 57 (2): 126–137.

0220 ■ 1998 Pal, S.K. *Drought*. Delhi: B R Publishing Corporation.

0221 ■ 2000 India Disaster Report Team. A hundred years of drought: An update. In S. Parasuraman and P.V. Unnikrishnan (eds). *India Disasters Report: Towards a Policy Initiative*. New Delhi: Oxford University Press. 177–178.

0222 ■ 2000 Sivasami, K. S. Drought and rainfall patterns: 1877–1999. *Economic and Political Weekly*, 35 (24): 1993–1994.

0223 ■ 2001 Special Correspondent. Chhattisgarh in grip of unprecedented drought. *New Age*, 49 (2): 14.

0224 ■ 2002 D' Souza, R. Colonialism, capitalism and nature: Debating the origins of Mahanadi delta's hydraulic crisis (1803–1928). *Economic and Political Weekly*, 35 (13): 1261–1272.

0225 ■ 2002 Das, B. Drought spreads its tentacles. *Down to Earth*, 11 (6): 13.

0226 ■ 2002 Das, H.P. and S.K. Banerjee. Incidence of large scale droughts in India with particular reference to classification, monitoring and its impact assessment. In R. Nagarajan (ed). *Proceedings of the National Workshop on Recent Trends in Drought Assessment Monitoring and Management*. Mumbai: Centre of Studies in Resources Engineering, Indian Institute of Technology, 1–7.

0227 ■ 2002 Editorial. Variability in drought incidents over districts of Maharashtra. *Mausam*, 53 (4): 533–542.

0228 ■ 2002 Nagarajan, R. *Proceedings of the National Workshop on Recent Trends in Drought Assessment Monitoring and Management*. Mumbai: Centre of Studies in Resources Engineering, Indian Institute of Technology.

0229 ■ 2002 Sikka, D.R. and S.M. Kulshrestha. *Indian Droughts in the Context of History and Climate*. Maryland: Centre for Ocean-Land-Atmosphere/ Cooperative for American Relief Everywhere, Technical Report No. 6.

0230 ■ 2003 Prasad, R. and H. Singh. Persistence, intensification and recession of meteorological drought over India during 1987. *Mausam*, 54 (2): 453–462.

0231 ■ 2003 Special Correspondent. All India drought. *Down to Earth*, 11 (18): 56.

0232 ■ 2004 Jat, M.L. *et al.* Spread of drought in Rajasthan during 1987 southwest monsoon. *Mausam*, 55 (2): 345–350.

Occurrence: Earthquake

0233 ■ 1830 Lyell, C. Earthquake of Kutch in Mumbai 1819. In C. Lyell (ed). *Principles of Geology.* London: John Murray Press/ Reprint (1998) Penguin.

0234 ■ 1834 Wright, B.W. Short notice of earthquake. *Chennai Journal of Literature Science,* 1: 104–111.

0235 ■ 1843 Hannay, W. Memoranda of earthquakes and other remarkable occurrences in Upper Assam from January 1839 to September 1843. *Journal of the Asiatic Society of Bengal,* 12: 907–909.

0236 ■ 1843 Smith, R.B. Memoirs on Indian Earthquakes. Part I. *Journal of the Asiatic Society of Bengal,* 12: 257–293.

0237 ■ 1843 Smith, R.B. Memoirs on Indian Earthquakes. Part II. *Journal of the Asiatic Society of Bengal,* 12: 1024–1056.

0238 ■ 1844 Smith, R.B. Memoirs on Indian Earthquakes. Part III. *Journal of the Asiatic Society of Bengal,* 13: 964–983.

0239 ■ 1849 Dalton, E.T. Earthquakes experiences in Assam in the later end of January 1849. *Journal of the Asiatic Society of Bengal,* 18: 173–174.

0240 ■ 1849 Hodgson, C.K. Memorandum on earthquakes in January 1849 at Burpetah (Assam). *Journal of the Asiatic Society of Bengal,* 18: 174–175.

0241 ■ 1849 Jenkins, F. Earthquakes in Assam. *Journal of the Asiatic Society of Bengal,* 18: 172–175.

0242 ■ 1856 Broun, J.A. On earthquakes in Travancore. *Chennai Journal of Literature Science,* 23: 109–114.

0243 ■ 1859 Dykes, J.W.B. Account of an earthquake in Guntoor, July 1859. *Chennai Journal of Literature Science,* 26: 165–166.

0244 ■ 1860 Davison, C. Earthquake in Ganjam 25 February 1860. *Chennai Journal of Literature Science,* 27: 341–342.

0245 ■ 1860 Knox, T.J. Earthquake of Southern India, 17 December 1859. *Chennai Journal of Literature Science,* 27: 340.

0246 ■ 1861 Brett, H.A. Earthquake in the Shevaroy Hills, 4 March 1861. *Chennai Journal of Literature Science,* 28: 139–141.

0247 ■ 1861 Thornhill, G. Earthquake at Nandigama, Kistna district, 24 July 1861. *Chennai Journal of Literature Science,* 28: 295.

0248 ■ 1869 Godwin-Austin, H.H. On an earthquake in the Cachar Hills, 10 January 1869. *Proceedings of the Royal Geographical Society, London,* 13: 370–372.

0249 ■ 1869 Leonard, H. Remarks on the earthquakes of 10 January 1869 at Cachar. *Proceedings of the Asiatic Society of Bengal,* 38: 102–103.

0250 ■ 1878 Blanford, W.T. Remarks on earthquakes in India. *Proceedings of the Asiatic Society of Bengal,* 47: 144–145.

0251 ■ 1878 Keatinge, R. H. Record of the occurrence of earthquake in Assam during 1877. *Proceedings of the Asiatic Society of Bengal,* 47 (2): 4–11.

0252 ■ 1879 Keatinge, R. H. Record of the occurrence of earthquakes in Assam during 1878. *Proceedings of the Asiatic Society of Bengal,* 48 (2): 48–55.

0253 ■ 1883 Walker, J. T. The earthquake of 31 December 1881: General report of the operations of the Survey of India 1881–1882. *Proceedings of the Asiatic Society of Bengal,* 52: 60–62.

0254 ■ 1885 Medlicott, H. B. Preliminary notice of the Bengal earthquake of 14 July 1885. *In Records of the Geological Survey of India,* 18 (3): 156–158.

0255 ■ 1891 Smith, H. E. Earthquake in the district of Salem, 4 March 1891. *Chennai Journal of Literature Science,* 58: 139.

0256 ■ 1897 Blaserna, P. Kolkata earthquake of 12 June 1897. *Nature,* 56: 346.

0257 ■ 1897 Heath, T. An Edinburgh record of the Indian earthquake (12 June, 1897). *Nature,* 56: 174–287.

0258 ■ 1897 Omori, F. Preliminary note on the earthquake of 12 June 1897. *Nature,* 57: 59.

0259 ■ 1898 Oldham, R. D. Great Indian earthquake of June 12, 1897. *Nature,* 58: 586.

0260 ■ 1901 Oldham, R. D. *List of Aftershocks of the Great Earthquake of 12 June 1897.* Memoirs of the Geological Survey of India, No. 30 (1). Kolkata.

0261 ■ 1902 Oldham, R. D. On tidal periodicity in the earthquakes of Assam. *Proceedings of the Asiatic Society of Bengal,* 71: 139–153.

0262 ■ 1903 Oldham, R. D. *The Diurnal Variation in Frequency of the Aftershock of the Great Earthquake of 12 June, 1897.* Memoirs of the Geological Survey of India, No. 35 (2). Kolkata.

0263 ■ 1905 Davison, C. Record of the Indian Earthquake of 4 April 1905 at Burmingham. *Nature,* 71: 589.

0264 ■ 1905 Davison, C. The Indian earthquake 4 April. *Geographical Journal,* 25 (5): 548–549.

0265 ■ 1905 Holland, T. H. Kangra earthquake of 4 April 1904. *Nature,* 72: 428–429.

0266 ■ 1905 Special Correspondent. Indian earthquake of 4 April 1905. *Nature,* 71: 563–564.

0267 ■ 1929 Coulson, A. L. The epicentre of the Northwest Himalayan earthquake of 1 February 1929. In *Records of the Geological Survey of India,* 62: 274–289.

0268 ■ 1934 Banerji, S. K. North Bihar earthquake of 15 January 1934. *Current Science,* 2 (9): 326–331.

0269 ■ 1934 Hunter, J. and G. De. Indian earthquake1934. *Nature,* 133 (3355): 236–237.

0270 ■ 1934 Krishan, M.S. The North Bihar earthquake of the 15 January 1934. *Current Science*, 2 (9): 323–326.

0271 ■ 1934 Roy, S.C. Focal region of the North Bihar earthquake of 15 January 1934. *Current Science*, 2 (11): 419–422.

0272 ■ 1934 Special Correspondent. The earthquake in North India. *Current Science*, 2 (8): 280.

0273 ■ 1935 Fermor, L.L. *et al.* North Bihar earthquake 1934. *Current science*, 3 (8): 372–76

0274 ■ 1937 Special Correspondent. Earthquakes in India in 1937. *Science and Culture*, 3 (12): 661–662.

0275 ■ 1937 West, W.D. Earthquake in India. In *Proceedings of the 24 Indian Science Congress*. Hyderabad. 289–337.

0276 ■ 1938 Crookshank, H. Earthquake shocks at Paliyad in Kathiwar. In *Records of the Geological Survey of India*, 73 (4): 579–591.

0277 ■ 1938 Roy, S.C. On a seismometric study of Bihar earthquake of 15 January 1934 and its aftershock of 19 January 1934. In *Proceedings of the 25 Indian Science Congress.* Kolkata. 111.

0278 ■ 1938 Special Correspondent. Earthquake in India. *Nature*, 142 (3587): 203.

0279 ■ 1939 Special Correspondent. Earthquakes recorded in India, January–March 1938. *Nature*, 143 (3617): 327.

0280 ■ 1939 Special Correspondent. Bihar-Nepal earthquake of 1934. *Science and Culture*, 4 (11): 628–629.

0281 ■ 1950 Sohoni, V.V. Great Assam earthquake of 15 August 1950. *Current Science*, 19 (9): 265–268.

0282 ■ 1950 Special Correspondent. The Northeast Assam earthquake. *Science and Culture*, 16 (3): 111.

0283 ■ 1950 Tandon, A.N. The very Great Earthquake of 15 August 1950. *Science and Culture*, 16 (4): 147–155.

0284 ■ 1950 Wadia, D.N. and J.B. Auden. Northeast Assam earthquake of 15 August 1950. *Journal of Scientific and Industrial Research*, 9A (10): 351–352.

0285 ■ 1951 Malurkar, S.L. Great Earthquake of Assam border of 15 August 1950. *Journal of Scientific and Industrial Research*, 10A (1): 21–25.

0286 ■ 1951 Tillosson, F. The Great Assam earthquake of 15 August 1950. *Nature*, 167, 128–130.

0287 ■ 1952 Bose, S.K. The Great Earthquake of Northeast Assam. *Geographical Review of India*, 14 (2): 33–41.

0288 ■ 1952 Gee, E.R. Assam earthquake of 1950. *Journal of Mumbai Natural History Society*, 50 (3): 629–635.

0289 ■ 1953 Corps, E. V. Assam earthquake of 15 August 1950. *Central Board of Geophysics, India,* 1: 43–49.

0290 ■ 1953 Mathur, L. P. Assam earthquake of 15 August 1950: A short note on factual observations. *Central Board of Geophysics, India,* 1: 56–60.

0291 ■ 1953 Poddar, M. C. Short note on the Assam earthquake of 15 August 1950. *Central Board of Geophysics, India,* 1: 38–42.

0292 ■ 1953 Pramanik, S. K. and Mukherjee, S. M. Assam earthquake of 1950. *Central Board of Geophysics, India,* 1: 26–34.

0293 ■ 1953 Seignior, C. H. T. Assam earthquake. *Central Board of Geophysics, India,* 1: 61–64.

0294 ■ 1953 Sohoni, V. V. Great Assam earthquake of 15 August 1950. *Central Board of Geophysics, India,* 1: 76–79.

0295 ■ 1953 Tillosson, F. Great Assam earthquake of 15 August 1950. *Central Board of Geophysics, India,* 1: 94–96.

0296 ■ 1953 Tandon, A. N. Very Great Earthquake of 15 August 1950. *Central Board of Geophysics, India,* 1: 80–89.

0297 ■ 1953 Ward, F. Assam earthquake of 1950. *Geographical Journal,* 119 (2): 169–182.

0298 ■ 1955 Bhan, S. N. Earthquakes felt at Srinagar during 1923–1952. *Indian Journal of Meteorology and Geophysics,* 6 (2): 197.

0299 ■ 1956 Ramamurthi, H. and G. M. Panchage. Succession law of aftershocks following Assam's Great Earthquake of 1950. *Journal of Irrigation and Power,* 13 (2): 217–222.

0300 ■ 1956 Tandon, A. N. and S. M. Mukherjee. The Rann of Kutch earthquake of 21 July 1956. *Indian Journal of Meteorology and Geophysics,* 7: 137–146.

0301 ■ 1956 Tandon, A. N. and S. M. Mukherjee. Manipur-Burma border earthquake of 22 March 1954. *Indian Journal of Meteorology and Geophysics,* 7 (1): 27–36.

0302 ■ 1960 Chakraborty, K. C. and D. P. Ghosh. Seismological studies of the crustal layers in Indian region from the data of near earthquakes. In *Proceedings of the World Conference on Earthquake Engineering.* Tokyo. 1633–1642.

0303 ■ 1964 Srivastava, L. S. *et al.* Badgam earthquake of 2 September 1963. *Bulletin of Indian Society of Earthquake Technology,* 1 (1): 79–94.

0304 ■ 1965 Heck, N. H. Description of Great Earthquakes: Rann of Cutch (India) earthquake of 16 June 1819. In N. H. Heck (ed). *Earthquakes.* New York and London: Hafner Publishing Company. 119.

0305 ■ 1965 Heck, N. H. Description of Great Earthquakes: Assam (India) 12 June 1897. In N. H. Heck (ed). *Earthquakes.* New York: Hafner Publishing. 120.

0306 ■ 1965 Heck, N.H. Description of Great Earthquakes: Kangra, (India) earthquake of 4 April 1905. In N.H. Heck (ed). *Earthquakes.* New York: Hafner Publishing Company. 121.

0307 ■ 1965 Heck, N.H. Description of Great Earthquakes: India earthquake of 15 January 1934. In N.H. Heck (ed). *Earthquakes.* New York and London: Hafner Publishing Company. 122.

0308 ■ 1965 Tandon, A.N. The Ratnagiri earthquake of 28 September 1962. *Indian Journal of Meteorology and Geophysics,* 16 (4): 127–128.

0309 ■ 1966 Singh, K. Earthquakes in India and neighbourhood. *Indian Journal of Meteorology and Geophysics,* 17 (4): 521–528.

0310 ■ 1967 Ray, S. The Koynanagar earthquake of December 1967. *Deccan Geographer,* 5: 89–93.

0311 ■ 1968 Bhaleha Atomic Research Centre. *Koyna Earthquake of 11 December 1967.* Mumbai.

0312 ■ 1968 Bhagavantam, S. Koyna earthquake. *Journal of Indian Geophysical Union,* 5 (1): 1–5.

0313 ■ 1968 Chauhan, R.K.S. *Studies of Seismicity of India: Analysis of Some Strong Earthquake Sequence.* Roorkee: University of Roorkee (Unpublished Ph.D thesis).

0314 ■ 1968 Lele, V.S. Earthquake tremors of Saurashtra. *Journal of Indian Geophysical Union,* 5 (2): 11–18.

0315 ■ 1968 Narain, H. and H.K. Gupta. Koyna earthquake. *Nature,* 217 (5134): 1138–1139.

0316 ■ 1968 Narian, H. and H.K. Gupta. Observations on Koyna earthquakes. *Journal of Indian Geophysical Union,* 5 (1): 30–34.

0317 ■ 1968 Tandon, A.N. and H.M. Choudhary. *Koyna Earthquake, December 1967.* Pune: India Meteorological Department, Scientific Report No. 59.

0318 ■ 1969 Gubin, I.E. *Earthquakes and Seismic Zoning of Indian Peninsula.* New Delhi: Central Water and Power Commission.

0319 ■ 1969 Gubin, I.E. *Seismic Zoning of the Western Margin of the Indian Peninsula in Maharashtra State.* Paris: United Nations Educational, Scientific and Cultural Organization.

0320 ■ 1969 Jhinrgran, A.G. *et al.* The Kolkata earthquakes of 15 April and 9 June, 1964. In *Records of the Geological Survey of India,* 97 (2): 1–29.

0321 ■ 1970 Biswas, S.K. and S.V. Deshpande. Geological and tectonic maps of Kutch. *Bulletin of the Natural Gas Commission,* 7: 115–116.

0322 ■ 1970 Guha, S.K. *et al.* Recent seismic disturbances in Western Maharashtra (1962–69). *Bulletin of Indian Society of Earthquake Technology,* 7 (4): 187–196.

0323 ■ 1970 Gupta, H.K. *et al.* Godavary valley earthquake sequence of April 1969. *Bulletin of the Seismological Society of America,* 60 (2): 601–615.

0324 ■ 1970 Varma, M.M. *et al.* Kothagudam (Andhra Pradesh) earthquake, 13 April 1969. Broach (Gujarat) earthquake, 23 March 1970 and seismicity of Peninsular India. *Bulletin of Indian Society of Earthquake Technology*, 7 (4): 207–221.

0325 ■ 1971 Gupta, H.K. *et al.* An earthquake cluster at Koyna. *Bulletin of the Seismological Society of America,* 61 (2): 297–315.

0326 ■ 1971 Ram, A. and H.S. Rathor. On frequency, magnitude and energy of significant Indian earthquakes. *Pure and Applied Geophysics,* 79: 26–32.

0327 ■ 1972 Ram, A. *et al.* Broach (Gujarat) earthquake of 23 March 1970. *Pure and Applied Geophysics,* 93 (1): 125–128.

0328 ■ 1972 Srivastava, H.N. and V.P. Kamble. Aftershock characteristics in Himalaya mountain belt and neighbourhood. *Indian Journal of Meteorology and Geophysics,* 23: 75–82.

0329 ■ 1975 Tandon, A.N. Some typical earthquake of north and west Uttar Pradesh. *Bulletin of Indian Society of Earthquake Technology,* 12 (2): 74–88.

0330 ■ 1976 Singh, S. *et al.* The Kinnaur earthquake of 19 January 1975: A field report. *Bulletin of the Seismological Society of America*, 66: 887–901.

0331 ■ 1978 Basu, S. *Statistical Analysis of Seismic Data and Seismic risk analysis of India Peninsula.* Kharagpur: Indian Institute of Technology, Kharagpur (Unpublished Ph.D thesis).

0332 ■ 1978 Khattri, K. and M.Wyss. Precursory variation of the seismicity rates in the Assam area. *Geology*, 6 (11): 685–688.

0333 ■ 1979 Kaila, K.L. and M.Rao. Seismic zoning maps of Indian subcontinent *Geophysics Research Bulletin,* 17: 293–301.

0334 ■ 1979 Srivastava, H.N. and H.M. Chaudhury. Seismicity and focal mechanism of earthquake in Andaman Islands. *Mausam,* 30 (2): 213–220.

0335 ■ 1979 Srivastava, H.N. *et al.* Precursory seismic observations in Himachal Pradesh and Shillong Plateau. *Mausam*, 30 (2): 289–296.

0336 ■ 1980 Rastogi, B.K. and P. Talwani. Relocation of Koyna earthquakes. *Bulletin of the Seismological Society of America*, 70: 1847–1868.

0337 ■ 1980 Srivastava, V.K. *Seismology of India with Special Reference to Seismotectonics.* Varanasi: Banaras Hindu University (Unpublished Ph.D thesis).

0338 ■ 1981 Chaudhary, H.M. Earthquake in the Himalayas. In J.S. Lall (ed). *The Himalayas: Aspects of Change.* Delhi: Oxford University Press, 99–119.

0339 ■ 1981 Kumar, A. *et al.* A macroseismic study of 14 June 1978 Dharamsala earthquake. *In Records of the Geological Survey of India,* 112 (8): 84–88.

0340 ■ 1981 Mohan, I. *et al.* Some recent earthquakes in Peninsular India. *Journal of Geological Society of India,* 22 (6): 292–298.

0341 ■ 1983 Banghar, A.R. Synthetic P wave seismograms for the Kinnaur (Himachal Pradesh) earthquake of 19, January 1975. *Physics of the Earth and Planetary Interiors,* 32 (4): 331–341.

0342 ■ 1983 Khattri, K.N. *et al.* Local seismic activity in the region of the Assam Gap, Northeast India. *Bulletin of the Seismological Society of America,* 73: 459–469.

0343 ■ 1984 Rao, B.R. *et al.* Historical seismicity of Peninsular India. *Bulletin of the Seismological Society of America,* 74 (6): 2519–2533.

0344 ■ 1984 Sarkar, I. *On Seismological Investigations of the Kumaon Himalayas Using Local Earthquake Data.* Roorkee: University of Roorkee (Unpublished Ph.D Thesis).

0345 ■ 1985 Goswami, D.C. *A Study of Seismic risk in the Northeast Indian region.* Guwahati: Guwahati University (Unpublished Ph.D Thesis).

0346 ■ 1986 Rastogi, B.K. *et al.* Microearthquake investigations near Sriramasagar reservoir, Andhra Pradesh State (India). *Physics of the Earth and Planetary Interiors,* 44 (2): 149–159.

0347 ■ 1986 Rastogi, B.K. *et al.* Microearthquakes near Osmansagar reservoir, Hyderabad (India). *Physics of the Earth and Planetary Interiors,* 44 (2): 134–141.

0348 ■ 1986 Srivastava, H.N. *et al.* Study of return periods of earthquakes in some selected Indian and adjoining regions. *Mausam,* 37 (3): 333–340.

0349 ■ 1987 Srivastava, H.N. *et al.* Space and time variation of seismicity pattern preceeding two earthquakes in Himachal Pradesh (India). *Tectonophysics,* 138 (1): 69–77.

0350 ■ 1989 Chakravorty, H.R. *Some Studies on Seismicity of Northeastern India Including Earthquake Spectral Analysis.* Imphal: Manipur University (Unpublished Ph.D Thesis).

0351 ■ 1989 Khattri, K. and K.Khanal. A perspective of seismic hazard in Indian region. *Science and Culture,* 55 (9): 325–331.

0352 ■ 1989 Ni, J.F. Active tectonics of the Himalayas. *Proceedings of Indian Academy of Sciences, Earth and Planetary Sciences,* 98 (1): 71–90.

0353 ■ 1990 Arora, B.R. and C.D. Reddy. Geomagnetic department sounding over seismically active area of South Gujarat. *Bulletin of Indian Society of Earthquake Technology,* 27: 105–112.

0354 ■ 1990 Sarna, H.K. *A Study of Seismic Hazard for Northeast India and Neighbourhood.* Guwahati: Guwahati University (Unpublished Ph.D Thesis).

0355 ■ 1990 Seth, A. *et al.* Sedimentary seismic activity in an immature passive margin basin (Lower member of Katrol formations, Upper Jurassic Kutch (India). *Sedimentary Geology,* 68 (4): 279–291.

0356 ■ 1990 Tewari, A.K. *Earthquake Hazard Zoning with Special Reference to Liquification Potential in Parts of Bihar and Uttar Pradesh Plains Using Remote Sensing Data.* Roorkee: Department of Earth Sciences, University of Roorkee (Unpublished Postgraduate Dissertation).

0357 ■ 1991 Kamble, V.P. Seismically vulnerable region hit again. *Current Science,* 61 (8): 503–505.

0358 ■ 1991 Rao, D.T. *et al.* Earthquake swarm activity in South Gujarat. *Mausam,* 42 (1): 89–98.

0359 ■ 1992 Kayal, J.R. *et al.* Aftershock sequence of Uttarkashi earthquake 20 October 1991. *Journal of Geological Society of India,* 39: 203–217.

0360 ■ 1992 Khattri, K.N. Seismic hazard in Indian region. *Current Science,* 62 (1–2): 109–116.

0361 ■ 1992 Special Correspondent. Uttarkashi earthquake, 20 October 1991. *Journal of Geological Society of India,* 39 (1): 83–88.

0362 ■ 1992 Vohra, C.P. *et al. Uttarkashi Earthquake of 20 October 1991.* Special Publication No. 30. Kolkata: Geological Survey of India.

0363 ■ 1993 Gupta, H.K. Earthquake exceeding magnitude 5 occur again in Koyna region. *Journal of Geological Society of India,* 42 (4): 413–416.

0364 ■ 1993 Gupta, H.K. and I. Mohan. A quick look at the Latur earthquake of 30 September 1993. *Current Science,* 65 (7): 517–521.

0365 ■ 1993 Kayal, J.R. *et al.* Microearthquakes at the main boundary thrust in Eastern Himalaya and the present day tectonic model. *Tectonophysics,* 218 (4): 375–381.

0366 ■ 1993 Nandy, D.R. *et al. Bihar-Nepal Earthquake, 20 August 1988.* Special Publication No. 31. Kolkata: Geological Survey of India.

0367 ■ 1993 Verma R.K. *et al.* Statistical analysis of seismicicity of Northeast India and Northern Burma during the period 1979–1990. *Journal of Himalayan Geology,* 4 (1): 71–80.

0368 ■ 1994 Gahalaut, V.K. Temporal variation of seismicity in a Himalayan tectonic block and 1991 Uttarkashi earthquake (India). *Physics of the Earth and Planetary Interiors,* 85 (3–4): 339–347.

0369 ■ 1994 Gaur, V.K. Evaluation of seismic hazard in India towards minimizing earthquake risk. *Current Science,* 67 (5): 324–329.

0370 ■ 1994 Gautam, A. *Earthquake: A Natural Disaster.* New Delhi: Ashish Publishing House.

0371 ■ 1994 Khattri, K.N. Earthquake hazard in Indian region. *Sadhana: Academy Proceedings in Engineering Sciences,* 19 (4): 609–614.

0372 ■ 1994 Levchenko, D.G. *et al.* Recording of ocean bottom seismic noise of a strong earthquake in the Himalayas by broad band digital ocean bottom seismograph installed on the Mid-Atlantic Ridge. *Physics of the Earth and Planetary Interiors,* 84 (1–4): 305–320.

0373 ■ 1994 Nandy, D.R. Earthquake hazard potential of Central and South Bengal basin. *Indian Journal of Earth Science,* 21 (2): 59–68.

0374 ■ 1994 Rao, P.P. *Geophysical Seismicity Mapping and Risk Evaluation Based on the Heuristic and Probabilistic Approach.* Vishakhapatnam: Andhra University, Waltair (Unpublished Ph.D Thesis).

0375 ■ 1994 Shamsher, P. Current trends in geotechnical earthquake engineering. *Indian Concrete Journal,* 68 (11).

0376 ■ 1995 Dattatrayam, R.S. and V.P. Kamble. A study of aftershocks of 20 October 1991 Uttarkashi earthquake. *Mausam,* 46 (4): 435–444.

0377 ■ 1995 Gupta, H.K. *et al.* Verification of magnitudes of Himalayan region earthquakes of 1903–1985 from Gottingen observatory. *Tectonophysics,* 244 (4): 267–284.

0378 ■ 1995 Gupta, P.R. *et al.* Remote sensing delineation of zones susceptiable to seismically induced liquefaction in the Ganga plains. *Journal of Geological Society of India,* 46: 75–82.

0379 ■ 1995 Kamble, V.P. *et al.* Spectra of seismic noise at selected Indian stations. *Mausam,* 46 (2): 169–174.

0380 ■ 1995 Kayal, J.R. The Killari earthquake of 30 September 1993, Latur district (Maharashtra). *Indian Journal of Geology,* 67 (2): 151–157.

0381 ■ 1995 Khattri, K.N. Earthquake hazard in the region of Doon Valley. *Journal of Himalayan Geology,* 6 (2): 127.

0382 ■ 1995 Kumar, S. *Seismological Studies of the Northwest Himalayas Based on Earthquake Data Recorded in Garhwal Himalayas.* Roorkee: University of Roorkee (Unpublished Ph.D thesis).

0383 ■ 1995 Kumar, S. *et al.* Seismotectonic observations at Tehri dam site for microzoning. *Himalayan Geology,* 16: 63–70.

0384 ■ 1995 Mahadevia, T.M. Seismicity of the Indian Peninsular shield. *Indian Journal of Geology,* 67 (1): 47–76.

0385 ■ 1995 Sarkar, I. *et al.* Evidence of small magnitude earthquakes on the active tectonics of Northwest Garhwal Himalaya. *Himalayan Geology,* 16: 279–284.

0386 ■ 1995 Sarkar, I. *et al.* On the aftershock sequence of a M 4.6 earthquake of the Garhwal Himalaya. *Proceedings of Indian Academy of Sciences, Earth and Planetary Sciences,* 104 (4): 683–691.

0387 ■ 1995 Verma, R.K. *et al.* Seismicity of Delhi and its surrounding region. *Journal of Himalayan Geology,* 6 (1): 75–68.

0388 ■ 1996 Cotton, F. *et al.* Rupture history and seismotectonics of 1991 Uttarkashi (Himalaya) earthquake. *Tectonophysics,* 258 (1–4): 35–51.

0389 ■ 1996 De, R. A microearthquake survey in the Himalayan foredeep region, North Bengal area. *Himalayan Geology*, 17: 71–80.

0390 ■ 1996 Narula, P.L. *et al* (eds). Killari earthquake 30 September 1993. *Geological Survey of India,* Special Publication No. 37. Kolkata.

0391 ■ 1996 Parvez, I. A. and A. Ram. Mapping seismic source of Himalayan earthquakes using body wave spectral estimates. *Himalayan Geology*, 17: 81–90.

0392 ■ 1996 Rajendran, C.P. and K. Rajendran. Low-moderate seismicity activity in the vicinity of Palghat gap and its implications. *Current Science,* 70 (4): 304–308.

0393 ■ 1996 Rajendran, C.P. and K. Rajendran. Earthquake recurrence in Peninsular India: Status and prospects. *Gondwana Geology Magazine*, 5: 107–124.

0394 ■ 1996 Ram, A. Application of array seismology for earthquake hazard studies: A review. *Himalayan Geology*, 17: 129–136.

0395 ■ 1996 Rao, C.V.R. and P.S. Raju. Study of the aftershock activity and source parameters of the Latur earthquake of 30 September 1993. *Journal of Geological Society of India*, 47 (2): 243–250.

0396 ■ 1996 Verma, R.K. *et al.* The Latur earthquake and intra-plate seismicity of Peninsular India. *Himalayan Geology,* 17: 119–128.

0397 ■ 1997 Gupta, H.K. *et al.* The Jabalpur earthquake of 22 May 1997. *Journal of Geological Society of India*, 50 (1): 85–91.

0398 ■ 1997 Parvez, I.A. *Earthquake Source Mapping and Hazards Assessment in Indian Subcontinent.* Varanasi: Banaras Hindu University (Unpublished Ph.D thesis).

0399 ■ 1997 Rajendran, C.P. Deformational features in the river bluffs at Ter, Osmanabad district (Maharashtra): Evidence for an ancient earthquake. *Current Science,* 72 (10): 750–755.

0400 ■ 1998 Acharyya, S.K. *et al.* Jabalpur earthquake of 22 May 1997: Constraint from aftershock study. *Journal of Geological Society of India*, 51 (3): 295–304.

0401 ■ 1998 Arya, A.S. *et al. Technolegal Aspects of Earthquake, Windstorm and Flood Hazards and Land use Zoning.* New Delhi: Ministry of Urban Affairs and Employment, Government of India, National Government Pulication.

0402 ■ 1998 Bansal, B.K. A glance through the seismicity of Peninsular India. *Journal of Geological Society of India,* 52 (1): 67–80.

0403 ■ 1998 Bhadauria, Y.S. *et al.* Seismicity of Delhi region from local earthquake data of Delhi seismic array. *Journal of Indian Geophysics Union*, 2: 7–21.

0404 ■ 1998 Bhattacharya, S.N. A perspective of historical earthquake in India and its neighbourhood upto 1900. *Mausam,* 49 (3): 375–382.

0405 ■ 1998 Dasgupta, S. *et al.* Quantitative assessment of seismic hazard in Northeastern India through Poisson possibility density function analysis. *Journal of Geological Society of India*, 52 (2): 181–194.

0406 ■ 1997 Iyengar, R.N. and D. Sharma. *Some Earthquakes of the Himalayan Region from Historical Sources.* Dehradun: Wadia Institute of Himalayan Geology.

0407 ■ 1998 Jain, S.K. Indian earthquakes: An overview. *Indian Concrete Journal*, 72 (11).

0408 ■ 1998 Kayal, J.R. Seismicity of Northeast India and surroundings: Development over the past 100 years. *Journal of Geophysics*, 19 (1): 9–34.

0409 ■ 1998 Mahajan, A.K. The 24 March 1995 Chamba earthquake (Northwest Himalaya): Field observations and seismotectonics. *Journal of Geological Society of India*, 51(2): 227–232.

0410 ■ 1998 Shanker, D. and M.L. Sharma. Estimation of seismic hazard parameters in the Himalayas and its vicinity from complete data files. *Pure and Applied Geophysics,* 152 (2): 267–279.

0411 ■ 1998 Sohoni, P.S. and J.N. Malik. Remnants of large magnitude earthquakes: Evidences from Great Rann sediments. *Current Science,* 74 (11): 985–989.

0412 ■ 1998 Special Correspondent. Earthquake strikes Garhwal Hills. *Journal of Geological Society of India,* 53 (5): 611.

0413 ■ 1999 Bhattacharya, S.N. A perspective of historical earthquakes in India and its neighbourhood upto 1900. *Himalayan Geology*, 20 (1): 117–122.

0414 ■ 1999 India Meteorological Department. *Chamoli Earthquake of 29 March 1999 and its Aftershocks.* New Delhi: Indian Meteorological Department.

0415 ■ 1999 Iyengar, R.N. Earthquakes in ancient India. *Current Science*, 77 (6): 827–838.

0416 ■ 1999 Iyengar, R.N. and D. Sharma. Some earthquakes of the Himalayan region. *Himalayan Geology*, 20 (1): 81–86.

0417 ■ 1999 Iyengar, R.N. *et al.* Earthquake history of India in medieval times. *Indian Journal of History of Science*, 34 (3): 181–237.

0418 ■ 1999 Jain, S.K. *et al. The Chamoli India Earthquake of 29 March 1999.* Oakland: Earthquake Engineering Research Institute.

0419 ■ 1999 Khattri, K.N. An evaluation of earthquake hazard and risk in Northern India. *Himalayan Geology,* 20 (1): 1–46.

0420 ■ 1999 Malik, J.N. *et al.* Modern and historic seismicity of the Kachchh peninsula, (Western India). *Journal of Geological Society of India,* 54 (5): 545–550.

0421 ■ 1999 Mukherjee, S. Microzonation of seismic and landslide prone areas for alternate highway alignment in a part of Western Coast of India using remote sensing techniques. *Photonirvachak, Journal of the Indian Society of Remote Sensing,* 27 (2): 81–90.

0422 ■ 1999 Rastogi, B.K. and P. Mandal. Foreshocks and nucleation of small-to moderate sized Koyna earthquakes (India). *Bulletin of the Seismological Society of America,* 89: 829–836.

0423 ■ 1999 Shanker, R. Seismic activity in India: The problem size. *Shelter,* special issue, October: 4.

0424 ■ 1999 Sukhija, B.S. *et al.* Paleoliquification evidences and periodicity of large pre-historic earthquakes in Shillong Plateau, (India). *Earth and Planetary Science Letters,* 167 (3–4): 269–282.

0425 ■ 2000 Gahalaut, V.K. and Kalpna. On the occurrence of small and moderate earthquakes of the Himalayan seismic belt. *Current Science,* 79 (7): 1004–1006.

0426 ■ 2000 Gupta, S.C. *Seimic Wave Attenuation Characteristics of Three Indian Regions.* Roorkee: University of Roorkee (Unpublished Ph.D thesis).

0427 ■ 2000 Khan, P.K. *Evolving Trends of Seismicity in India and its Neighbourhood.* Dhanbad: Indian School of Mines (Unpublished Ph.D thesis).

0428 ■ 2000 Lakhera, R.C. *Remote Sensing and Geographic Information System Based Approach for Seismotectonic Hazard Zonation: Case Studies from Garhwal Himalaya.* Dehradun: Indian Institute of Remote Sensing, National Remote Sensing Agency.

0429 ■ 2000 Mandal, P. *et al.* Recent India earthquake. *Current Science,* 79 (9): 1334–1346.

0430 ■ 2000 Rajendran, K. and C.P. Rajendran. Earthquake history at selected sites in the Himalaya: Recent initiatives. In *Proceedings of the International Symposium and School on Active Faulting.* Japan. 387–391.

0431 ■ 2000 Rastogi, B.K. Chamoli earthquake of magnitude 6.6 on 29 March 1999. *Journal of Geological Society of India,* 55 (5): 505–514.

0432 ■ 2001 Banerjee, P.K. *et al.* A qualitative assessment of seismic risk along the Peninsular Coast of India (South of 19° North). *Journal of Geodynamics,* 31 (5): 481–498.

0433 ■ 2001 Bilham, R. *et al.* Himalayan seismic hazard. *Indian Concrete Journal,* 75 (10).

0434 ■ 2001 Eidinger, J. M. (ed). *Gujarat (Kutch) India M 7.7 Earthquake of 26 January 2001 and Napa M 5.2 Earthquake of 3 September 2000: Lifeline Performance.* Virginia: Technical Council on Life Line Earthquake Engineering, American Society of Civil Engineers.

0435 ■ 2001 England P. and R. Bitham. Plateau 'pop-up' in the Great 1897 Assam earthquake. *Nature*, 410 (April 12): 806–809.

0436 ■ 2001 Gaur, V. K. The Rann of Kutch earthquake 26 January 2001. *Current Science*, 80 (3): 338–340.

0437 ■ 2001 Geological Survey of India. *Chamoli Earthquake of 29 March 1999.* Bulletin No. 53. Kolkata.

0438 ■ 2001 Mandal, P. *et al.* Aftershock activity and frequency dependent low coda Qc in the epicentral region of the 1999 Chamoli earthquake of M 6.4. *Pure and Applied Geophysics*, 158 (9–10): 1719–1735.

0439 ■ 2001 Rajendran, C. P. Earthquakes in India. In Ganesh *et al* (eds). *Proceedings of the Workshop on Popular Lectures on Facts and Myths of Earthquake, Indian Context.* Tiruchirapalli: Bharathidasan University. 31–34.

0440 ■ 2001 Rastogi, B. K. Erattupetta earthquake of 12 December 2000 and seismicity of Kerala. *Journal of Geological Society of India,* 57 (3): 273–274.

0441 ■ 2001 Thakur, V. C. *et al.* Microseismicity and seismotectonics of 1905 Kangra earth-quake affected region in Dhauladhar-Chamba (Northwest Himalaya). In O. P. Verma (ed). *Focus on Seismicity.* New Delhi: Department of Science and Technology, (Special Volume–2). Government of India. 65–73.

0442 ■ 2002 Bhattacharya, S. N. and R. S. Dattatrayan Earthquake sequence in Kerala during December 2000 and January 2001. *Current Science,* 82: 1275–1278.

0443 ■ 2002 Narayan, J. P. *et al.* A seismological report on 26 January 2001 Bhuj, Indian earthquake. *Seismological Research Letters,* 73 (3): 343–355.

0444 ■ 2002 Rajendran, K. *et al.* Sand blows from the 2001 Bhuj earthquake reveal clues on past seismicity. *Current Science,* 83: 603–610.

0445 ■ 2002 Susan, E. *et al.* The 26 January 2001 M 7.6 Bhuj, India earthquake: Observed and predicted ground motions. *Bulletin of the Seismological Society of America*, 92: 2061–2079.

0446 ■ 2003 Ambraseys, N. and R. Bilham. Revaluated intensities for the Great Earthquake of 12 June 1897 Shillong (India). *Bulletin of the Seismological Society of America*, 93 (2): 655–673.

0447 ■ 2003 Gupta, H. K. Major and great earthquakes in the Himalaya region: An overview. In H. K. Gupta (ed). *Disaster Management.* Hyderabad: University Press (India) Private Limited. 100–109.

0448 ■ 2003 Jayachandran, G. *et al.* A quantitative study of the energy release in the aftershocks of the Bhuj earthquake 2001 (India) using Lg phase. *Physics of the Earth and Planetary Interiors*, 138 (2): 131–145.

0449 ■ 2003 Kumar, A. *et al.* Aftershock activity of Bhuj earthquake of 26 January 2001. Proceedings of Indian Academy of Sciences, *Earth and Planetary Sciences*, 112 (3): 391–396.

0450 ■ 2003 Mishra, P.S. Application of geographic information systems in seismic hazard and risk assessment microzonation of Jabalpur urban area (Madhya Pradesh). In M. Mohanty and S. Aggarwal (eds). *Applications of Geographic Information Systems in Earthquake and Landslide Hazard Assessment.* New Delhi: Department of Science and Technology, Government of India. 10–11.

0451 ■ 2003 Misra, K.S. *et al.* Aftershocks of 26 January 2001 Bhuj earthquake and seismotectonics of the Kutch region. *Proceedings of Indian Academy of Sciences, Earth and Planetary Sciences,* 112 (3): 385–390.

0452 ■ 2003 Nath, S.K. Seismic microzonation and hazard mapping of Sikkim Himalaya through geographic information systems integration of geological and strong motion features. In M. Mohanty and S. Aggarwal (eds). *Application of Geographic Information Systems in Earthquake and Landslide Hazard Assessment.* New Delhi: Department of Science and Technology. 5–9.

0453 ■ 2003 Rajendran, K. *et al. Seismic Hazard Assessment of the Kumaun Himalaya using Seismological, Paleoseismological and Neotectonic Methods.* New Delhi: Department of Science and Technology, Government of India.

0454 ■ 2003 Shah, S. Application of mapping and geographic information systems technologies after the Bhuj earthquake. In *Proceedings of the Map India 2003.* New Delhi: GIS Development, Survey of India and National Consortium on Remote Sensing in Transportation.

0455 ■ 2004 Ambraseys, N. and J. Douglas. Magnitude calibration of North India earthquakes. *Geophysical Journal International,* 159 (1): 165–206.

0456 ■ 2004 Lay, T. *et al.* The Great Sumatra-Andaman earthquake of 26 December 2004. *Science,* 306: 1123.

0457 ■ 2004 Mahajan, A.K. *et al.* Macroseismic field observations of 26 January 2001 Kachchh earthquake and its seismotectonics. *Journal of Asian Earth Sciences,* 23 (1): 17–23.

0458 ■ 2004 Mukhopadhyay, P. and P. Bormann. Low cost seismic microzonation using microtremor data: An example from Delhi (India). *Journal of Asian Earth Sciences,* 24 (3): 271–280.

0459 ■ 2004 Naithani, A.K. *et al.* Chamoli earthquake of 29 March 1999 (Garhwal Himalaya, India): An observation. *Science and Culture,* 70 (1–2): 21–31.

0460 ■ 2004 Nath, S.K. Seismic hazard mapping and microzonation in Sikkim Himalaya through geographic information systems integration of site effects and strong ground motion attributes. *Natural Hazard Journal,* 31 (2): 319–342.

0461 ■ 2004 Pande, P. and J.R. Kayal (eds). *Kutch (Bhuj) Earthquake 26 January 2001*. Special Publication No. 76. Kolkata: Geological Survey of India.

0462 ■ 2004 Prakash, R. and H.V. Srivastava. Spatio-temporal seismicity variation in the earthquakes of Uttaranchal region. *Mausam,* 55 (4): 681–690.

0463 ■ 2004 Rajendran, C.P. and K. Rajendran. Towards better seismic hazard assessment: Need for an integrated approach. In K.S. Valdiya (ed). *Coping with Natural Hazards: Indian Context.* Hyderabad: Orient Longman. 59–70.

0464 ■ 2004 Rajendran, C.P. and K. Rajendran. Studying earthquake recurrence in the Kachchh region (India). *EOS American Geophysical Union Transaction,* 84 (48): 529–536.

0465 ■ 2004 Rajendran, C.P. and K. Rajendran. Understanding earthquake in space and time: Application of new investigation tools. In *Silver Jubilee Compendium Earth System Science and Natural Resources Management*. Tiruvananthpuram: Centre of Earth Science Studies. 249–262.

0466 ■ 2004 Rajendran, K. and C.P. Rajendran. Study of earthquake in Kerala: Status and prospects. In *Silver Jubilee Compendium Earth System Science and Natural Resources Management.* Tiruvananthpuram: Centre of Earth Science Studies. 231–248.

0467 ■ 2005 Nath, S.K. An initial model of seismic microzonation of Sikkim Himalaya through thematic mapping and geographic information systems integration of geological and strong features. *Journal of Asian Earth Sciences,* 25 (2): 329–343.

0468 ■ 2005 Nath, S.K. *et al.* A seismic hazard scenario in the Sikkim Himalaya from seismotectonics, spectral amplification, source parameterization and spectral attenuation laws using strong motion seismometry. *Journal of Geophysical Research,* 110 (B01301): 1–24.

0469 ■ 2005 Rajendran, C.P. *Earthquake Recurrence in the Andaman-Nicobar Arc.* Delhi: Department of Science and Technology, Government of India.

0470 ■ 2005 Rajendran, K. and Rajendran, C.P. Study of earthquakes in Kerala: Status and prospectus. In *CESS Jubliee Compendium.* Tiruvananthapuram: Centre for Earth Science Study. 231–247.

0471 ■ 2005 Rajendran, K. and C.P. Rajendran. The Great Sumatra-Andaman earthquake of 26 December 2004. *Current Science,* 88 (1): 11–12.

0472 ■ 2005 Raju, P.S. *et al.* A note on the 26 December 2004 Great Sumatra earthquake. *Journal of Geological Society of India*, 65 (2): 249–251.

0473 ■ 2005 Rastogi, B.K. *et al.* Improved earthquake locations and estimation of Pn and Sn path anomalies for India, using multiple event relocations and reference events. *Current Science,* 88 (10): 1586–1591.

0474 ■ 2005 Seih. K. Aceh-Andaman earthquake: What happened and what's next? *Nature,* 434 (March 31): 573–574.

0475 ■ 2005 Singh, H.N. *et al.* Occurrence of anomalous seismic activity preceding large to Great Earthquake in Northeast India region, with special reference to 6 August 1988. *Physics of the Earth and Planetary Interiors,* 148 (2–4): 261–284.

0476 ■ 2005 Srinivasan, C. Sumatra earthquake of December 2004 and its aftershocks as recorded at the Kolar Gold Field Observatory (India). *Journal of Geological Society of India*, 65 (3): 370–372.

0477 ■ 2006 Sitharam, T.G. *et al.* Use of remote sensing and seismotectonic parameters for seismic hazard analysis of Bangalore. *Natural Hazard and Earth System Sciences*, 6 (6): 927–939.

0478 ■ 2006 Thakur, V.C. Reassessment of earthquake hazard in the Himalaya and implications from the 2004 Sumatra-Andaman earthquake. *Current Science,* 90 (8): 1070–1072.

Occurrence: Flood

0479 ■ 1923 Sharma, S.R. Floods in South Kanara. *Modern Review*, 34 (4): 419–424.

0480 ■ 1924 Editorial. Floods in South India. *Indian Review*, 25 (8): 497.

0481 ■ 1935 Andrews, C.F. Floods in Orissa. *Modern Review*, 57 (4): 413–414.

0482 ■ 1935 Saha, M. Great flood in Northern Bengal. *Modern Review*, 32 (5): 605–611.

0483 ■ 1935 Saha, M. Damodar flood of 1935. *Modern Review*, 58 (5): 527–529.

0484 ■ 1937 Prasad, R. Floods. *Modern Review*, 62 (4): 431–435.

0485 ■ 1938 Agarwala, K.S. Floods in India. *Indian Review*, 39 (1): 15–16.

0486 ■ 1953 Sircar, P.K. Floods of the Rupnarayan, the Kansai and the Subarnrekha. *Geographical Review of India*, 15 (3): 6–19.

0487 ■ 1954 Special Correspondent. Floods in India: 1954. *Indian Journal of Power and River Valley Development*, 4 (9): 21–26.

0488 ■ 1954 Special Correspondent. Flood of 15 August 1950. *Science and Culture,* 20 (4): 181–182.

0489 ■ 1955 Bagchi, K. Floods in West Bengal rivers. *Geographical Review of India*, 17 (2): 38–41.

0490 ■ 1955 Parthasarathy, K. Flood in Bihar, Bengal and Assam during July and August 1954. *Journal of Scientific and Industrial Research*, 14 A: 115–119.

0491 ■ 1955 Special Relief Department. *Orissa Floods 1955.* Cuttack: Government of Orissa.

0492 ■ 1956 Sinha, B.N. Flood frequency in the major rivers of Orissa. *Indian Journal of Power and River Valley Development,* 6 (7): 5–10.

0493 ■ 1956 Sinha, B.N. The 1955 floods in Orissa. *Geographical Review of India*, 18 (2): 22–37.

0494 ■ 1957 Mehta, D. Floods in India. *Indian Review*, 58 (10): 446–448.

0495 ■ 1958 Bose, U.K. A study of the floods in West Bengal and West Uttar Pradesh during September-October 1956. *Indian Journal of Meteorology and Geophysics*, 9: 23–34.

0496 ■ 1961 Chatterjee, S.P. and K. Bagchi. Floods in West Bengal: A regional study. *Geographical Review of India,* 23 (3): 1–5.

0497 ■ 1961 Editorial. The flood situation. *Bhagirath,* 8 (October): 149–151.

0498 ■ 1961 Editorial. The flood affected states. *Bhagirath*, 8 (December): 221–223.

0499 ■ 1961 Editorial. Season of floods. *Indian Journal of Power and River Valley Development*, 11(7): 31–32.

0500 ■ 1961 Ramchander, M.S. Flood picture in states. *Bhagirath,* 8 (November): 209–212.

0501 ■ 1961 Special Correspondent. Flood picture in states. *Bhagirath,* 8 (Dec.): 246–248.

0502 ■ 1962 Dhawan, S.K. Flood picture in states. *Bhagirath,* 9 (November): 202–205.

0503 ■ 1962 Jain, R.K. Flood picture in states. *Bhagirath*, 9 (October): 166–168.

0504 ■ 1964 Bose, N.K. and A.K. Sinha.1959 October floods of Damodar river. *Indian Journal of Power and River Valley Development*, 14 (8): 1–4.

0505 ■ 1964 Shah, R.B. Flood situation in India during 1963. *Bhagirath*, 11 (Jan.): 26–27.

0506 ■ 1964 Special Correspondent. Flood in Assam. *Bhagirath,* 11 (October): 130–133.

0507 ■ 1965 Ramesh Rao, K. Flood picture in states. *Bhagirath,* 12 (January): 19–21.

0508 ■ 1966 Banerjee, S. *et al.* Flood producing storms in lower Bihar, with particular reference to the storm of October 1961. *Indian Journal of Meteorology and Geophysics*, 17 (4).

0509 ■ 1966 Rao, K.R. Flood picture in states. *Bhagirath*, 13 (January): 22–24.

0510 ■ 1968 Special Correspondent. Flood picture in states. *Bhagirath*, 15 (January): 24–28.

0511 ■ 1969 Banerjee, B. Floods in West Bengal. *Indian Journal of Power and River Valley Development*, 19 (10).

0512 ■ 1969 Sanyal, C.C. The flood in 1968. *Geographical Review of India,* 31(3): 47–53.

0513 ■ 1969 Special Correspondent. Flood picture in states. *Bhagirath,* 16 (April-September): 56–59.

0514 ■ 1970 Bose, N.K. The challenge of floods in North Bengal. *Indian Journal of Power and River Valley Development,* 20 (8): 347–348.

0515 ■ 1971 Special Correspondent. Flood picture in states. *Bhagirath,* 18 (January): 26–29.

0516 ■ 1972 Goswami, A. Regional flood frequency analysis of Brahmaputra basin in Northeastern area. *Journal of Institution of Engineers, India,* 52: 169–174.

0517 ■ 1973 Special Correspondent. Flood picture in states. *Bhagirath,* 20 (January): 19–21.

0518 ■ 1974 Ramachandran, R. and S.C. Thakur. India and Ganga floodplains. In G. White (ed). *Natural Hazards: Local, National and Global.* New York: Oxford University Press. 36–42.

0519 ■ 1974 Special Correspondent. Flood picture in states. *Bhagirath,* 21 (January): 18–25.

0520 ■ 1975 Das, P. An appraisal of flood incidence of the Kaliaghai river basin. *Geographical Review of India,* 37 (1): 73–80.

0521 ■ 1975 Dhar, O.N. *et al.* A study of extreme rainfall over flash flood prone region of Himalayan foothills. *Hydrological Sciences Bulletin,* 20 (1): 61–67.

0522 ■ 1975 Special Correspondent. Flood picture in states. *Bhagirath,* 22 (April): 19–24.

0523 ■ 1975 Special Correspondent. Flood picture in states. *Bhagirath,* 22 (Oct.): 154–158.

0524 ■ 1976 Special Correspondent. Flood picture in states. *Bhagirath,* 23 (Oct.): 147–153.

0525 ■ 1977 Special Correspondent. Flood picture in states. *Bhagirath,* 24 (Oct.): 159–164.

0526 ■ 1978 Mukherjee, A.K. Indian floods. *Modern Review,* 144 (3): 161–163.

0527 ■ 1978 Special Correspondent. Flood picture in states. *Bhagirath,* 25 (Oct.): 169–179.

0528 ■ 1978 Subramanium, A.R and A.S.R.A.S. Sastri. Analysis of floods and droughts in Andhra Pradesh. *Deccan Geographer,* 16 (2): 470–478.

0529 ■ 1979 Special Correspondent. Flood picture in states. *Bhagirath,* 26 (Oct.): 154–158.

0530 ■ 1980 National Remote Sensing Agency. *Brahmaputra Flood Mapping and River Migration Studies (Airborne Scanner Survey).* Hyderabad.

0531 ■ 1981 Girglani, J.M. Flash floods in Vamsadhara river, Srikulam district (Andhra Pradesh): A Case Study. In *Proceedings of the International Conference on Flood Disaster.* New Delhi: Indian National Science Academy.

0532 ■ 1981 Special Correspondent. Flood picture in states. *Bhagirath,* 28 (Oct.): 175–180.

0533 ■ 1982 Sharma, K.D. and N.S. Vangani. Unprecedented floods in Luni basin, Western Rajasthan, during July 1979: A case study. *Mausam,* 33 (3): 377–384.

0534 ■ 1983 Kayastha, S.L. Floods in India: A study of their occurrence, causes, forecasting and control. *National Geographical Journal of India,* 29 (3–4): 121–141.

0535 ■ 1983 Nagi, R. S. and B. B. Koti Reddy. Flood in Sahibi river *Photonirvachak, Journal of the Indian Society of Remote Sensing,* 11(2): 59–62.

0536 ■ 1983 Special Correspondent. Flood picture in states. *Bhagirath,* 30 (Oct.): 203–207.

0537 ■ 1984 Naidu, E.M. Floods and cyclones in Andhra Pradesh. In B.R. Naidu (ed). *Disaster Management.* Tirupati: Sri Venkateswara University. 71–77.

0538 ■ 1984 Pandharinath, N. A case study of 1981 flood in the Godavari river. *Mausam,* 35 (2): 195–198.

0539 ■ 1985 Murty, K.L.N. Flood prone area: A case study of river Vamsadhara, Srikakulam district (Andhra Pradesh). In *Proceedings of the Sixth Indian Geographical Congress.* Department of Architecture and Regional Planning Kharagpur: Indian Institute of Technology.

0540 ■ 1985 Ramaswamy, C. *Review of Floods in India during the Past 75 Years.* New Delhi: Indian National Science Academy.

0541 ■ 1985 Special Correspondent. Flood picture in states. *Bhagirath,* 32 (January): 29–33.

0542 ■ 1986 Dhar, O.N. *et al.* The highest ever recorded floods in the major Indian rivers: A brief appraisal. *Transactions, Institute of Indian Geographers,* 8 (1): 13–26.

0543 ■ 1986 Government of Assam. *Assam Floods October 1986: Report of the Expert Committee.* Guwahati.

0544 ■ 1986 Khare, S.B. Challenges of Ganga-Brahmaputra systems. *Indian Journal of Power and River Valley Development,* 36 (5 and 6): 162–164.

0545 ■ 1986 Special Correspondent. Flood picture in states. *Bhagirath,* 33 (April): 94–102.

0546 ■ 1987 Special Correspondent. Floods in Bihar: 1986. *Bhagirath,* 34 (January): 40–41.

0547 ■ 1987 Tewari, D.N. Floods in India. In D.N. Tewari (ed). *Victims of Environmental Crisis.* Dehradun: Educational Private Limited, 131–150.

0548 ■ 1988 Subrahmanyam, V.P. Hazards of floods and droughts in India. In M.I. El–Sabh and T.S. Murty (eds). *Natural and Man Made Hazards, Proceedings of the International Symposium.* Holland: D. Riedel Publisher, 337–356.

0549 ■ 1989 Dhar, O.N. and S.S. Nandargi. Floods in Indian rivers during contrasting monsoon seasons of 1987 and 1988. *Hydrology,* 12 (1–2): 21–34.

0550 ■ 1989 Kumar, A. and R.P.S. Chhonkar. Flood inundation mapping in part of Mahanadi river basin. *Photonirvachak, Journal of the Indian Society of Remote Sensing,* 17: 13–16.

0551 ■ 1990 Govind Rao, P. Frequency analysis of annual peak floods in the monthly maximum flows in Mahanadi basin. *Indian Journal of Power and River Valley Development,* 40 (1): 12–14.

0552 ■ 1990 Kar, M. *Cyclone and Floods in Orissa: A Geographical Analysis.* Bhubaneswar: Department of Geography, Utkal University (Unpublished Master of Philosophy Dissertation).

0553 ■ 1990 Prasad, M.C. and G. Prasad. Disastrous floods in Bihar during August to September 1987. *Vayumandal,* 20 (3–4): 102–107.

0554 ■ 1991 Chowdhury, and S.V. Mhasawade. Variations in meteorological floods during monsoon over India. *Mausam,* 42 (2): 167–170.

0555 ■ 1992 Singh, A.K. *Flood Hazard in North Bihar.* Delhi: Department of Geography, Delhi School of Economics, University of Delhi (Unpublished Master of Philosophy Dissertation).

0556 ■ 1993 Dhar, O.N. and S.S. Nandargi. Worst flood-prone rivers and sites in India. *Vayumandal,* 23 (1–2): 86–92.

0557 ■ 1993 Kale, V.S. *et al.* Prehistoric flood deposits on the Choral River, Central Narmada basin. *Current Science,* 65: 877–878.

0558 ■ 1993 Kar, M. and D.C. Goswami. Flood hazard zonation of Nagaon district (Assam). In S.M. Rashid (ed). *Remote Sensing in Geography,* New Delhi: Manak Publications. 116–29.

0559 ■ 1993 Sinha, R. and V. Jain. Flood hazards in North Bihar rivers (Indo-Gangetic plains). *Journal of Geological Society of India,* 41: 27–52.

0560 ■ 1994 Dhar, O.N. and S.S. Nandargi. Floods in Indian rivers. *Indian Journal of Power and River Valley Development,* 44 (7 and 8): 228– 236.

0561 ■ 1995 Das, C. *et al.* Flood zoning using geographic information systems: A case study from the Barak basin (Assam). *Indian Cartographer,* 15: 188–191.

0562 ■ 1996 Sharma, C.K. Catastrophic floods of 21 July 1993 in the Bagmati basin of Nepal. *Himalayan Geology,* 17: 169–17.

0563 ■ 1997 Kale, V.S. and Gadgil, A.S. *Flood hydrology in the Upper Tapi Basin using Palaeoflood and Historical Record.* New Delhi: Department of Science and Technology, Government of India (Unpublished report).

0564 ■ 1997 Kale, V. S. *et al.* A 2000 year paleoflood record from Sakarghat on Narmada (Central India). *Journal of Geological Society of India,* 50: 285–288.

0565 ■ 1997 Iqbal, L. Flood in the Gomati basin. In T. Iwasaki, *et al.* (eds). *Proceedings of Group Seminar on Natural Disaster Reduction for Roads.* 148–149.

0566 ■ 1997 Rao, P. V. R. and G. S. Rao. Flash flood over South Coastal Andhra Pradesh and Rayalseema in October 1996. *Vayumandal,* 27 (3–4): 45–50.

0567 ■ 1997 Srivastava, S. S. Flood plain zoning in the Brahmaputra valley, around South Salmana, Goalpara district (Assam). In *Proceedings of the Symposium on the Earth Sciences in Environmental Assessment and Management.* Special Publication No. 48 (2). Kolkata: Geological Survey of India. 177–180.

0568 ■ 1998 Baker, V. R. Future prospects of past floods in India. In V. S. Kale (ed). *Flood Studies in India.* Bangalore: Memoirs of the Geological Society of India, No. 41. 219–228.

0569 ■ 1998 Kurothe, R. S. *Development of Physically Based Flood Frequency Models.* Roorkee: University of Roorkee (Unpublished Ph.D thesis).

0570 ■ 1998 Nandargi, S. and O. N. Dhar. *High Frequency Floods and their Magnitudes in the Indian Rivers.* Pune: Indian Institute of Tropical Meteorology.

0571 ■ 1998 Nandargi, S. and O. N. Dhar. A study of floods in Ganga and its sub-basin. *Indian Journal of Power and River Valley Development,* 48 (5 and 6): 85–90.

0572 ■ 1998 Sinha, R. and V. Jain. Flood hazards of North Bihar rivers (Indo-Gangetic plains). In V. S. Kale (ed). *Flood Studies in India.* Bangalore: Memoirs of the Geological Society of India, No. 41. 27–52.

0573 ■ 1998 Wohl, E. and D. Cenderelli. Flooding in the Himalaya mountains. In V. S. Kale (ed). *Flood Studies in India.* Bangalore: Memoirs of the Geological Society of India, No. 41, 77–100.

0574 ■ 1999 Dhar, O. N. and S. S. Nandargi. Summer monsoon of 1998 and associated floods: A brief appraisal. *Journal of Geological Society of India,* 54 (2): 193–194.

0575 ■ 1999 Fard, A. F. *Regionalisation of Flood Extremes Using Pattern Analysis.* New Delhi: Indian Institute of Technology (Unpublished Ph.D thesis).

0576 ■ 1999 Kale, V. S. Long period fluctuations in monsoon floods in the Deccan Peninsula. *Journal of Geological Society of India,* 53 (1): 5–16.

0577 ■ 2000 Rao, A. R. and K. H. Hamed. *Flood Frequency Analysis.* London: CRC Press.

0578 ■ 2000 Special Correspondent. Study of floods in Assam. *Bhagirath,* 47 (July-September): 174–175.

0579 ■ 2000 United Nations Children's Fund. *West Bengal Flood Emergency: Situation Report.* New Delhi.

0580 ■ 2001 Bhan, S.K. *et al.* Study of floods in West Bengal during September 2000 using Indian remote sensing satellite data. *Photonirvachak, Journal of the Indian Society of Remote Sensing,* 29 (1–2): 1–2.

0581 ■ 2001 Chatterjee, S. *Mapping of Inundated Area in Sub-Mahanadi Basin.* Dehradun: Indian Institute of Remote Sensing (Unpublished Postgraduate Dissertation).

0582 ■ 2001 Dhar, O.N. and S.S. Nandargi. Flood study of Central Indian rivers. *Indian Journal of Power and River Valley Development,* 51: 161–166.

0583 ■ 2001 Dhar, O.N. and S.S. Nandargi. A comparative flood frequency study of Ganga and Brahmaputra river systems of North India. *Water Policy,* 3 (1): 101–107.

0584 ■ 2001 Mirza, M.M.Q. *et al.* Are floods getting worse in the Ganges, Brahmaputra and Meghna basins? *Environmental Hazards,* 3 (2): 37–48.

0585 ■ 2001 Rudra, K. *Floods in West Bengal.* Kolkata: National Alliance of Peoples' Movement.

0586 ■ 2001 Shukla, S. and M. Chatterjee. *Mapping of Inundated Area in Sub-Mahanadi Basin.* Dehradun: Indian Institute of Remote Sensing.

0587 ■ 2001 Special Correspondent. Floods in India. *Down to Earth,* 10 (12).

0588 ■ 2002 Dhar, O.N. and S.S. Nandargi. Flood study of the Himalaya tributaries of the Ganga river. *Meteorological Applications,* 9 (1): 63–68.

0589 ■ 2002 Rakhecha, P. Highest floods in India. In A. Snorrason *et al.* (eds). *Proceedings of Symposium on the Extremes: Extraordinary Floods.* United Kingdom: International Association of Hydrological Sciences Press. 167–171.

0590 ■ 2003 Kale, V.S. The spatio-temporal aspects of monsoon floods in India: Implication for flood hazard management. In H.K. Gupta (ed). *Disaster Management.* Hyderabad: University Press (India) Private Limited. 22–47.

0591 ■ 2003 Kumar, R. and C. Chatterjee. Use of L-movements in development of regional flood frequency relationship for gauged and ungauged catchment. *Indian Journal of Power and River Valley Development,* 53 (5 and 6): 86–92.

0592 ■ 2003 Nandargi, S. and O.N. Dhar. High frequency floods and their magnitudes in the Indian rivers. *Journal of the Geophysical Society of India,* 61: 90–96.

0593 ■ 2004 Dhar, O.N. and S.S. Nandargi. Floods in North India river systems. In K.S. Valdiya (ed). *Coping with Natural Hazards: Indian Context.* Hyderabad: Orient Longman. 104–123.

0594 ■ 2004 Dhar, O.N. and S.S. Nandargi. Floods in India during the drought year of 2004. *Indian Journal of Meteorology, Hydrology, and Geophysics,* 28 (281): 249–255.

0595 ■ 2004 Kale, V.S. Floods in India: Their frequency and pattern. In K.S. Valdiya (ed). *Coping with Natural Hazards: Indian Context.* Hyderabad: Orient Longman. 91–103.

0596 ■ 2005 Jain, S.K. *et al.* Delineation of flood prone areas using remote sensing techniques. *Water Resources Management,* 19 (4): 333–347.

0597 ■ 2005 Kalwar, S.C. *et al.* Drought and flood in Rajasthan. *Annals of National Association of Geographers, India,* 20 (1): 81–93.

0598 ■ 2006 Kumar, D. *et al.* Geo-informatics in flood mapping in a part of River Ramganga. In K.K. Sharma, *et al* (eds). *Environmental Geohazards: Science and Society.* New Delhi: Research India Press. 253–261.

0599 ■ 2006 Sanyal, J. and X.X. Lu. Geographic information systems based flood hazard mapping at different administrative scales: A case study in Gangetic West Bengal (India). *Singapore Journal of Tropical Geography,* 27(2): 207–220.

Occurrence: Heat and Cold Wave

0600 ■ 1938 Agarwala, K.S. Cold waves in India. *Indian Review,* 39 (4): 222–224.

0601 ■ 2000 India Disaster Report Team. Heat and cold waves. In S. Parasuraman and P.V. Unnikrishnan (eds). *India Disasters Report: Towards a Policy Initiative.* New Delhi: Oxford University Press. 177–178.

Occurrence: Heavy Rain

0602 ■ 1989 Dhar, O.N. and S.S. Nandargi. A study of heavy rainfall distribution over the intermediate catchment of the Narmada basin between Narmada Sagar and Sardar Sarovar dam sites. *Indian Journal of Power and River Valley Development,* 39 (9): 46–50.

0603 ■ 1989 Nageshkumar, S. Beyond rain gauge. *Frontline,* 22 (August 19): 110.

0604 ■ 1990 Gupta, G.R. A meso-scale study of unusually heavy rainfall over South Kerala. *Mausam,* 41 (1): 109–114.

0605 ■ 1992 Dubey, D.P. and T.K. Balakrishanan. A study of heavy to very heavy rainfall over Madhya Pradesh for the period 1977–1987. *Mausam,* 43 (4): 326–329.

0606 ■ 1993 Das, A.T. Exceptionally heavy rain over north Tamil Nadu during first half of November 1985. *Vayumandal,* 23 (3–4): 77–81.

0607 ■ 1993 Dhar, O.N. and S.S. Nandargi. Spatial distribution of severe rainstorms over India and their associated area rainfall depths. *Mausam,* 44 (4): 373–380.

0608 ■ 1993 Sharma, S.N. Cloudbursts in Rajasthan. In Pal, I. *et al* (eds). *Changing Contours of Arid Ecology: Physical and Cultural Heritage,* Volume I. Jaipur: Asian Environmental Council.

0609 ■ 1995 Dhar, O.N. and S.S. Nandargi. Zones of severe rainfall activity over India: Some refinements. *International Journal of Climatology,* 15: 811–819.

0610 ■ 2001 Rao, N.G. Occurrence of heavy rainfall around the confluence line in monsoon disturbances and its importance in causing floods. *Proceedings of Indian Academy of Sciences, Earth and Planetary Sciences,* 110 (1): 87–94.

0611 ■ 2004 Basu, G.C. and U. Bhattacharjee. Statistical analysis of rainfall distribution and trend of rainfall anomalies district wise. *Mausam,* 55 (3): 409–418.

0612 ■ 2004 Bhan, S.C. *et al.* Cloudbursts in Himachal Pradesh. *Mausam,* 55 (4): 712–713.

0613 ■ 2006 Goswami, B.N. *et al.* Increasing trend of extreme rain events over India in a warming environment. *Science,* 314: 1442–1445.

Occurrence: Landslide

0614 ■ 1894 Strachey, R. The landslip at Gohna in British Garhwal. *Geographical Journal,* 4: 162–170.

0615 ■ 1950 Ray, K.B. Landslides in the Himalayas. *Modern Review,* 88 (6): 493–494.

0616 ■ 1958 Gupta, R.S. A note on serious erosion and landslips beyond Kalsi on Dehradun-Chakrata road. *Indian Forester,* 84: 678–683.

0617 ■ 1962 Dey, A.K. Minutang landslide, Mishmi Hills, North East Frontier Area (Assam). In *Records of the Geological Survey of India,* 87 (3): 485–494.

0618 ■ 1974 Sen Sharma, S.B. and S.E. Hassan. Landslide at Nimu village along the Srinagar-Leh road, Ladakh (Jammu and Kashmir). *Indian Minerals,* 28 (4): 120–123.

0619 ■ 1975 Krishnaswamyan, V.S. and M.S. Jain. A review of some of the major landslides in Northern and Northeastern Himalayas. In *Proceedings of the Seminar on Landslides and Toe Erosion Problems with Special Reference to Himalayan Region.* Kolkata: Indian Society of Engineering Geology. 3–40.

0620 ■ 1975 Sinha, B.N. *et al.* Landslides in Darjeeling district and adjoining areas. *Bulletin of the Geological Survey of India,* Series B, No. 36. Kolkata.

0621 ■ 1979 Basu, P. Soil erosion and landslides in Darjeeling Himalayas. *Geographical Review of India*, 41 (1): 50–57.

0622 ■ 1979 Chaudhury, A.K. *A Note on the Study of Slides around the Bomdila Town, Kameng district (Arunachal Pradesh)*. Kolkata: Geological Survey of India (Unpublished Report).

0623 ■ 1980 Didwal, R.S. Occurrences of landslides in Jammu Province of State and their control. In *Proceedings of the International Symposium on Landslides*. New Delhi: Sarita Pradhan. 37–40.

0624 ■ 1980 Natrajan, T.K. *et al.* Some case record of landslides in Sikkim. In *Proceedings of the International Symposium on Landslides*. Delhi: Sarita Pradhan. 45–60.

0625 ■ 1980 Singhal, M.K. *et al.* Slips at Chila Power House site (Hardwar). In *Proceedings of the International Symposium on Landslides*. New Delhi: Sarita Pradhan. 221–226.

0626 ■ 1982 Seshagiri, N. *et al. The Nilgiri Landslides*. Miscellaneous Publication No. 57. Kolkata: Geological Survey of India.

0627 ■ 1983 Chaudan, R.P.S. and B.B. Mullick. *Progress Report on the Systematic Geological Mapping Investigation of Landslides in Parts of Udampur (Jammu and Kashmir)*. Kolkata: Geological Survey of India.

0628 ■ 1984 Bhandari, R.K. and G.S. Mehrotra. *Landslides in Garhwal Himalayas*. Roorkee Central Building Research Institute.

0629 ■ 1986 Valdiya, K.S. Accelerated erosion and landslide prone zones in the Central Himalayan region. In J.S. Singh (ed). *Environmental Regeneration in Himalaya, Concepts and Strategies*. Nainital: Central Himalayan Environment Association and Gyanodaya Prakashan. 12–38.

0630 ■ 1988 Bhattacharya, A. and D.K. Sengupta. Landslide hazard zonation along National Highway 31 from Sevok to Rangpo: An evaluation study. In *Proceedings of the Workshop on Geohazards and Related Societal Issues*. Lucknow: Indian Society of Engineering Geology.

0631 ■ 1988 Gupta, S.K. Landslide zonation: An input for geoenvironmental assessment of parts of Satluj valley . *Journal of Engineering Geology*, 17 (3–4): 53–66.

0632 ■ 1988 Jayakumar, M. and P.K. Vinayar. Case studies on landslides in Wynad Region of Western Ghats. In C. Bonnard (ed). *Proceedings of the Fifth International Symposium on Landslides*. Rotterdam: Balkema. 433–438.

0633 ■ 1988 Kumar, H. Landslide hazard near village Banswara near Rudraprayag-Kedarnath Road. In *Proceedings of the Workshop on Geohazards and Related Societal Issues*. Lucknow: Indian Society of Engineering Geology.

0634 ■ 1988 Paul, C. and K. Guha. Landslides hazards in Northeastern India with particular reference to Aizwal town. In *Proceedings of the Workshop on Geohazards and Related Societal Issues*. Lucknow: Indian Society of Engineering Geology.

0635 ■ 1989 Bartarya, S.K. and K.S. Valdiya. Landslides and erosion in the catchment of Gaula river, Kumaon. *Mountain Research and Development*, 9 (4): 405–419.

0636 ■ 1989 Joshi, K.C. *et al.* Landslide hazard zonation and its correlation with coseismic slides during Uttarkashi earthquake of 20 October 1991. In E. Brabb and B.L. Harrod (eds). *Landslides: Extent and Economic Significance*. Rotterdam: Balkema. 977–981.

0637 ■ 1990 Gairola, B.M. and S.D. Shukla. Preparation of landslide hazard zonation map of part of Uttar Pradesh Himalayas. *In Records of the Geological Survey of India,* 123 (8): 184–186.

0638 ■ 1990 Gupta, R.P. and B.C. Joshi. Landslide hazard zoning using geographic information system approach: A case study from Ramganga catchment (Himalaya). *Engineering Geology,* 28 (1–2): 119–131.

0639 ■ 1991 Froehlich, W. *et al.* Ambootia landslide valley in Darjeeling Himalaya active since 1968. *Bulletin of the Seismological Society of America,* 81: 111–128.

0640 ■ 1992 Chaubey, V.D. and P.K. Litoria *et al.* Landslide hazard zonation in Uttarkashi and Tehri districts (Uttar Pradesh Himalayas). In D.H. Bell (ed). *Proceedings of the Sixth International Symposium on Landslides.* New Zealand. Rotterdam: Balkema. 911–917.

0641 ■ 1992 Haigh, M.J. Analysis of landslide sites: Kilbury Road, Kumaon Himalaya. *Current Science*, 65 (7): 518–522.

0642 ■ 1992 Pauchari, A.K. and M. Pant. Landslide hazard mapping based on geological attributes. *Engineering Geology*, 32 (1–2): 81–100.

0643 ■ 1992 Rao, P.J. and D. Mukherjee. Nainital Highway: A case study in landslide hazard zonation. In D.H. Bell (ed). *Proceedings of the Sixth International Symposium on Landslide.* Rotterdam: Balkema. 1051–1055.

0644 ■ 1992 Sharan, R.B. Landslide hazard zonation in parts of Chenab basin (Jammu and Kashmir). *In Records of the Geological Survey of India,* 125 (8): 182–183.

0645 ■ 1992 Sharma, A. A study of landslide hazards from newspapers. *Geographical Review of India,* 54 (3): 56–77.

0646 ■ 1993 Sati, S.P. *et al.* Landslide hazard zones between the valleys of Alak nanda and Mandakini rivers. In G.S. Rajwar (ed). *Garhwal Himalaya: Ecology and Environment.* Delhi: Ashish Publishing House. 89–98.

0647 ■ 1993 Sharma, V.K. and G.C. Kandpal. *Report on the Preparation of Landslide Zonation Map of Bhagirathi Valley in Parts of Uttarkashi and Tehri Garhwal (Uttar Pradesh)* Kolkata: Geological Survey of India (Unpublished Report).

0648 ■ 1994 Chandra, P. and P.S. Bhatnagar. Landslide hazard zonation in parts of Beas catchment (Himachal Pradesh). In *Records of the Geological Survey of India,* 127 (8): 105–110.

0649 ■ 1994 Chandra, P. and P.S. Bhatnagar. *A Report on Preparation of Landslide Zonation Map of Part of Beas Valley Area, Districts Chamba and Kangra (Himachal Pradesh).* Kolkata: Geological Survey of India.

0650 ■ 1994 Joshi, K.C. and P. Pandey. Post earthquake landslide hazard evaluation, Uttarkashi district (Uttar Pradesh). In *Records of the Geological Survey of India,* 127 (8): 284–287.

0651 ■ 1994 Kandpal, G.C. Preparation of landslide zonation map of part of Uttarkashi. In *Records of the Geological Survey of India,* 127 (8): 235–237.

0652 ■ 1994 Sharda, Y.P. Landslides hazard zonation in parts of Chenab basin (Jammu and Kashmir). In *Records of the Geological Survey of India,* 127 (8): 110–114.

0653 ■ 1994 Sharma, V.K. Preparation of landslide zonation map of parts of Tehri and Uttarkashi districts (Uttar Pradesh). In *Records of the Geological Survey of India,* 127 (8): 232–234.

0654 ■ 1995 Bhatnagar, P.S. Landslide hazard zonation and detailed study of landslides around Bharmour and Thibra hydel project, Chamba district, Himachal Pradesh. In *Records of the Geological Survey of India,* 128(8): 241–241.

0655 ■ 1995 Chandra, P. Landslide hazard zonation and detailed study of ground behaviour and Thibra Hydel Projects, Chamba district. In *Records of the Geological Survey of India*, 128 (8): 239–240.

0656 ■ 1995 Gupta, P. *et al.* Landslide hazard zonation mapping around Shivpuri, Garhwal Himalaya (Uttar Pradesh). *Himalayan Geology*, 16: 95–102.

0657 ■ 1995 Gupta, V. *et al.* Landslide hazard zonation in the upper Satluj valley, Kinnaur district (Himachal Pradesh). *Himalayan Geology*, 16: 81–94.

0658 ■ 1995 Kandpal, G.C. Preparation of landslides zonation map of Bhagirathi valley in parts of Uttarkashi and Tehri districts. In *Records of the Geological Survey of India,* 128 (8): 242–244.

0659 ■ 1995 Karlekar, S.N. and U. Keskar. The quantitative assessment of a few landslides around Chipluan and Sangameshwar (Maharashtra). *Geographical Review of India,* 57 (1): 1–19.

0660 ■ 1995 Sarkar, I. *et al. Landslides Hazard Zonation in Parts of Sikkim (Northeast Himalaya).* Roorkee: Central Building Research Institute. (Unpublished report).

0661 ■ 1995 Sarkar, S. *et al.* Landslides hazard zonation: A case study in Garhwal Himalaya. *Mountain Research and Development,* 15 (4): 301–309.

0662 ■ 1996 Anbalagan, R. and S. Bhavani. Landslide hazard and risk assessment mapping of mountainous terrain: A case study from Kumaon Himalaya (India). *Engineering Geology*, 43 (4): 237–246.

0663 ■ 1996 Balachandran, V. *et al.* Marappalam landslide, the Nilgiri district (Tamil Nadu): A case study. In A.P. Santhakumar, *et al* (eds). *Proceedings of the International Conference on Disasters and Mitigation,* (A-4). Chennai: Anna University. 21–29.

0664 ■ 1996 Bartarya, S.K. *et al.* Landslides hazards: Some case studies from the Satluj valley. *Himalayan Geology,* 17: 193–208.

0665 ■ 1996 Champati Ray, P.K. *Landslides Hazard Zonation Using Fuzzy Logic and Probability Analysis in Western Himalayas.* Netherlands: Indian Institute of Remote Sensing-International Institute for Geo-Information Science and Earth Observation Programme.

0666 ■ 1996 Jhingran, B. and D. Mukharjee. Engineering geological approach for assessment of landslides hazards in Himalayan terrain for highways. In K. Senneset (ed). *Proceedings of Seventh International Symposium on Landslides.* Rotterdam: Balkema. 309–314.

0667 ■ 1996 Nagarajan, R. and M.V. Khire. Debris slides of Varandh Ghat, West Coast of India. *Bulletin of Engineering Geology and the Environment,* 57 (1): 59–63.

0668 ■ 1996 Naithani, A.K. and C. Prasad. Studies of landslides and mass movements along main central thrust zone in part of Garhwal Himalayas. In K. Senneset (ed). *Proceedings of the Seventh International Symposium on Landslides.* Rotterdam: Balkema. 1947–54.

0669 ■ 1996 Naithani, A.K. *et al.* Landslide zonation and geoenvironmental appraisal along main central thrust zone in Mandakini valley (Garhwal Himalaya). *Himalayan Geology,* 18: 135–144.

0670 ■ 1996 Ramaswamy, S.M. Frequent landslides in Nilgiris (India): A phenomenon related to Pleistocene tectonics. In K. Senneset (ed). *Proceedings of Seventh International Symposium on Landslides.* Rotterdam: Balkema.

0671 ■ 1996 Sengupta, C.K. and S.S. Ghosh. *Landslide Hazard Zonation along National Highway 31 A.* Kolkata: Geological Survey of India (Unpublished Report).

0672 ■ 1996 Sharma, V.K. A probabilistic approach for landslide zonation in part of Garhwal Himalaya. In K. Senneset (ed). *Proceedings of Seventh International Symposium on Landslides.* Rotterdam: Balkema. 381–386.

0673 ■ 1996 Sharma, V.K. and G.C. Kandpal. Landslide zonation mapping: An impact for geoenvironmental hazard assessment of a part of Garhwal Himalaya. In *Proceedings of the Symposium on Recent Advances in Geological Studies of Northwest Himalayas and Foredeep*. Special Publication No. 21. Kolkata: Geological Survey of India, 81–89.

0674 ■ 1996 Tiwari, R.S. and Snehmani. Use of remote sensing and geographic information systems technology of landslide hazard evaluation. In R.B. Singh(ed). *Disasters Environment and Development.* New Delhi: Oxford and IBH Publishing Co. Pvt. Ltd. 157–168.

0675 ■ 1997 Deva, Y. Slope failure along Chamba-Kharamukh road: A testimony to the fragility of the Himalayan slopes. In *Proceedings of the Symposium on the Earth Sciences in Environmental Assessment and Management, Lucknow*, 9–11 April 1996. Special Publication No. 48 (2). Kolkata: Geological Survey of India. 143–146.

0676 ■ 1997 Naithani, A.K. and C. Prasad. Landslide hazard zonation mapping in the Okhimath, Kedarnath area. In *Proceedings of the Symposium on the Earth Sciences in Environmental Assessment and Management.* Special Publication No. 48 (2). Kolkata: Geological Survey of India. 37–42.

0677 ■ 1997 Sambhu, V. *et al.* Landslide hazard analysis of the area around Dehradun and Mussoorie (Uttar Pradesh). *Current Science,* 73 (12):1117–1123.

0678 ■ 1998 Gupta, P. *Landslide Hazard Zonation Mapping Considering Geoenvironment Conditions of Parts of Bhagirathi River Valley, (Uttar Pradesh).* Roorkee: University of Roorkee (Unpublished Ph.D thesis).

0679 ■ 1998 Nagarajan, R. *et al.* Technical note on temporal remote sensing data and geographic information systems application in landslide hazard zonation of part of Western Ghat. *International Journal of Remote Sensing,* 19 (4): 573–585.

0680 ■ 1998 Prakash, C. Landslide inventory: A prerequisite for geohazard management. *Journal of Engineering Geology*, 27.

0681 ■ 1998 Raju, M. Landslide hazard zonation studies in southern part of Mizoram state (India). *Journal of Engineering Geology*, 27: 63–65.

0682 ■ 1998 Sidharthan, S. *et al.* Relevance of socio-economic investigations in landslides delineation: A case study in part of the Western Ghats of Kerala. *Indian Cartographer,* 17: 281–289.

0683 ■ 1998 Valdiya, K.S. Catastrophic landslides in Uttaranchal (Central Himalaya). *Journal of Geological Society of India,* 52 (4): 483–486.

0684 ■ 1999 Gupta, R.P. *et al.* Landslide hazard zonation in a part of the Bhagirathi valley (Garhwal Himalayas), using remote sensing and geographic information systems. *Himalayan Geology,* 20 (2): 71–85.

0685 ■ 1999 Joshi, D.D. *A Report on Landslide Hazard Zonation Studies around Itanagar Capital Complex, (Arunachal Pradesh).* Kolkata: Geological Survey of India.

0686 ■ 1999 Pant, P.D. and K. Luirei. Malpa rockfalls of 18 August 1998 in the Northeastern Kumaon Himalaya. *Journal of Geological Society of India,* 54 (4): 415–420.

0687 ■ 1999 Raju, N. *et al. A Comprehensive Report on Landslide Hazard Zonation of South Mizoram.* Kolkata: Geological Survey of India (Unpublished Report).

0688 ■ 1999 Thampi, P.K. *et al.* Landslide hazard zonation: A case study from Western Ghat. *Journal of Geological Society of India,* 53.

0689 ■ 2000 Acharya, S.K. and P.L. Narula. Landslides and seismic hazard zonation. In P.S. Roy, *et al* (eds). *Natural Disasters and their Mitigation: A Remote Sensing and Geographic Information Systems Perspective.* Dehradun: Indian Institute of Remote Sensing. 77–88.

0690 ■ 2000 Champati Ray, P.K. Operationalisation of cost-effective methodology for landslides hazard zonation using remote sensing and geographic information systems: Indian Institute of Remote Sensing initiative. In P.S. Roy, *et al* (eds). *Natural Disasters and their Mitigation: A Remote Sensing and Geographic Information Systems Perspective.* Dehradun: Indian Institute of Remote Sensing. 95– 101.

0691 ■ 2000 Champati Ray, P.K. *Operationalisation of Cost Effective Methodology for Landslide Hazard Zonation using Remote Sensing and Geographic Systems: Indian Institure of Remote Sensing Initiative.* Dehradun: Indian Institute of Remote Sensing and National Remote Sensing Agency.

0692 ■ 2000 India Disasters Report Team. Landslides and floods: An update. In S. Parasuraman and P.V. Unnikrishnan (eds). *India Disasters Report: Towards a Policy Initiative.* New Delhi: Oxford University Press. 164.

0693 ■ 2000 Mukhopadhyay, S.C. Landslides in the Sikkim, Darjeeling Himalaya: An environmental hazard. *Science and Culture,* 66 (1–2): 26–39.

0694 ■ 2000 Nagarajan, R. *et al.* Landslide hazard susceptibility mapping based on terrain and climatic factors for tropical monsoon regions. *Bulletin of Engineering Geology and the Environment,* 58 (4): 275–287.

0695 ■ 2000 Sarkar, S. and D.P. Kanungo. Landslides: Hazard zonation and risk assessment. In P.S. Roy, *et al* (eds). *Natural Disasters and their Mitigation: A Remote Sensing and Geographic Information Systems Perspective.* Dehradun: Indian Institute of Remote Sensing. 102–109.

0696 ■ 2000 Singh, R.B. Landslide hazard in Himachal Himalaya. In V. Subramanian and R. Jayakumar (eds). *Proceedings of the International Workshop on Geoenvironmental Hazards.* Delhi: United Nations Educational, Scientific and Cultural Organisation and Jawaharlal Nehru University. 71–84.

0697 ■ 2001 Ghosh, D. and A. Mitra. Landslide hazard zonation mapping in East Sikkim district.In *Proceedings of National Seminar on Geotechniques and Geological Hazards in the Indian Context.* Kolkata: Geological Survey of India.

0698 ■ 2001 Khullar, V.K. Landslides in Mizoram (Northeast India): An opinion. In *Proceedings of the National Seminar on Geotechniques and Geological Hazards in the Indian Context.* Kolkata: Geological Survey of India.

0699 ■ 2001 Malbarna, B.D. and G.S. Manhas. Development of landslide at Udaipur end portal of tunnel: Udaipur-Katra gauge railway line. In *Proceedings of the Role of Earth Sciences in Integrated Development and Related Societal Issues.* Special Publication No. 65 (2). Kolkata: Geological Survey of India. 116–121.

0700 ■ 2001 Nainwal, H.C. and C. Prasad. Micro-zonation of slope instability using slope mass rating approach: A case study from Garhwal Himalaya. *Himalayan Geology,* 22 (2): 135.

0701 ■ 2001 Pandey, B.W. Landslides hazard: Typology, mechanism and mapping in Western Himalaya. *Annals of National Association of Geographers, India,* 21 (1): 23–35.

0702 ■ 2001 Raju, M. Rapid assessment of landslide hazard zonation: A fast track approach. In *Proceedings of the Role of Earth Sciences in Integrated Development and Related Societal Issues.* Special Publication No. 65 (2). Kolkata: Geological Survey of India. 51–59.

0703 ■ 2001 Sidharthan, S. *et al.* Application of geographic information systems in landslide hazard zonation. *Indian Cartographer,* 20: 149–158.

0704 ■ 2001 Singh I. and S.K. Kulshrestha. Landslide hazard zonation along Upper Beas valley, Kullu district. In *Proceedings of the Role of Earth Sciences in Integrated Development and Related Societal Issues.* Special Publication No. 65 (2). Kolkata: Geological Survey of India. 77–83

0705 ■ 2001 Singh, L.P. *et al.* Micro landslide hazard zonation studies for urban planning around Gangtok (Sikkim State). In *Proceedings of the Role of Earth Sciences in Integrated Development and Related Societal Issues.* Special Publication No. 65 (1). Kolkata: Geological Survey of India. 145–154.

0706 ■ 2002 Saha, A.K. *et al.* Geographic information systems based landslide hazard zonation in the Bhagirathi valley (Himalayas). *International Journal of Remote Sensing,* 23 (3): 357–369.

0707 ■ 2002 Sharma, V.K. and P. Jaiswal. *Landslide Hazards Zonation Studies along Almora-Mansarovar Pilgrim Route and Hazard Assessment in the Vicinity of Some Important Cultural Heritage Sites, Kumaon Himalaya.* Kolkata: Geological Survey of India (Unpublished Report).

0708 ■ 2002 Singh, R.B and B.W. Pandey. Landslides in Himachal Himalaya. In V. Subramaniam (ed). *Environmental Hazards in South Asia.* New Delhi: Concept Publishing Company. 55–64.

0709 ■ 2002 Soman, K. *et al.* Need for multiparametric landslide hazard zonation. *Journal of Geological Society of India,* 60 (2): 232.

0710 ■ 2003 M. Mohanty and S. Aggarwal (eds). *Applications of Geographic Information Systems in Earthquake and Landslide Hazard Assessment.* New Delhi: Department of Science and Technology, Government of India.

0711 ■ 2003 Pal, D.J. and A. Kaushal. Application of geomatics technology in landslide hazard zonation. In M. Mohanty and S. Aggarwal (eds). *Applications of Geographic Information Systems in Earthquake and Landslide Hazard Assessment.* New Delhi: Department of Science and Technology, Government of India. 22–24.

0712 ■ 2003 Sharma, R. *A Study of Landslide Hazard along Chamba-Brahmapur Road in District Chamba, Himachal Pradesh (India).* Shimla: Department of Geography, University of Himachal Pradesh (Unpublished Master of Philosophy Dissertation).

0713 ■ 2004 Arora, M.K *et al.* An artificial neural network approach for landslide hazard zonation in the Bhagirathi, Ganga valley, (Himalaya). *International Journal of Remote Sensing,* 25 (6): 559–572.

0714 ■ 2004 Bhandari, R.K. Landslide hazards zonation: Some thoughts. In K.S. Valdiya (ed). *Coping with Natural Hazards: Indian Context.* Hyderabad: Orient Longman. 134–152.

0715 ■ 2004 Bhattachraya, A. Remote sensing and landslides hazard zonation of road network in parts of Chamba valley. In R. Nagarajan (ed). *Landslide Disaster: Assessment and Monitoring.* New Delhi: Anmol Publications. 71–80.

0716 ■ 2004 Kulkarni, M.N. Global positioning system and its application to landslide. In R. Nagarajan (ed). *Landslide Disaster: Assessment and Monitoring.* New Delhi: Anmol Publications. 141–155.

0717 ■ 2004 Kundu, S. and A.P. Krishna. Landslides hazard assessment along the selected highways in Sikkim Himalaya: A remote sensing and geographic information systems based approach. In R. Nagarajan (ed). *Landslide Disaster: Assessment and Monitoring.* New Delhi: Anmol Publications. 104–112.

0718 ■ 2004 Mathur, V.K. and S. Sarkar. Recent trends in landslide hazard zonation and risk assessment: An overview. In R. Nagarajan (ed). *Landslide Disaster: Assessment and Monitoring.* New Delhi: Anmol Publications. 97–103.

0719 ■ 2004 Nagarajan, R. Landslide hazard mapping of seismic areas. In R. Nagarajan (ed). *Landslide Disaster: Assessment and Monitoring.* New Delhi: Anmol Publications. 222–240.

0720 ■ 2004 Ray, C. Landslides hazard zonation using spatial models in geographic information systems. In R. Nagarajan (ed). *Landslide Disaster: Assessment and Monitoring.* New Delhi: Anmol Publications. 81–96.

0721 ■ 2005 Geological Survey of India. *Inventory of Landslides of Northwest Himalaya.* Special Publication No. 71. Kolkata.

0722 ■ 2005 Pant, P.D. and K. Luirei. Amiya landslide in the catchment of Gaula river Southern Kumaon (Uttaranchal). *Journal of Geological Society of India,* 65 (3): 291–295.

0723 ■ 2005 Sarkar, S. and P.K. Gupta. Techniques for landslide hazard zonation: Application to Srinagar-Rudraprayag area of Garhwal Himalaya. *Journal of Geological Society of India,* 65 (2): 217–230.

0724 ■ 2006 Gupta, S.S. and V.K. Verma. Landslide hazards in Indian hilly areas. In K.K. Sharma, *et al* (eds). *Environmental Geohazards: Science and Society.* New Delhi: Research India Press. 162–170.

0725 ■ 2006 Gupta,V. Landslides hazard zonation mapping in the Himalayas: A case study from Satluj Valley. In K.K. Sharma, *et al* (eds). *Environmental Geohazards: Science and Society.* New Delhi: Research India Press. 102–116.

0726 ■ 2006 Kanungo, D.P. *et al.* A comparative study of concentional, artificial neural network black box, fuzzy and combined neural and fuzzy weighting procedures for landslide susceptibility zonation in Darjeeling Himalayas. *Engineering Geology,* 85 (3–4): 347–366.

0727 ■ 2006 Singh, T. and M. Devi. Landslide occurrences and risk assessment in Itanagar capital complex (Arunachal Pradesh). *Himalayan Geology,* 27 (1): 145–162.

Occurrence: Storm

0728 ■ 1937 Agarwala, K.S. Hail storms in India. *Indian Review,* 38 (10): 615–616.

0729 ■ 1945 Agarwala, K.S. Storms in the Indian seas. *Indian Review,* 46 (12): 701–705.

0730 ■ 1946 Agarwala, K.S. Dust storms in India. *Indian Review,* 47 (10): 533–535.

0731 ■ 1974 Kumar, L. *On Thunderstorms and Hailstorms in the Extreme Northeast Indian Region.* Varanasi: Banaras Hindu University (Unpublished Ph.D thesis).

0732 ■ 1980 Gupta, H.N. and S.K. Ghosh. North Delhi tornado of 17 March 1978. *Mausam,* 31: 93–100.

0733 ■ 1989 Chatterjee, R.N. and P. Prakash. Radar study of severe thunderstorms around Delhi during the monsoon season. *Atmospheric Research,* 22 (4): 373–384.

0734 ■ 1990 Chatterjee, R.N. and P. Prakash. Radar study of thunderstorms around Delhi during monsoon season. *Mausam,* 41 (2): 161–165.

0735 ■ 1993 Dhar, O.N. and S.S. Nandargi. The zones of severe rainstorm activity over India. *International Journal of Climatology,* 13.

0736 ■ 1996 Yadav, S.C. and G.R. Chary. Hailstorm: A recurrent phenomenon in Vidharbha. *Geographical Review of India,* 58 (1): 33–40.

0737 ■ 2000 India Disaster Report Team. Tornadoes in the East Coast. In S. Parasuraman and P.V. Unnikrishnan (eds). *India Disasters Report: Towards a Policy Initiative.* New Delhi: Oxford University Press. 198.

0738 ■ 2003 Mandke, S.K. and U.V. Bhide. Study of decreasing storm frequency over Bay of Bengal. *Journal of Indian Geophysical Union,* 7 (2): 53–58.

0739 ■ 2005 Basu, G.C. Diurnal variation of thunderstorm activities over Kolkata during pre-monsoon season. *Mausam,* 56 (3): 707–710.

Occurrence: Tsunami

0740 ■ 1999 Murty, T.S. *et al.* Tsunamis on the coastlines of India. *Science of Tsunami Hazards,* 17 (3).

0741 ■ 2004 Charaborty, S.C. Tsunami of 26 December 2004. *Geographical Review of India,* 66 (4): 311–320.

0742 ■ 2005 Ballukraya, P.N. Tsunami of 26 December 2004: Observations at Chennai and East Coast of India. *Journal of Geological Society of India,* 65 (3): 368–369.

0743 ■ 2005 Gupta, H.K. A note on the 26 December 2004 tsunami in the Indian Ocean. *Journal of Geological Society of India,* 65 (2): 247–248.

0744 ■ 2005 Jagkumar, S. and D. Ilangovan. Run up and inundation limits along southeast coast of India during the 26 December 2004 Indian Ocean tsunami. *Current Science,* 88 (11): 1741–1743.

0745 ■ 2005 Mahadevan, T.N. General path of the tsunami of 26 December 2004. *Journal of Geological Society of India,* 65 (3): 374.

0746 ■ 2005 Mishra, D.C. and R.P. Rajasekhar. Tsunami of 26 December 2004 and related tectonic setting. *Current Science,* 88 (5): 680–682.

0747 ■ 2005 Ramachandran, R. Waves from the past. *Frontline,* 22 (2): 116–120.

0748 ■ 2005 Sadhuram, Y. Tsunami of 26 December 2004. *Science and Culture,* 88 (10): 1530–1531.

0749 ■ 2005 Sekhar, R.M. Natural calamities: Tsunami waves. *Yojna,* 46: 20–25.

0750 ■ 2006 Rajendran, N. History of tsunami. In S.M. Ramaswamy *et al* (eds). *Geomatics in Tsunami.* New Delhi: New India Publishing Agency. 1–9.

Occurrence: Assorted

0751 ■ 1972 Special Correspondent. Disastrous weather events of 1971. *Indian Journal of Meteorology and Geophysics,* 23 (1): 110–111.

0752 ■ 1974 Special Correspondent. Disastrous weather events of 1973. *Indian Journal of Meteorology and Geophysics,* 25 (3–4): 515–518.

0753 ■ 1976 Singh, D.J. *Natural Calamities in Jammu and Kashmir State from the Middle of 19 Century to 1925 A.D.* Jammu: Jammu University (Unpublished Ph.D thesis).

0754 ■ 1983 Aggarwal, C.M. *Natural Calamities and the Great Moghuls.* Patna: Bodh Gaya Kanchan Publications.

0755 ■ 1989 Das, P.K. The Norwesters as a natural hazard. *Science and Culture,* 55 (8): 299–305.

0756 ■ 1992 Raghavandran, K. A natural disaster century. In *Proceedings of the World Congress on Natural Hazards Reduction,* 307–311.

0757 ■ 1993 International Federation of Red Cross and Red Crescent Society. *World Disasters Reports* (Annual Publications: Latest volume 2006).Geneva.

0758 ■ 1995 Singh, M. *History of Natural Calamities in Bihar since Independence (1947–1990): A Case Study*. Ranchi: Ranchi University (Unpublished Ph.D thesis).

0759 ■ 1996 Singh, R.B. and B.W. Pandey. Hazard zone mapping and risk assessment analysis in the upper Beas basin. In R.B. Singh (ed). *Disasters, Environment and Development*. New Delhi: Oxford and IBH Publishing Co. Ltd. 121–136.

0760 ■ 1997 Bhatta, B.B. Natural calamities in Orissa during the British rule. In N.R. Patina (ed). *Economic History of Orissa*. New Delhi: Indus Publishing Company. 356–365.

0761 ■ 1997 Bhatta, B.B. *The Natural Calamities in Orissa in the 19 Century*. New Delhi: Commonwealth Publishers.

0762 ■ 1997 Singh, R.B. *et al.* Risk from natural hazards in the Manali area. In F. Berkes and J.S. Gardner (eds). *Sustainability of Mountain Environments in India and Canada* Winnipeg: University of Manitoba. 237–254.

0763 ■ 1998 Lok Soochna Evam Sahayata Kendra. *The Calamity Prone Central Himalayas*. Chamoli: Dasholi Gram Swaraj Mandal.

0764 • 2000 Singh, R.B. and D. Tingal. Man-made and natural geohazards in Kullu valley. In R.B. Singh (ed). *Disaster Management*. Jaipur: Rawat Publications. 259–286.

0765 ■ 2000 Subramanian, V. and R. Jayakumar (eds). *Proceedings of the International Workshop on Geoenvironmental Hazards*. New Delhi: United Nations Educational, Scientific and Cultural Organisations and Jawaharlal Nehru University.

0766 ■ 2001 De. U.S. and M.M. Dandekar. Natural disasters in urban areas. *Deccan Geographer*, 39: 1–13.

0767 ■ 2002 Gardener, J.S. Natural hazard risk in the Kullu district (Himachal Pradesh). *Geographical Review of India*, 92 (1): 282–302.

0768 ■ 2002 Pandey, B.W. *Geoenvironmental Hazards in Himalayas: Assessment and Mapping (the upper Beas basin)*. New Delhi: Mittal Publication.

0769 ■ 2005 De, U.S. *et al.* Extreme weather events over India in the last 100 years. *Journal of Indian Geophysical Union*, 9 (3): 173–187.

0770 ■ 2006 Pande, R.K. Hazard zonation in Uttaranchal (India): A tool to understand the problem. *Disaster Prevention and Management*, 15 (4): 598–607.

0771 ■ 2006 Pandey, B.W. Hydrogeomorphic hazards assessment and mapping in the marginal region of Western Himalayas. In K.K. Sharma, *et al* (eds). *Environmental Geohazards: Science and Society*. New Delhi: Research India Press, 262–284.

Characteristics

A characteristic refers to the distinguishing features or quality of any object, event, place or time. By attempting to answer the curiosity of what a characteristic signifies references that evaluate, estimate or even describe the attributes of disaster are grouped under the umbrella of characteristics. These could pertain to geological, meteorological, hydrological or any of the anthropological dimensions. Characteristics also refers to measures of their magnitude, intensity and frequency. Since overviews, case studies, observations, and notes detail on characteristics these are drawn within its catchment. Within the 4004 research references those that fall into the class of characteristics are 292. This is 8 per cent of the total (Tables 6.10, 6.11 and 6.12).

Table 6.10 References for each Category

'Natural' Disaster	References	Number
Cyclone	0772–0822	51
Drought	0823–0849	27
Earthquake	0850–0971	122
Floods	0972–1009	38
Heat and Cold Wave	1010–1011	2
Heavy Rain	1012–1017	6
Landslide	1018–1028	11
Storm	1029–1042	14
Tsunami	1043–1052	10
Assorted	1053–1060	8

Table 6.11 Research References on Characteristics

Book	Report	Thesis	Article[1]	Chapter[2]	Feature[3]	Total	Book
Cyclone	0	2	2	45	1	1	51
Drought	1	2	5	14	5	0	27
Earthquake	1	10	4	96	5	6	122
Floods	1	8	5	14	9	1	38
Heat and Cold Wave	0	0	0	2	0	0	2
Heavy Rain	0	0	0	4	2	0	6
Landslide	0	0	2	4	5	0	11
Storm	0	1	0	12	1	0	14
Tsunami	2	3	0	5	0	0	10
General Hazard	1	1	1	0	5	0	8
Total	**6**	**27**	**19**	**196**	**33**	**8**	**289**

[1] *in a journal;* [2] *in a book/report/proceedings;* [3] *in a magazine.*

Table 6.12 Dominance of Natural over Human in Research References on Characteristics

Phase	Natural	Human	Ecological	Total
1830–1947	18	0	0	18
1948–1989	84	6	7	97
1990–2006	159	8	10	177
Total	**261**	**14**	**17**	**292**

Characteristics: Cyclone

0772 ■ 1937 Sen, S.N. Mechanism of cyclones in Bay of Bengal. *Science and Culture,* 3 (9): 458–461.

0773 ■ 1953 Jordan, C.L. On the formation of Masulipatnam cyclone of 1949. *Indian Journal of Meteorology and Geophysics*, 4 (4): 339–342.

0774 ■ 1965 Bhattacharjee, P. and A.C. De. Radar study of the cyclonic storm of 21 September 1962 in the Bay of Bengal. *Indian Journal of Meteorology and Geophysics,* 16 (1): 81–84.

0775 ■ 1965 Das Gupta, S.C. Severe cyclonic storm of Bay of Bengal on 11–12 May 1965: Some noteworthy features as observed and felt at Agartala. *Indian Journal of Meteorology and Geophysics*, 16 (4): 713–722.

0776 ■ 1965 Raman, S. The severe cyclonic storm of 11–12 June 1964 in the Arabian Sea and some of its surface characteristics. *Indian Journal of Meteorology and Geophysics,* 16 (4): 623–630.

0777 ■ 1967 Special Correspondent. Cyclonic storm of 13 November. *Indian Journal of Meteorology and Geophysics,* 18 (1): 163–169.

0778 ■ 1968 Mukharjee, A.K. and P.K. Mishra. Satellite study of the Bay cyclone. *Indian Journal of Meteorology and Geophysics,* 19 (3): 295–304.

0779 ■ 1976 Arunachalam, G. and R. Murlikrishna. A radar study of the Gopalpur cyclone of 20–25 September 1972. *Indian Journal of Meteorology, Hydrology, and Geophysics,* 27 (3): 299–302.

0780 ■ 1976 Mishra, D.K. *et al.* Estimation of maximum wind speed in tropical cyclones occurring in Indian seas. *Indian Journal of Meteorology, Hydrology and Geophysics,* 27 (3): 285–290.

0781 ■ 1977 Mukherji, A.K. *et al.* Some features of Porbander cyclone, October 1975. *Indian Journal of Meteorology, Hydrology, and Geophysics,* 28 (4): 439–446.

0782 ▲ 1979 Government of Andhra Pradesh. *Note on the Recent Cyclone.* Hyderabad.

0783 ■ 1980 Johns, B. and A. Ali. The numerical modelling of storm surges in the Bay of Bengal. *Quarterly Journal of Royal Meteorological Society,* 106: 1–18.

0784 ■ 1980 Raghavan, S. *et al.* Radar study of the Bay of Bengal cyclone of 19 November 1977. *Mausam,* 31 (2): 229–240.

0785 ■ 1980 Raghavan, S. *et al.* The Andhra cyclone of 12 May 1979: The radar meteorologist's point of view. *Mausam,* 31 (4): 573–580.

0786 ■ 1980 Rao, B.M. *et al.* A preliminary study of the Saurashtra cyclone from television infrared operational satellite-next generation. *Mausam,* 31(2): 225–228.

0787 ■ 1981 Indian Meteorological Department. The intensity of Andhra cyclone of 1977. *Mausam,* 32 (3): 321–328.

0788 ■ 1981 Mandal, G.S. *et al.* Characteristics of an Arabian Sea cyclone. *Mausam,* 32 (2): 139–144.

0789 ■ 1981 Subbaramayya, I. and S.R.M. Rao. Cyclone climatology of North Indian Ocean. *Indian Journal of Marine Sciences,* 10: 366–68.

0790 ■ 1981 Subramanian, D.V. Radar observations of cyclones in the Arabian Sea and the Bay of Bengal. *Mausam,* 32 (2): 189–194.

0791 ■ 1984 Lakshminarayanan, S. Cyclone eye. *Mausam,* 35 (4): 425–434.

0792 ■ 1986 Mooley, D.A. and C.M. Mohile. Some aspects of the rainfall associated with cyclonic storms of the Bay of Bengal. *Journal of Climatology,* 6 (2): 149–160.

0793 ▲ 1987 Prasad, K.N. *et al.* Studies on November 1977 and May 1979 cyclones of the Southern Andhra coastal areas. In *Records of the Geological Survey of India,* 115(5): 97–115.

0794 ■ 1989 Kalsi, S.R. *et al.* On aspects of marginal cyclones. *Mausam,* 40 (1): 47–50.

0795 ■ 1989 Pandey, M.K. *et al.* Satellite and radar study of Sriharikota severe cyclonic storm of November 1984. *Mausam,* 40 (4): 425–430.

0796 ■ 1989 Verghese, P.A. *Characteristics of 76 mm Dense Medium Cyclone.* Dhanbad: Indian School of Mines (Unpublished Ph.D thesis).

0797 ■ 1990 Lakshminarayan, S. Fujiwara effect in cyclone pairs of southern tropics. *Vayumandal,* 20 (1–2).

0798 ■ 1990 Mandal, G.S. and Chakraborty. Composite sounding of a monsoon depression in the Bay of Bengal and a cyclonic storm in the Arabian Sea. *Vayumandal,* 20 (1–2): 25–32.

0799 ■ 1990 Raghavan, S. Structure of tropical cyclones in the Bay of Bengal. *Mausam,* 41 (2): 325–328.

0800 ■ 1991 Mukherjee, A.K. Some less known facts about tropical cyclones. *Vayumandal,* 21 (1–2): 8–15.

0801 ■ 1992 Subramaniam, S.K. and G. Mohnan. Some features of the Bay cyclone of May 1990. *Vayumandal,* 22 (3–4): 97–102.

0802 ■ 1993 Baneerjee, S. *et al.* Micro-computer based modelling and design of cyclone separators. *Journal of Institution of Engineers India,* 73: 65–68.

0803 ■ 1994 Kesari, A.R. *The Role of Surface Layer in Cyclogenesis and Some Structural Aspects of Bay Cyclones.* Vishakhapatnam: Andhra University, Waltair (Unpublished Ph.D thesis).

0804 ■ 1996 Sen Sarma, A.K. Reconstructing the Great Bengal cyclone of 1737. *Mausam,* 47 (1): 67–72.

0805 ■ 1998 Behera, S.K. *et al.* Investigation of mixed layer response to Bay of Bengal cyclone using a simple ocean model. *Meteorology and Atmospheric Physics*, 65 (1–2): 77–91.

0806 ■ 1999 Indian Meteorological Department. *A Report on Cyclonic Disturbances over North Indian Ocean during 1998*. Delhi: Indian Meteorological Department.

0807 ■ 1999 Sharma, R.V. Structure of the Kakinada cyclone of 6 November 1996. *Mausam,* 50 (2): 129–134.

0808 ■ 2000 Gadgil, A.S. Some aspects of Orissa Super Cyclone of October 1999. *Transactions, Institute of Indian Geographers,* 22 (2): 52–55.

0809 ■ 2001 Alam, M.M. *et al.* Study of some meteorological aspect of November 1992 cyclone: A case study. *Mausam,* 52 (3): 602–605.

0810 ■ 2003 Kalsi, S.R. *et al.* Decaying nature of Super Cyclone of Orissa after landfall. *Mausam,* 54 (2): 393–396.

0811 ■ 2004 Desai, D.S. and M.G. Huparikar. Comparison of two Super Cyclone storms over the Bay of Bengal: 14–20 November 1977 and 2–31 October 1999. *Mausam,* 55 (4): 715–720.

0812 ■ 2004 Dube, S.K. *et al.* Numerical modelling of storm surge in the head of Bay of Bengal using location specific model. *Natural Hazard Journal,* 31 (2): 437–453.

0813 ■ 2005 Bhagat, D.K.U.R. Balance of kinetic energy in the field of mid-tropospheric cyclone. *Mausam,* 56 (2): 473–476.

0814 ■ 2005 Bhagat, D.K.U.R. Absolute angular momentum generation and its vertical transport in the field of mid-tropospheric cyclones. *Mausam,* 56 (4): 759–764.

0815 ■ 2005 Kumar, P. Mathematical approach on size, maximum wind, radius of maximum wind and inflow angle in tropical cyclone. *Mausam,* 56 (4): 893–899.

0816 ■ 2005 Singh, C. and B.K. Bandhopadhyay. Decaying characteristics of severe cyclonic storms after rainfall over East Coast of India. *Mausam,* 56 (2): 395–400.

0817 ■ 2006 Kalsi, S.R. Orissa Super Cyclone: A synopsis. *Mausam,* 57 (1): 1–20.

0818 ■ 2006 Kalsi, S. R. and K. B. Srivastava. Characteristic features of Orissa Super Cyclone of 29 October 1999 as observed through cyclone detection radar at Paradip. *Mausam*, 57 (1): 21–30.

0819 ■ 2006 Loe, B. R. *et al.* Comparative analysis of subjective and advanced objective techniques of tropical cyclone intensity estimation. *Mausam,* 57 (1): 159–164.

0820 ■ 2006 Loe, B. R. *et al.* Recent very severe tropical cyclone over the Bay of Bengal: Analysis with satellite data. *Mausam,* 57 (1): 37–46.

0821 ■ 2006 Varma, A. K. *et al.* Study of geophysical parameters associated with the Orissa Super Cyclone, using active and passive microwave remote sensing measurements. *International Journal of Remote Sensing*, 27 (18): 3753–3765.

0822 ■ 2006 Venkatesh, T. N. Meso-scale interactions during the genesis and intensification of October 1999 Orissa Super Cyclone. *Mausam,* 57 (1): 31–36.

Characteristics: Drought

0823 ■ 1964 Subramanyam, V. P. *et al.* Application of water balance concept for a climatic study of drought in South India. *Indian Journal of Meteorology and Geophysics,* 15 (4): 393–402.

0824 ■ 1971 Subrahmanyam, V. P. and C. V. S. Sastry. Studies in drought climatology: Dry climates of South India. *Tropical Ecology,* 12 (2): 250–256.

0825 ■ 1971 Subramanyam, V. *Studies in Drought Climatology of Dry Climatic Zones of South India.* Vishakhapatnam: Andhra University, Waltair (Unpublished Ph.D thesis).

0826 ▲ 1973 Bhasin, M. G. Drought: A geographical appraisal. *Deccan Geographer,* 11 (1 and 3): 95–96.

0827 • 1973 Ladenjinsky, W. Drought in Maharashtra: Not in hundred years. *Economic and Political Weekly*, 8 (46): 383–396.

0828 ■ 1973 Ratnam, B. P. *et al.* Drought climatology at Dharwar. *Indian Journal of Agricultural Science*, 28: 799–801.

0829 ▲ 1973 Srivastava, R. K. *A Geographical Study of Recent Drought Affected Areas of Uttar Pradesh and Bihar.* Allahabad: University of Allahabad (Unpublished Ph.D thesis).

0830 ■ 1973 Subrahmanyam, V. P. Studies in drought climatology: Moist climates of South India. *Tropical Ecology,* 14.

0831 ▲ 1974 Reddy, N. B. K. (ed). *Drought Prone Areas of India.* Tirupati: Sri Venkateshwara University.

0832 ■ 1975 Ramasastry, C. V. *Studies in Drought Climatology of the Different Climatic Zones of North India.* Vishakhapatnam: Andhra University, Waltair (Unpublished Ph.D thesis).

0833 • 1979 Despande, C.D. Drought prone areas and the human resource. In N.B.K Reddy (ed). *Proceedings of the All India Symposium on Drought Prone Areas of India.* Tirupati: Rayalseema Geographical Society. 43–46.

0834 • 1979 Reddy, N.K.B. Drought prone areas of India. In N.B.K Reddy (ed). *Proceedings of the All India Symposium on Drought Prone Areas of India.* Tirupati: Rayalseema Geographical Society.

0835 • 1979 Venugopal, B. and R. Geetha. Drought and famines in Rayalseema: A study of drought prone areas. In N.B.K Reddy (ed). *Proceedings of the All India Symposium on Drought Prone Areas of India.* Tirupati: Rayalseema Geographical Society. 93–104.

0836 ■ 1980 Ram Mohan, H.S. and V.P. Subrahmanyam. Drought climatology of the semi-arid zone of Tamil Nadu. *Indian Geographical Journal,* 55 (1): 12–21.

0837 ■ 1981 Subrahmanyam, V.P. Drought climatology. *Annals of National Association of Geographers, India,* 1 (2): 11–16.

0838 ■ 1984 Chowdhury, A. and V.P. Abhyankar. On some climatological aspects of drought in India. *Mausam,* 35 (3): 375–378.

0839 ■ 1984 Ravindranath, M. *Drought Climatology of the Central South India Region (Andhra Pradesh).* Vishakhapatnam: Andhra University, Waltair (Unpublished Ph.D thesis).

0840 ■ 1986 Diwedi, D.K. Climatology of drought of North India. *Geographical Review of India,* 48 (3): 27–33.

0841 ■ 1986 Gregory, S. and B. Parthasarathy. Extreme monsoonal rainfall deficits in India. *Theoretical and Applied Climatology*, 37 (4): 194–204.

0842 ■ 1992 Mishra, S. Drought: A case study of West Bengal (1901–1990). *Geographical Review of India*, 54 (4): 71–80.

0843 ■ 1992 Mohan, S. and S. Udanti. An optimization model for drought conditions. *Indian Journal of Power and River Valley Development*, 43 (4): 9–16.

0844 ■ 1995 Chaudhuary, J.L. Drought classifaction study of Bastar district of Chattisgarh region. *Geographical Review of India*, 57 (2): 126–137.

0845 ■ 1997 Samuel, R.B. *Drought Climatology: Water Balance and Development of Land and Water Resources of Kurnool District (Andhra Pradesh).* Anantapur: Sri Krishna Devaraya University (Unpublished Ph.D thesis).

0846 ▲ 2001 Gupta, L.C. and V.K. Sharma. *Drought in Rajasthan: 1999–2000 and 2000–2001*. New Delhi: National Centre for Disaster Management.

0847 • 2001 Gupta, L.C. and V.K. Sharma. *Drought in Gujarat: 1999–2000 and 2000–2001*. New Delhi: National Centre for Disaster Management.

0848 ■ 2001 Gupta, L.C. and V.K. Sharma. Drought in Rajasthan and Gujarat 1999–2000 and 2000–01. In V.K. Sharma (ed). *Disaster Management.* New Delhi: Indian Institute of Public Administration. 108–125.

0849 ■ 2002 Singh, R.S. Drought assessment in arid Western Rajasthan. In R. Nagarajan (ed). *National Workshop Proceedings on Recent Trends in Drought Assessment Monitoring and Management.* Mumbai: Centre of Studies in Resources Engineering, Indian Institute of Technology. 109–114.

Characteristics: Earthquake

0850 ■ 1837 Wright, B.W. Note on an earthquake in Ganjam 29 June 1837 *Chennai Journal of Literature Science*, 2: 246.

0851 ■ 1845 Blagrave. Note on earthquakes in Kutch, 19 June and 25, 1845. *Journal of the Asiatic Society of Bengal*, 14: 18.

0852 ■ 1845 Hannay, H. Note on earthquake in Assam 26 July 1845. *Journal of the Asiatic Society of Bengal,* 14: 118–119.

0853 ■ 1849 Butler, J. Note on an earthquake at Guwahati, Assam, 22 January 1849. *Journal of the Asiatic Society of Bengal*, 18: 172–173.

0854 ■ 1852 Burgess. Note on a severe earthquake at Pillibheet, 31 March 1852. *Journal of the Asiatic Society of Bengal*, 21: 432.

0855 ■ 1858 Saxton, G.H. Note on an earthquake at Cuttack. *Journal of the Asiatic Society of Bengal*, 27: 366.

0856 ■ 1869 Godwin-Austin, H.H. Notes from Assaloo North Cachar on Great Earthquake of January 1869. *Proceedings of the Asiatic Society of Bengal,* 38: 91–103.

0857 ■ 1870 Steel, F.H. Note on earthquake in Assam, 11 April 1870. *Proceedings of the Asiatic Society of Bengal,* 39: 222–223.

0858 ■ 1873 Lamb. Note on the earthquake in Kamrup, 19 December 1872. *Proceedings of the Asiatic Society of Bengal,* 42: 65–66.

0859 ■ 1878 Wynne, A.B. Notes on the earthquake in Punjab of 2 March 1878. *Proceedings of the Asiatic Society of Bengal*, 47 (2): 131–140.

0860 ■ 1882 Rogers, R.E. Memorandum on the earthquake of the 31 December 1881 and the great sea waves resulting there from as shown on the diagrams of the tidal observatories in the Bay of Bengal. *Proceedings of the Asiatic Society of Bengal*, 51: 63–66.

0861 ■ 1897 Heath, T. Note on the Kolkata earthquake (12 June 1897) as recorded by the Bifilar pendulum at the Edinburgh Royal Observatory. *Proceedings of the Royal Society, Edinburgh,* 21: 481–488.

0862 ■ 1897 Inglis, J.W. Notes on some earthquakes in India. *Proceedings of the Royal Society, Edinburgh,* 21: 506–509.

0863 ■ 1898 Oldham, R.D. *A Note on the Allah Bund in the Northwest of Rann of Kutch*. Memoirs of the Geological Survey of India, No. 28. Kolkata.

0864 ■ 1942 Mukherjee, S.M. Seismological features of the Satpura earthquake of 14 March 1938. *Proceedings of Indian Academy of Sciences, Earth and Planetary Sciences,* 16: 167–175.

0865 ■ 1953 Akima, T. On dispersion curves of surface waves from the Great Assam earthquake of 15 August 1950. *Central Board of Geophysics, India*, 1: 108.

0866 ■ 1953 Gulati, B.L. Geodetic and geophysical aspects of the earthquakes in Assam. *Central Board of Geophysics,* India, 1: 16–25.

0867 ■ 1953 Ray, S. Isoseismals for the Great Assam earthquake of 15 August 1950. *Central Board of Geophysics,* India, 1: 35–37.

0868 ■ 1953 Ward, F. Notes on the Assam earthquake. *Central Board of Geophysics,* India, 1: 97–100.

0869 ■ 1962 Saha, B.P. On the depth of focus of Delhi earthquake of 27 August 1960. *Indian Journal of Meteorology and Geophysics*, 13 (1): 137–139.

0870 ■ 1964 Basu, K.L. Note on the Coimbatore earthquake of 8 February 1900. *Indian Journal of Meteorology and Geophysics,* 15 (2): 281–286.

0871 ■ 1965 Special Correspondent. Study on earthquake. *Science and Culture,* 31 (12).

0872 ■ 1969 Bhaskar Rao, V. *et al.* Some geophysical and geological aspects of the Koyna earthquake, December 1967. *Tectonophysics,* 7 (3): 265–271.

0873 ■ 1969 Jai Krishna *et al.* Analysis of Koyna acceleration of 11 December 1967. *Bulletin of the Seismological Society of America*, 59 (4): 1719–1731.

0874 ■ 1969 Lee, W.H.K. *et al.* Fault-plane solution of the Koyna (India) earthquake. *Nature,* 223 (July 12): 172–173.

0875 ■ 1969 Ray, S. Paper on Koyna earthquake. *Journal of Geological Society of India*, 10 (1): 103–105.

0876 ■ 1969 Special Correspondent. Expert report on Koyna earthquake. *Indian Journal of Power River Valley Development*, 19 (6): 247–252.

0877 ■ 1970 Agarwal, P.N. *Evaluation of Earthquake Parameters in Koyna and Dakpathar Regions*. Roorkee:University of Roorkee (Unpublished Ph.D thesis).

0878 ■ 1970 Arora, S.K. *et al.* Some aspects of the structure of Southern India based on recent Bhadrachalam earthquakes. *Nature,* 225: 261–262.

0879 ■ 1970 Fitch, T.J. Earthquake mechanisms in the Himalayan, Burmese and Andaman regions and continental tectonics in Asia. *Journal of Geophysical Research,* 75 (14): 2699–2709.

0880 ■ 1970 Murty, G.S. Surface reflected phase from Koyna aftershocks. *Indian Journal of Meteorology and Geophysics,* 21 (3): 479–484.

0881 ■ 1971 Karunakaran, C. *et al.* Some aspects of the seismicity of the Peninsular shield. *Indian Journal of Power and River Valley Development,* 21: 15–24.

0882 ■ 1971 Ram, A. *et al.* Note on the seismicity of the Koyna region (Maharashtra). *Bulletin of the Seismological Society of America,* 61 (2): 457–471.

0883 ■ 1972 Gupta, H.K. *et al.* Some discriminatory characteristics of earthquakes near Kariba Kremasta and Koyna artificial lakes. *Bulletin of the Seismological Society of America,* 62 (1): 481–492.

0884 ■ 1975 Singh, D.D. *et al.* Surface wave radiation pattern and source parameters of Koyna earthquake of 10 December 1967. *Bulletin of the Seismological Society of America,* 65: 711–731.

0885 ■ 1978 Sinvhal, A. *et al.* Neotectonics and time space seismicity of the Andaman-Nicobar region. *Bulletin of the Seismological Society of America,* 68: 399–409.

0886 ■ 1981 Langston, C.A. Source inversion of seismic waveforms: The Koyna (India) earthquake of 13 September 1967. *Bulletin of the Seismological Society of America,* 71: 1–24.

0887 ■ 1984 Molnar, P. Structure and tectonics of the Himalaya: Constraints and implications of geophysical data. *Annual Review of Earth and Planetary Sciences,* 12.

0888 ■ 1984 Srivastava, H.N. *et al.* Crustal studies of Koyna region using expansion data from deep seismic sounding. *Tectonophysics,* 110 (1–2): 61–72.

0889 • 1985 Raul, L.J. Social and political aspects of the South India earthquake of 1980. *Disasters: The Journal of Disaster Studies, Policy and Management,* 9 (3): 206–212.

0890 ■ 1988 Chander, R. Rupture propagation direction in major earthquakes along the Himalayan convergence zone. *Current Science,* 57 (14): 805–806.

0891 ■ 1989 George, J. *Travel Times for P Waves of the Earthquakes in and around Northeast India.* Jorhat: Jorhat University (Unpublished Ph.D thesis).

0892 ■ 1989 Khattri, K.N *et al.* New seismological results on the tectonics of the Garhwal Himalaya. *Proceedings of Indian Academy of Sciences, Earth and Planetary Sciences,* 98 (1): 91–109.

0893 ■ 1989 Molnar, P. and M.R. Pandey. Rupture zone of Great Earthquake in the Himalayan region. *Proceedings of Indian Academy of Sciences, Earth and Planetary Sciences,* 98 (1): 61–70.

0894 ■ 1989 Rajendran, K. and H.K. Gupta. Seismicity and tectonic stress field of a part of the Burma-Andaman-Nicobar arc. *Bulletin of the Seismological Society of America,* 79: 989–1005.

0895 ■ 1990 Chandrasekaran, A.R. and J.D. Das. Strong motion arrays in India and characteristics of recent recorded events. *Bulletin of Indian Society of Earthquake Technology,* 27: 1–66.

0896 ■ 1990 Kumar, S. and A.K. Mahajan. Study of intensities of 26 April 1986 Dharamshala (Himachal Pradesh) earthquake and associated tectonics. *Journal of Geological Society of India,* 35 (2): 213–219.

0897 ■ 1990 Mukhopadhyay, S. *Seismic Velocity Structure and Seismotectonics of the Shillong Massif (Northeastern India).* Roorkee: University of Roorkee (Unpublished Ph.D thesis).

0898 ■ 1991 Kayal, J.R. and R. De. Microseismicity and tectonics in Northeast India. *Bulletin of the Seismological Society of America,* 81: 131–138.

0899 ■ 1991 Shanker. D. *Characteristic Studies of Tectonics, Siesmicity and Occurrence of Major Earthquakes in Northeast India.* Varanasi: Banaras Hindu University (Unpublished Ph.D thesis).

0900 ■ 1991 Talwani, P. and K. Rajendran. Some seismological and geometric aspects of intraplate earthquake. *Tectonophysics,* 186 (1–2): 19–41.

0901 ■ 1992 Chandra, U. Seismotectonics of Himalaya. *Current Science,* 62 (1): 40–71.

0902 ■ 1992 Chandrasekharan, A.R. and J.D. Das. Analysis of strong motion accelerograms of Uttarkashi earthquake of 20 October 1991. *Bulletin of Indian Society of Earthquake Technology,* 29 (1): 35–55.

0903 ■ 1992 Rai, S.S. *et al.* Seismic tomography of the South Indian shield. *Current Science,* 62 (1–2): 213–226.

0904 ■ 1992 Rastogi, B.K. Seismotectonics inferred from earthquakes and earthquake sequences in India during 1980s *Current Science,* 62 (1–2): 101–108.

0905 ■ 1992 Singh, V.P. and D. Shankar. On the seismicity and tectonic activity of the Bengal basin. *Mausam,* 43 (4): 371–378.

0906 • 1993 Government of Maharashtra. *A Preliminary Report on the 30 September 1993 Earthquake.* Mumbai.

0907 ■ 1993 Gupta, H.K. Patterns preceding major earthquakes in Northeast India. *Current Science,* 65 (11 and 12): 889–893.

0908 ■ 1993 Kailasam, L.N. Geophysical and geodynamical aspects of Maharashtra earthquakes of 30 September 1993. *Current Science,* 65: 736–739.

0909 ■ 1993 Kumar, S. and A.K. Mahajan. The Uttarkashi earthquake of 20 October 1991: Field observations. *Terra Nova,* 6 (2): 95–99.

0910 • 1993 Satpute, S. The Latur earthquake. *Mausam,* 44 (1): 18–20.

0911 ■ 1993 Thakur, V.C. and S. Kumar. Details of the 20 October 1991 Uttarkashi earthquake in Garhwal Himalaya and its seismotectonic significance. *Himalayan Geology,* 14: 229–236.

0912 ■ 1994 Das, P.K. and P. Phukan. *Earthquakes and Volcanoes.* New Delhi: Indus Publishing Company.

0913 ■ 1994 Indian Meteorological Department. *Latur Earthquake of 30 September 1993 and its Aftershocks: A Consolidated Report.* New Delhi.

0914 ■ 1994 Jain, S.K. *et al.* The *29 September 1993 M 6.4 Killari Maharashtra, earthquake in Central India.* Oakland: Earthquake Engineering Research Institute.

0915 ■ 1995 Dattatrayam, R.S. *et al.* Characteristics of some foreshocks and aftershocks of 20 October 1991 Uttarkashi earthquake. *Journal of Geological Society of India,* 45 (1): 51–64.

0916 ■ 1995 Ramachandran, B. Waves of shock: Anatomy of earthquakes. *Frontline,* 28 (June 30): 78–82.

0917 ■ 1996 Rajendran, C.P. and K. Rajendran. Mechanism of microseismic activity in Kerala: A suggestion. *Journal of Geological Society of India,* 47 (4): 467–476.

0918 ■ 1996 Rajendran, C.P. *et al.* Palaeoseismic indicators in the rupture zone of the 1993 Killari (Latur) earthquake. *Current Science,* 70 (50): 385–390.

0919 ■ 1997 Kumar, D. The representation of attenuation characteristic of strong ground motion observed in 1986 Dharmashala and 1991 Uttarkashi earthquakes by available empirical relations. *Current Science,* 73: 543–548.

0920 ■ 1998 Kayal, J.R. Three dimensional seismic structure beneath Shillong Plateau and Assam Valley, (Northeast India). *Bulletin of the Seismological Society of America,* 88: 667–676.

0921 • 1998 Maharashtra Emergency Earthquake Rehabilitation Programme. *Risk Assessment and Vulnerability Analysis.* Mumbai: Government of Maharashtra.

0922 ■ 1998 Paul, A. *et al.* Estimation of focal parameters for Uttarkashi earthquake using peak horizontal accelerations. *ISET Journal of Earthquake Technology,* 35 (1–3): 1–8.

0923 ■ 1998 Rajendran, K. and C.P. Rajendran. Characteristics of 1997 Jabalpur earthquake and their bearing on its mechanism. *Current Science,* 74 (2): 168–174.

0924 ■ 1999 Gahalaut, V.K. and R. Chander. Geodetic evidence for accumulation of earthquake generating strains in the Northwest Himalayas near 75.5° E. *Bulletin of the Seismological Society of America,* 89 (4): 837–843.

0925 ■ 1999 Gupta, H.K. *et al.* Anatomy of surface rupture zones of two stable continental region earthquakes, 1967 Koyna and 1993 Latur (India). *Geophysical Research Letters,* 26 (13): 1985–1988.

0926 ■ 1999 Singh, S.K. *et al.* A spectral analysis of the 21 May 1997, Jabalpur (India) earthquake (M 5.8) and estimation of ground motion from future earthquakes in the Indian shield region. *Bulletin of the Seismological Society of America,* 89, 1620–1630.

0927 ■ 1999 Singh, S.K. *et al.* Crustal and upper mantle structure of Peninsular India source parameters of the 21 May 1997, Jabalpur earthquake (M 5.8): Results from a new regional broadband network. *Bulletin of the Seismological Society of America,* 89: 1631–1641.

0928 ■ 2000 Ambraseys, N. and R. Bilham. A note on the Kangra M 7.8 earthquake of 4 April 1905. *Current Science,* 79 (1): 45–50.

0929 ■ 2000 Narula, P.L. *et al.* Rupture mechanism of Chamoli earthquake on 29 March 1999 and its implications for seismotectonics of Garhwal Himalaya. *Journal of Geological Society of India,* 55 (5): 493–504.

0930 ■ 2000 Rajendran, K. *et al.* The Chamoli earthquake:Field observations and implications for seismic hazard. *Current Science,* 79 (1): 45–51.

0931 ■ 2001 Bhattacharya, P.M. Application of fractals in seismology: A case study in Northeast India. In *Proceedings of the Role of Earth Sciences in Integrated Development and Related Societal Issues.*Special Publication No. 65 (2) Kolkata: Geological Survey of India, 1.

0932 ■ 2001 Bilham, R. *et al.* Himalayan seismic hazard. *Science,* 293: 1442–1444.

0933 ■ 2001 Goel, R. Earthquake in Gujarat: A case of Kutch. In S.L. Goel and R. Kumar (eds). *Disaster Management.* Delhi: Deep and Deep Publications. 219–237.

0934 ■ 2001 Hough, S.E. and S. Martin. Intensity distribution of the seismological report on 26 January 2001 Bhuj earthquake. *Seismological Research Letters,* 72 (3).

0935 ■ 2001 Indian Meteorological Department. *Bhuj Earthquake of 26 January 2001:A Consolidated Preliminary Report.* Delhi: Indian Meteorological Department.

0936 ■ 2001 Joshi, A. Strong motion envelope modelling of the source of the Chamoli earthquake of 29 March 1999 in the Garhwal Himalaya. *Journal of Seismology,* 5 (4): 499–518.

0937 ■ 2001 Karanth, R.V. and P.S. Sohoni. Geological observations of 26 January 2001 Bhuj earthquake. *Journal of Geological Society of India,* 58 (3): 193–202.

0938 ▲2001 Mistry, R. *et al. Interdisciplinary Observations on the 26 January 2001 Bhuj Earthquake.* USA World Seismic Safety Initiative and Earthquakes and Megacities Initiatives.

0939 ■ 2001 Rajendran, C.P. and K. Rajendran. The 1993 Killari (Latur) earthquake: A paleoseismological perspective. *Indian Geological Journal,* 71: 185–200.

0940 ■ 2001 Rajendran, C.P. and K. Rajendran. Characteristics of deformation and past seismicity with the 1819 Kutch earthquake. *Bulletin of the Seismological Society of America,* 91 (3): 407–426.

0941 ■ 2001 Rao, K.S. *et al.* Study of Bhuj earthquake: A remote sensing approach. In H.R.Wason *et al* (eds). *Proceedings of the Workshop on Recent Earthquakes of Chamoli and Bhuj.* Roorkee.

0942 ■ 2001 Shankar, R. and P. Pande. Geoseismological studies of Kutch (Bhuj) earthquake 26 January 2001. *Journal of Geological Society of India,* 58 (3): 203–208.

0943 ■ 2001 Talwani, P. and A.Gangopadhyay.Tectonic framework of Kachchh earthquake of 26 January 2001. *Seismological Research Letters,* 72 (3): 336–345.

0944 ■ 2001 Tiwari, S.C. Earthquake observation. *Current Science,* 80 (4): 485–486.

0945 ■ 2001 Wesnousky, S. Eight days in Bhuj: Field report bearing on surface rupture and genesis of the 26 January 2001 earthquake in India. *Seismological Research Letters,* 72 (5): 514–522.

0946 ■ 2002 Earthquake Engineering Research Institute. *Bhuj, India, Earthquake of 26 January 2001.* (Reconnaissance Report Supplement to *Earthquake Spectra* Volume 18). Oakland.

0947 ■ 2002 Gupta, L.C. *et al. Gujarat Earthquake 26 January 2001.* New Delhi: National Centre for Disaster Management, Indian Institute of Public Administration.

0948 ■ 2002 Kayal, J.R. and S. Mukhopadhyay. Seismic tomography structure of the 1993 Killari earthquake source area. *Bulletin of the Seismological Society of America,* 92 (5): 2036–2039.

0949 • 2002 Khera, P.D. Ethnography of an earthquake. *Economic and Political Weekly,* 37 (11): 1021–1023.

0950 ■ 2002 Mandal, P. *et al.* Magnitude estimation of Koyna-Warna earthquakes. *Journal of Seismology,* 6 (4): 447–458.

0951 ■ 2002 Nath, S.K. *et al.* Determination of S wave site response in the Garhwal Himalaya from the aftershock sequence of the 1999 Chamoli earthquake. *Bulletin of the Seismological Society of America,* 92 (3): 1072–1081.

0952 ■ 2003 Gaonkar, S.G. *et al.* Study of the epicentral trends and depth sections for aftershocks of the 26 January 2001 Bhuj earthquake in western India. *Proceedings of Indian Academy of Sciences, Earth and Planetary Sciences,* 112 (3): 401–412.

0953 ■ 2003 Kumar, A. Seismological studies in Manipur and adjoining areas. In M. Mohanty and S. Aggarwal (eds). *Applications of Geographic Information Systems in Earthquake and Landslide Hazard Assessment.* New Delhi: Department of Science and Technology, Government of India. 20–21.

0954 ■ 2003 Mandal, P. and B.K. Rastogi. Results from local earthquake velocity tomography: Implications towards the source process involved in segregating the 2001 Bhuj earthquake in the lower crust beneath Kuchchh. *Bulletin of the Seismological Society of America,* 94 (2): 6333–649.

0955 ■ 2003 Mishra, O.P. *et al.* Crack density, saturation rate and porosity at the 2001 Bhuj earthquake hypocenter: A fluid driven earthquake. *Earth and Planetary Science Letters,* 212 (3–4): 393–405.

0956 ■ 2003 Mukhopadhyay, S. and J.R. Kayal. Seismic tomography structure of the 1999 Chamoli earthquake source area in the Garhwal Himalaya. *Bulletin of the Seismological Society of America,* 93(4): 1854–1861.

0957 ■ 2003 Shaik, M.A. and S. Srivastava. Statistical parameters of Bhuj earthquake sequence of 26 January 2001. *Proceedings of Indian Academy of Sciences, Earth and Planetary Sciences*, 112 (3): 397–400.

0958 ■ 2004 Jain, R. *et al.* Precursory changes in source parameter for the Koyna-Warna (India) earthquake. *Geophysical Journal International,* 158 (3): 915–921.

0959 ■ 2004 Mandal, P. *et al.* Characterisation of the causative fault system for the 2001 Bhuj earthquake of M 7.7. *Tectonophysics,* 378 (1–2): 105–122.

0960 ■ 2004 Mandal, P. and R. Narsaiah. Coda duration magnitude scale of 2001 Bhuj aftershocks. *Current Science,* 87 (4): 520–527.

0961 ■ 2004 Mandal, P. *et al.* Low coda Qc in the epicentral region of the 2001 Bhuj earthquake of M 7.7. *Pure and Applied Geophysics,* 161 (8): 1635–1654.

0962 ■ 2004 Padhy, S. Intermittent criticality on a regional scale in Bhuj. *Geophysical Journal International,* 158 (2): 676–680.

0963 ■ 2005 Cervone, G. *et al.* Wavelength maxima curves of surface latent heat flux anomalies associated with Indian earthquakes. *Natural Hazard and Earth System Sciences,* 5 (1): 87–99.

0964 ■ 2005 Geological Survey of India. *Contributions to the Kangra Earthquake Centenary Seminar 2005.* Special Publication No. 85. Kolkata.

0965 ■ 2005 Menon, A.K. Earthquake. *Indian Journal for Practicing Doctors*, 2 (4).

0966 ■ 2005 Mishra, D.C. *et al.* Tectonic and crustal structures related to Bhuj earthquake of 26 January 2001: Based on gravity and magnetic surveys constrained from seismic and seismological studies. *Tectonophysics,* 396 (3–4): 195–207.

0967 ■ 2005 Neetu, S. *et al.* Comment on the Great Sumatra-Andaman earthquake of 26 December 2004. *Science,* 310 (5753): 1431.

0968 ■ 2005 Tolstoy, M. and D.R. Bohnenstiehl. Hydroacoustic constraint on the rupture duration length and speed of the Great Sumatra-Andaman earthquake. *Seismological Research Letters,* 76 (4): 419–425.

0969 ■ 2006 Gahalaut, V.K. *et al.* Constraints on 2004 Sumatra-Andaman earthquake rupture from global positioning system measurements in Andaman and Nicobar Islands. *Earth and Planetary Science Letters,* 242 (3–4): 365–374.

0970 ■ 2006 Joshi, A. Analysis of strong motion data of the Uttarkashi earthquake of 20 October 1991 and the Chamoli earthquake of 28 March 1999 for determining the Q value and source parameters. *ISET Journal of Earthquake Technology,* 43 (1–2): 11–30.

0971 ■ 2006 Malagnini, L. *et al.* Unbiased moment-rate spectra and absolute site effects in the Kachchh basin from the analysis of the aftershocks of the 2001 M 7.6 Bhuj earthquake. *Bulletin of the Seismological Society of America,* 96: 456–466.

Characteristics: Flood

0972 ■ 1887 Highams, T. *Report on Floods.* Patiala: Punjab State Archives Department.

0973 ■ 1893 Jaffery, T.J.P. *Report on Floods.* Patiala: Punjab State Archives Department.

0974 ■ 1957 Dutt, G.N. Geological aspects of North Bengal floods. *Indian Journal of Power and River Valley Development,* 7 (9).

0975 ■ 1959 Special Correspondent. Floods. *Science and Culture,* 25 (5): 279–280.

0976 ▲ 1968 Katiyar, V.S. *A Geographical Study of Floods in Uttar Pradesh with Special Reference to Ghaghra Flood Plains.* Agra: Agra University (Unpublished Ph.D thesis).

0977 ■ 1970 Jagannathan, P. Climatology of flood and heavy rainfall in India. In *Proceedings of the Symposium on Flood Forecasting and Flood Damage Protection.* Publication. No. 107, Central Board of Irrigation and Power.

0978 ■ 1971 Central Board of Irrigation and Power. *Manual on River Behaviour: Control and Training.* New Delhi: Government of India.

0979 ■ 1975 Das, S.P. The floodplain of the Ganga. In S.P. Dasgupta (ed). *The Upper Gangetic Flood-Plain: A Regional Survey.* Kolkata: National Atlas and Thematic Mapping Organization. 1–6.

0980 ■ 1979 Guha, M.K. *et al.* A non-linear model on flood flow. *Mausam,* 30 (1): 92–96.

0981 ■ 1980 Dhar, O.N. *et al.* A brief appraisal of India's worst recorded floods. *Journal of Indian Association of Hydrologists,* 4 (1–2): 1–11.

0982 • 1981 Chandrasekeran, K.S. Orissa flood-problems and solutions. In *Proceedings of the International Conference on Flood Disaster.* New Delhi: Indian National Science Academy, 46–56.

0983 ■ 1981 Chaudhury, A.K. *et al.* An analytical study of severe floods of July 1978 in Bagmati and Adhwara group of rivers. *Mausam,* 32 (4): 405–410.

0984 ■ 1982 Central Water Commission. *The Kosi River: Its Morphology and Mechanics in Retrospect and Prospect,* Volume 1. New Delhi: Government of India.

0985 ■ 1985 Goswami, A. and Kumar. Studies on the nature of floods in the depressed belt of southern West Bengal. *Geographical Review of India,* 47 (2): 189–201.

0986 ■ 1985 Joshi, K.N. *A Geographical Study of Flood Hazard in Desert and Semi-desert Environment: A Case Study of Luni Basin with Satellite Remote Sensing Technique.* Jodhpur: University of Jodhpur (Unpublished Ph.D thesis).

0987 ■ 1987 Ramaswamy, C. *Meteorological Aspects of Severe Floods in India, 1923–1979.* New Delhi: Indian Meteorological Department.

0988 ▲ 1989 Department of Relief and Rehabilitation. *Report on Flood in 1989.* Patna: Government of Bihar.

0989 ■ 1990 Desai, B.N. and S.G. Rao. Flood hydrology at Sardar Sarovar. *Indian Journal of Power and River Valley Development,* 40 (9): 171–179.

0990 ■ 1990 Dey, S. Design flood of the river Teesta in North Bengal. *Indian Journal of Earth Science,* 17: 172–179.

0991 ■ 1991 Rao, G.N. and D.I. Subbaramayya. A note on the hydro-meteorology of August 1990 floods in Godavari. *Indian Journal of Power and River Valley Development,* 41 (4): 75–80.

0992 • 1992 De, N.K. and H.K. Sarkar. Flood hazard: A study of natural hazards in India. In O. Singh *et al* (eds). *Frontiers in Environmental Geography.* New Delhi: Concept Publishing Company. 229–238.

0993 ■ 1992 Sharma, K.D. Characteristics of a flash flood in the Luni basin, Northwest India. *Indian Journal of Power and River Valley Development,* 42 (4): 31–41.

0994 ■ 1993 Dogra, B. *Floods in South Asia: A Report on the Ganga-Brahmaputra Region.* Delhi: Publishers for South Asian Peoples Environment Network.

0995 ■ 1994 Special Correspondent. Punjab floods: A case study. *Disasters: The Journal of Disaster Studies, Policy and Management,* 18: 80–85.

0996 ▲1995 Kar, M. *Flood Hazard in Nagaon and Marigaon Districts of Assam: A Geographical Perspective.* Guwahati: Guwahati University (Unpublished Ph.D thesis).

0997 ■ 1996 Kale, V.S. Introduction to flood Hydrology. In V.S. Kale (ed). *Fluvial Geomorphology,* Pune: Department of Geography, University of Pune, 24–39.

0998 ■ 1996 Khatoon, R. *Floods in Uttar Pradesh: A Geographical Perspective.* Aligarh: Aligarh Muslim University (Unpublished Ph.D thesis).

0999 ■ 1996 Sharma, A.K. *Floods in Rajasthan: A Case Study of 1995 Floods in Alwar District.* New Delhi: Advanced Professional Programme in Public Administration, Indian Institute of Public Administration (Unpublished Master of Philosophy Dissertation).

1000 ■ 1996 Singh, V.P. (ed). *Hydrology of Disasters.* London: Kluwer Academic Publishers Dordrecht.

1001 ■ 1997 Kale, V.S. *et al.* Flood hydrology and geomorphology of monsoon dominated rivers: The Indian Peninsula. *Water International,* 22: 259–265.

1002 ■ 1998 Goswami, D.C. Flood regime and flood hydrology of the Brahmaputra river (Assam). In V.S. Kale (ed). *Flood Studies in India.* Bangalore: Memoirs of the Geological Society of India, No. 41. 53–76.

1003 ■ 1998 Kadarmandalgi, S.G. Ecology and flood problem of the Gangavali river basin, Uttara Kannada district (Karnataka). In B. Sinha (ed). *Ecosystem Degradation in India.* New Delhi: Ashish Publishing House. 132–140.

1004 ■ 1998 Kale, V.S. (eds). *Flood Studies in India.* Bangalore: Memoirs of the Geological Society of India, No 41.

1005 ■ 1998 Panda, G.K. and B.N. Sinha. Flood hazards of the coastal and deltaic environment of Orissa: A study of their nature and characteristics. In B. Sinha (ed). *Ecosystem Degradation in India.* New Delhi: Ashish Publishing House. 46–61.

1006 ▲1999 Sangwan, R.S. Flood hazard in Rohtak city: A geographical perspective. *Transactions, Institute of Indian Geographers,* 21(2): 65–83.

1007 • 2000 Winchester, P. The political economy of riverine and coastal floods in South India. In D.J. Parker (ed). *Flood. Volume-I* (Hazard and Disaster Series). New York: Routledge. 56–68.

1008 ■ 2001 Verma, R.K. and M. Ram. A case study of heavy flood in river Ken during September 1992. *Mausam,* 52 (3): 597–601.

1009 ■ 2002 Betal, H.R. Flood problem of Maldah: A geographical analysis. *Geographical Review of India,* 64 (4): 337–345.

Characteristics: Heat and Cold Wave

1010 ■ 1966 Raghavan, K. A climatological study of severe heat wave in India. *Indian Journal of Meteorology and Geophysics,* 17.

1011 ■ 1967 Raghavan, K. A climatological study of severe cold waves in India. *Indian Journal of Meteorology and Geophysics,* 18.

Characteristics: Heavy Rain

1012 ■ 2002 Prasad, C. *et al.* Studies in mechanism of cloudbursts in Himalaya: An analysis. In S.K. Tandon and B. Thakur (eds). *Recent Advances in Geomorphology, Quaternary Geology and Environmental Geosciences: Indian Case Studies.* New Delhi: Manisha Publications. 403–408.

1013 ■ 2003 Krishnamurty, T.N. and T.S.V. Vidyakumar: Current thrust on tropical rainfall measuring mission and special sensor microwave imager based modelling studies on heavy rains and flooding episodes. *Mausam*, 54 (1): 121–140.

1014 ■ 2003 Laxminarayanan, R. and R. Rajamohan. A study of usually heavy to very heavy rainfall over Kerala in October 1999. *Mausam,* 54 (4): 861–866.

1015 ■ 2005 Guhathakurta, P. *et al.* Extreme rainfall analysis of Andhra Pradesh using a probability distribution model: A regional estimate. *Mausam,* 56 (4): 785–794.

1016 ■ 2005 Mahapatra, M. *et al.* Some characteristics of very heavy rainfall over Orissa during summer monsoon season. *Journal of Earth System Science,* 114 (1): 17–36.

1017 ■ 2006 Sati, V.P. and M.S.S. Rawat: Natural hazards in Himalayas: A case study of cloudburst hit areas of Uttaranchal. In K.K. Sharma *et al* (eds). *Environmental Geohazards: Science and Society.* Delhi: Research India Press. 215–228.

Characteristics: Landslide

1018 ■ 1976 Prasad, C. and H.N. Mathur. Study of landslips along the main boundary fault near Rajpur, Dehradun. *Himalayan Geology*, 6: 517–525.

1019 ■ 1992 Mehrotra, G.S. and S. Sarkar *et al.* Landslide hazard assessment in Rishikesh-Tehri area (Garhwal Himalayas). In D.H. Bell (ed). *Proceedings of the Sixth International Symposium on Landslides.* Rotterdam: Balkema. 1001–1007.

1020 ■ 1996 Anbalagan, R. An overview of landslides hazards in Himalaya: Available knowledge based gaps and recommendations for future research. *Himalayan Geology*, 17: 165–168.

1021 ■ 1996 Mehrotra, G.S. Geomorphic assessment of hill slopes instability of Matli landslides in Uttarkashi district . In K. Senneset (ed). *Proceedings of Seventh International Symposium on Landslides.* Rotterdam: Balkema. 327–332.

1022 ■ 1996 Singh, R. B. Landslide hazard in Indian Himalaya and Canadian Rockies: A comparative analysis. In J. Chacon (ed). *Proceedings of the Eighth International Conference on Landslides.* Rotterdam: Balkema. 63–69.

1023 ■ 1997 Tewari, R.A. and S.K. Singh. Geophysical characteristics of Kandon landslide area, Manali limestone quarry, Sirmur district (Himachal Pradesh). In *Proceedings of the Symposium on the Earth Sciences in Environmental Assessment and Management,* Special Publication No. 48 (2) Kolkata: Geological Survey of India. 5154.

1024 ■ 2000 Nainwal, H.C. *Geological and Morphostructural Studies of Some Landslide Zones along Alaknanda Valley (Garhwal Himalayas).* Srinagar: H.N.B. Garhwal University (Unpublished Ph.D thesis).

1025 ■ 2001 Ray, T.K. A study of landslides in Assam hills. *Mausam,* 52 (3): 547–550.

1026 ■ 2002 Sharma, L.R. *A Study of Landslide Hazard along the Sainj-Wangtu National Highway in Himachal Pradesh (India).* Shimla: Department of Geography, University of Himachal Pradesh (Unpublished Master of Philosophy Dissertation).

1027 ■ 2004 Nachiappan, R.P. *et al.* Hydrochemical characteristics of groundwater in the Nainital lake basin: Implication on landslide hazards. *Journal of Geological Society of India*, 64 (2): 139–145.

1028 ■ 2004 Shah, S.D. and R. Nagarajan. Geological, structural and tectonic aspects of landslides. In R. Nagarajan (ed). *Landslide Disaster: Assessment and Monitoring.* New Delhi: Anmol Publications. 22–36.

Characteristics: Storm

1029 ■ 1955 Bhaskar Rao, N.S. and K.P. Ramakrishnan. Some features of Kakinada storm May 1955. *Indian Journal of Meteorology and Geophysics*, 6 (4): 367–370.

1030 ■ 1969 Swaminathan, D.R. The extra-ordinary path of the Bay of Bengal storm of 7–15 December 1965 in relation to the Tiros–10 satellite observation and the upper tropospheric wind field. *Indian Journal of Meteorology and Geophysics,* 20 (4): 357–360.

1031 ■ 1980 Singh, R. *et al.* On the formation, structure and decay of a monsoon cyclonic storm near Delhi: A case study. *Mausam,* 31 (1): 65–74.

1032 ■ 1984 Biswas, B. and C.V. Bahdram. A study of major rainstorms of the Teesta basin. *Mausam,* 35: 187–190.

1033 ■ 1989 Sinha, D.K. *et al* (eds). *On Aspects of Study of Norwesters and Local Severe Storms.* Kolkata: Kolkata University.

1034 ■ 1990 Manohar, G.K. *et al.* Some characteristics of point discharge current during two pre-monsoon season thunderstorms at Pune. *Current Science,* 59 (7): 367–370.

1035 ■ 1990 Singh, N. and K.K. Kumar. Extreme vaule distribution for modeling extreme rain events at Mumbai. *Mausam,* 41 (3): 427–432.

1036 ■ 1992 Kalsi, S.R. and R.C. Bhatia. Satellite observations of development of thunderstorm complexes in weakly forced environment. *Vayumandal,* 22 (3–4): 65–76.

1037 ■ 1995 Dhar, O.N. and S.S. Nandargi. On some characteristics of severe rainstorms of India. *Theoretical and Applied Climatology*, 50 (3–4): 205–212.

1038 ■ 2001 Santosh, K. Some climatological features of thunderstorm at Thiruvananthapuram, Kochi-Kozhikode airports. *Mausam,* 52 (2): 357–364.

1039 ■ 2001 Sharma, S. and U.K. De. Differentiation between pre-monsoon thunderstorm and non-thunderstorm days at Dumdum, Kolkata. *Mausam,* 52 (3): 606–608.

1040 ■ 2003 Manohar, G.K. and A.P. Kesarkar. Climatology of thunderstorm activity over Indian region: A case study of west-east contrast. *Mausam,* 54 (4): 819–828.

1041 ■ 2005 Pattanaik, D.R. Variability of oceanic and atmospheric conditions during active and inactive period of storms over the Indian region. *International Journal of Climatology,* 25 (11): 1523–1530.

1042 ■ 2006 Kumar, A. *et al.* Unsteady state modelling on convective thunderstorm. In N.C. Mahanti *et al* (eds). *Disaster Management.* Delhi: Narosa Publishing House.

Characteristics: Tsunami

1043 ■ 2004 Earthquake Engineering Research Institute. *Earthquake Report of 23–26 December 2004: Earthquake and Tsunami.* Oakland.

1044 ■ 2005 Kanakasabai, V. and Rajendran, M. Trend of micro-meteorological parameters during tsunami on the coast of India. *Science of Tsunami Hazard,* 23 (1).

1045 ■ 2005 Murthy, T.S. *et al.* Far field characteristics of the tsunami of December 2004. *ISET Journal of Earthquake Technology,* 42 (4): 213–218.

1046 ■ 2005 Narayan, A.C. Tsunami of 26 December 2004: Observations on Kerala coast. *Journal of Geological Society of India*, 65 (2): 239–246.

1047 ■ 2005 Nath, S.K. *et al.* Great Sumatra-Andaman earthquake of 26 December 2004 as perceived from broadband seismograms of Indian Institute of Technology, Kharagpur, seismological observatory. *Current Science,* 88 (9): 1381–1383.

1048 ■ 2005 National Institute of Science Communication and Information Resources. *Tsunami.* New Delhi.

1049 ▲2005 Satake, K. (ed). *Tsunamis: Case Studies and Recent Development.* Netherlands: Springer.

1050 ■ 2006 Horrillo, J. *et al.* Wave dispersion study in the Indian Ocean tsunami of 26 December 2004. *Science of Tsunami Hazard,* 25 (2).

1051 ■ 2006 Ramaswamy, S.M. *et al* (eds). *Geomatics in Tsunami.* New Delhi: New India Publishing Agency.

1052 ■ 2006 Singarasubramanian, S.R. *et al.* Sediment characteristics of the M–9 tsunami event between Rameswaram and Thoothukudi, Gulf of Mannar, Southeast Coast of India. *Science of Tsunami Hazard,* 25 (3).

Characteristics: Assorted

1053 ▲1994 Verghese, B.G *et al.* Natural hazards. In B.G. Verghese and R.R. Iyer (eds). *Harnessing Eastern Himalayan Rivers: Regional Cooperation in South Asia.* New Delhi: Konark Publishers. 87–103.

1054 ■ 1999 Arasu, S. *Risk Assessment and Natural Hazards: A Comparative Study of India and United States of America.* New Delhi: Jawaharlal Nehru University (Unpublished Ph.D thesis).

1055 ▲1999 National Centre for Disaster Management. *Natural Disasters in India: Some Recent Case Studies.* New Delhi: Indian Institute of Public Administration.

1056 ▲2001 Mandal, G.S. Natural disasters. In V.K. Sharma (ed). *Disaster Management.* New Delhi: Indian Institute of Public Administration. 222–233.

1057 ▲2001 Tahir, M. Natural hazards. In M. Taher and P. Ahmed (eds). *Geography of North-East India.* Guwahati: Manik Prakash. 55–66.

1058 ● 2005 Sharma, S.L. Social dimensions of disaster and its management. In R.K. Sharma and G. Sharma (eds). *Natural Disaster: Institutional Building for Mitigating the Impact on Vulnerable Sections.* New Delhi: APH Publishing Corporation. 223–230.

1059 ▲2006 Husain, A. *Natural Disaster.* New Delhi: Sumit Enterprises.

1060 ■ 2006 Krishnan, P. A. *et al.* Natural hazard assessment through remote sensing and geographic information systems: A case study from Sikkim State, India. In N. C. Mahanti *et al* (eds). *Disaster Management.* New Delhi: Narosa Publishing House.

Cause

Research that attempts to answer the question 'Why' forms a class apart called cause. Derived from the Latin word *causa* which means 'case' these seek the form of an explanation which is integral to any manifest effect. It describes the factors which lead to the occurrence of any event or emergence of a situation, usually of a grave concern. The search for cause is fundamental to any scientific inquiry. A research that explores and explains the determinants of any disaster is grouped into the category called, 'cause'. Adopting different routes and varied methodologies research that engages in searching for the cause be it in the domain of the natural or human or ecological is bracketed as 'cause'. Tracing the genesis of disasters are references that seek cause within relationships. Be it rainfall and flood or dams and seismcity or even deforestation and cyclones, all such references are bracketed within the group called 'cause'. Within the 4004, 596 references belong to the group of 'cause'. This is 15 per cent of the total (Tables 6.13, 6.14 and 6.15).

Table 6.13 References for each Category

Natural Disaster	References	Number
Cyclone	1061–1087	27
Drought	1088–1171	84
Earthquake	1172–1410	239
Floods	1411–1512	102
Heat and Cold Wave	–	–
Heavy Rain	1513–1523	11
Landslide	1524–1611	88
Storm	1612–1622	11
Tsunami	1623–1631	9
Assorted	1632–1653	22

Table 6.14 Research References in Cause

'Natural' disaster	Book	Report	Thesis	Article[1]	Chapter[2]	Feature[3]	Total
Cyclone	0	2	0	21	2	2	27
Drought	2	4	3	47	13	15	84
Earthquake	3	17	15	148	32	24	239
Floods	0	4	13	49	16	20	102
Heat and Cold Wave	0	0	0	0	0	0	0
Heavy Rain	0	0	0	9	2	0	11
Landslide	0	30	8	19	29	2	88
Storm	0	0	0	10	0	1	11
Tsunami	1	1	0	5	1	1	9
General Hazard	4	1	1	5	5	6	22
Total	**10**	**59**	**40**	**313**	**100**	**71**	**593**

[1] *in a journal;* [2] *in a book/report/proceedings;* [3] *in a magazine.*

Table 6.15 Dominance of Natural over Human in Research References on Cause

Phase	Natural	Human	Ecological	Total
1830–1947	27	0	0	27
1948–1989	181	45	0	226
1990–2006	234	106	0	340
Total	**442**	**151**	**0**	**593**

Cause: Cyclone

1061 ■ 1952 Sen, S.N. and C.A. George. An analysis of the Masulipatnam cyclone of October 1949. *Indian Journal of Meteorology and Geophysics,* 3 (4): 264–277.

1062 ■ 1968 Special Correspondent. Orissa cyclone. *Indian Journal of Meteorology and Geophysics,* 19 (1): 129–131.

1063 ■ 1977 Mukherjee A.K. and T.R. Sivaramkrishnan. Surface winds and sea waves in a hurricane field. *Nature,* 267 (May 19) 236–237.

1064 ■ 1981 Eliot, J. *Handbook of Cyclonic Storms in the Bay of Bengal (Volume 1, Second Edition).* Pune: Indian Meteorological Department.

1065 ■ 1981 Mooley, D.A. Increase in annual frequency of severe cyclonic storms of the Bay of Bengal after 1964: Possible causes. *Mausam,* 32 (1): 35–40.

1066 ■ 1982 Dube, S.K. *et al.* The effect of coastal geometry on the location of peak surge. *Mausam,* 33 (4): 445–450.

1067 ■ 1986 Basu, B.K. Storm surge associated with the Paradip cyclone of 1982. *Mausam,* 37: 343–346.

1068 ■ 1987 Singh, S.V. *et al.* Relationships of southern oscillation and other large scale features with Bay of Bengal cyclones during the post-monsoon season. *Advanced International Atmospheric Sciences,* 4 (2): 169–174.

1069 ■ 1989 Maurice, D. and T.S. Murty. Tropical cyclones in the Bay of Bengal and carbon dioxide warming. *Natural Hazard Journal,* 2 (3–4): 387–390.

1070 • 1989 Patel, S. Agni, cyclone and Baliapal agitation. *Economic and Political Weekly,* 24 (29): 1381–1382.

1071 ■ 1989 Rajeevan, M. Post monsoon tropical cyclone activity in the North Indian ocean in relation to the El-Nino and southern oscillation phenomenon. *Mausam,* 41 (1): 43–46.

1072 ■ 1989 Singh, R. On the development of the cyclonic storms and rainbands during October–December 1987. *Mausam,* 40 (4): 403–408.

1073 ■ 1991 Gupta, A. *et al.* El–Nino and tropical storm tracks over Bay of Bengal during post-monsoon season. *Mausam,* 42 (3): 257–260.

1074 ■ 1992 Murty, T.S. and V.R. Neralla. On the impossibility of the co-existence of monsoons and severe storm surges in the Bay of Bengal. *Natural Hazard Journal,* 6 (3): 281–285.

1075 ■ 1994 Ganeshan. Formation of tropical cyclones in the Bay. *Mausam,* 45 (3): 235–242.

1076 ■ 1996 Murty, T.S. and V.R. Narella. El Nino and storm surges in Bay of Bengal. In M.I. El–Sabh *et al* (eds). *Land Based and Marine Hazard: Scientific and Management Issues.* Netherlands: Kluwer Academic Publishers. 97–112

1077 • 1998 Jayanthi, N. Cyclone hazard, coastal vulnerability and disaster risk assessment along the Indian coasts. *Vayumandal,* 28 (3–4): 115–119.

1078 ■ 1998 Kumar, V. Coastal morphological influence for tropical cyclone track deviation along Andhra Coast: Geographic information systems and remote sensing based approach. *Current Science,* 75 (9): 955–958.

1079 • 2000 Shiva, V. and K. Emani. *Climate Change, Deforestation and the Orissa Super Cyclone: Ecological Cost of Globalization.* Dehradun: Research Foundation for Science, Technology and Ecology.

1080 • 2001 Agarwal, A. Cyclones, drought and monsoon. *Down to Earth,* 10: 23.

1081 ■ 2001 Naidu, J. V. M. and D. V. Rao. Tornadoes associated with Bay of Bengal cyclone of 8–10 November 1995. *Mausam,* 52 (2): 427–430.

1082 ■ 2001 Shyamala, B. and S. Sudevan. Behaviour of recurring cyclonic storms in Arabian Sea as a response to atmospheric interaction. *Mausam,* 52 (3): 469–479.

1083 ■ 2001 Singh, O. P. *et al.* Tropical cyclone frequency in North Indian Ocean in relation to southern oscillation phenomenon. *Mausam,* 52 (3): 511–514.

1084 ■ 2001 Subba Rao, M. V. and D. R. Kameshwar. Geohazards: A study of the November 1996 cyclone of Godavari delta, Andhra coast. In *Proceedings of the Role of Earth Sciences in Integrated Development and Related Societal Issues.* Publication No. 65 (2). Kolkata: Geological Survey of India. 59–67.

1085 ■ 2002 Nandakar, P. N. and G. Srinivasan. Wind and wave regime over the Mumbai High area under the influence of cyclonic storm. *Mausam,* 53 (2): 187–196.

1086 ■ 2003 Bhowmick, S. K. R. An evaluation of cyclone genesis parameters over the Bay of Bengal using model analysis. *Mausam,* 54 (2): 351–358.

1087 ■ 2005 Muthuchami, A. and S. Sridharan. Role of cyclonic systems in excess and deficit monsoon year in India. *Mausam,* 56 (4): 851–856.

Cause: Drought

1088 ■ 1884 Blanford, H. F. On the connection of the Himalyan snowfall with the dry winds and seasons of drought in India. In *Proceedings of Royal Society, London,* (37): 3–22.

1089 ■ 1913 Eliot, J. *Droughts and Famines in India (Report of the Chicago Meteorological Congress 1913).* Washington D.C.: Government Printing Office.

1090 ■ 1963 Subramaniam, B. P. and A. R. Subramaniam. A climatic study of drought in Peninsular India. *Geographical Journal,* 129.

1091 ■ 1964 Subrahmanyam, V. P. and A. R. Subramaniam. Application of water balance concepts for the climatic study of drought in South India. *Annals of Arid Zone,* 3 (4): 46–55.

1092 ■ 1968 Special Correspondent. Andhra: Rains play truant again. *Economic and Political Weekly*, 3 (25): 1351–1356.

1093 ■ 1969 Subrahmanyam, V. P. and C. V. S. Sastri. Some aspects of drought climatology of the dry subhumid zone of South India. *Journal of Meteorological Society, Japan,* 47: 239–244.

1094 ■ 1969 Subrahmanyam, V. P. and C. V. S. Sastri. A study of aridity and drought in Vishakhapatnam. *Annals of Arid Zone,* 8: 18–22.

1095 ■ 1971 Subrahmanyam, V.P. and C.V.S. Sastry. New techniques in drought analysis. *Annals of Arid Zone,* 10: 120–135.

1096 ■ 1973 Rao, Y.P. *et al.* Conditions for normal summer monsoon rainfall and causes of droughts over India. *Indian Journal of Meteorology and Geophysics,* 24 (4): 131–136.

1097 ■ 1974 Rakhecha, P. The study of droughts by water budget method over Andhra Pradesh. *Indian Journal of Meteorology and Geophysics,* 25: 411–16.

1098 • 1974 Special Correspondent. West Bengal: 1943 being reenacted. *Economic and Political Weekly,* 9 (20): 1696–709.

1099 ■ 1975 Subramanyam, V.P. Water balance techniques in drought climatology. *Vayumandal,* 5.

1100 ■ 1977 Bora, L.N. *A Study on Water Balance and Drought Climatology of Assam and the Vicinity.* Vishakhapatnam: Andhra University, Waltair (Unpublished Ph.D thesis).

1101 ■ 1977 Prajapati, M.C. *et al.* Rainfall features and agricultural droughts at Agra. *Annals of Arid Zone,* 16 (2): 176–184.

1102 • 1977 Special Correspondent. Maharashtra: Who is to blame? *Economic and Political Weekly,* 12 (18): 827.

1103 ■ 1979 Rao, R. *A Study on the Water Balance and Drought Climatology of Tamil Nadu.* Vishakhapatnam: Andhra University,(Unpublished Ph.D thesis).

1104 ■ 1979 Reddy, N.K.B. and U. Prabhu. Desertification and drought prone area. In N.B.K Reddy (ed). *Proceedings of the All India Symposium on Drought Prone Areas of India.* Tirupati: Rayalseema Geographical Society. 107–113.

1105 ■ 1979 Roopali. Rainfall as a conceptual aspect of the drought prone area: A case of district Bankura (West Bengal). In N.B.K Reddy (ed). *Proceedings of the All India Symposium on Drought Prone Areas of India.* Tirupati: Rayalseema Geographical Society. 225–232.

1106 • 1979 Sharma, M.L. *et al.* Socio-demographic factors accentuating drought conditions in India. *In Proceedings of the International Symposium on Hydrological Aspects of Drought.* Delhi: Indian Institute of Technology. 453–459.

1107 ■ 1979 Subramanyam, M.J. and K.S. Reddy. Telengana rainfall variability as an aspect of drought. In N.B.K Reddy (ed). *Proceedings of the All India Symposium on Drought Prone Areas of India.* Tirupati: Rayalseema Geographical Society. 127–132.

1108 ■ 1979 Teli, B.L. Rainfall trends in the drought prone areas of Rajasthan. In N.B.K Reddy (ed). *Proceedings of the All India Symposium on Drought Prone Areas of India.* Tirupati: Rayalseema Geographical Society. 201–204.

1109 ■ 1979 Varma, B.L. and R.B. Singh. Periodic variations in water balance and drought. In N.B.K Reddy (ed). In *Proceedings of the All India Symposium on Drought Prone Areas of India.* Tirupati: Rayalseema Geographical Society. 205–214.

1110 ■ 1980 Bhalme, H.N. and D.A. Mooley. Large scale droughts and floods and monsoon circulation. *Monthly Weather Review,* 108 (8): 1197–1211.

1111 ■ 1980 Subramaniam, A.R. and G.S.L.H.V.P. Rao. Climatic study of water balance, aridity and drought in Rajasthan state. *Annals of Arid Zone,* 19 (4): 371–377.

1112 ■ 1981 Raman C.R.V. and Y.P. Rao. Blocking highs over Asia and monsoon droughts over India. *Nature,* 289 (January 22) 271–273.

1113 ■ 1981 Sastri, A.S.R.S.A. and A.R. Malakar. Climatological analysis of drought over Northwest India during the year 1979. *Mausam,* 32 (3): 259–262.

1114 ■ 1981 Sastri, A.S.R.S.A. and Y.S. Rama Krishna. Drought during 1979 over Western Rajasthan: A climatic perspective. *Annals of Arid Zone,* 21(1): 41–47.

1115 ■ 1982 Sastri, A.S.R.S.A. *et al.* Drought climatology of Guhiya catchment in Rajasthan. *National Geographical Journal of India,* 28 (3–4): 146–148.

1116 ■ 1983 Subramaniam, A.R. and M.V. Rao. A climatic study of drought types in Rayalseema. *Geographical Review of India,* 45 (2): 45–54.

1117 ■ 1984 Bora, L.N. Climatic study of droughts in Northeast India. *National Geographer,* 19 (1): 11–18.

1118 • 1986 Nair, J. Karnataka: Many facets of drought. *Economic and Political Weekly,* 21 (18): 767–769.

1119 ■ 1986 Subramaniam, A.R. and P.A. Narasimha Raju. Droughts in North Coastal Andhra: A climatic study. *Annals of National Association of Geographers, India,* 6 (1): 9–13.

1120 ■ 1987 Kumar, A. Consequences of the 1987 monsoon failure: A preliminary assessment. *Economic and Political Weekly,* 22 (39): 1635–1640.

1121 • 1987 Olsen, K.W. Man-made drought in Rayalseema. *Economic and Political Weekly,* 22 (11): 441–43.

1122 • 1987 Ray, T. Droughts and deforestation. *Frontline,* (January 10): 94.

1123 • 1987 Sathe, M.D. Maharashtra: Causes of recurrent drought in Ahamadnagar. *Economic and Political Weekly*, 22 (24): 924–927.

1124 ■ 1988 Bandyopadhyay, J. *The Indian Drought 1987–88: The Ecological Causes of Water Crisis. What to do?* Dehradun: Research Foundation for Science, Technology and Ecology.

1125 ■ 1988 Homji, V.M.M. Deforestation, ecological destabilisation and drought. *Social Action,* 38 (2): 115–128.

1126 • 1989 Bandhopadhyay, J. Riskful confusion of drought and man-induced water scarcity. *Ambio-A Journal of the Human Environment,* 18 (5): 284–292.

1127 • 1992 Bandhopadhyay, J. and V. Shiva. Green Revolution and increased vulnerability to drought and desertification in arid regions. In S. Bahuguna (ed). *Environmental Crisis and Sustainable Development.* Dehradun: Natraj Publication, 217–239.

1128 • 1992 Bhalla, S. S. *et al.* The politics and economics of drought in India. In K. Basu and P. Nayak (eds). *Development Policy and Economic Theory.* New Delhi: Oxford University Press.

1129 • 1992 Bhatia, B. Lush fields and parched throats: Political economy of groundwater in Gujarat. *Economic and Political Weekly,* 27 (51 and 52): A142–A170.

1130 • 1993 Nayak, P. K. Anthropology and natural calamities. In P. K. Nayak and A. Mahajan (eds). *Kalahandi-Human Encounter with Drought.* Delhi: Reliance Publishing House. 1–10.

1131 • 1993 Pradhan, J. Drought in Kalahandi: The real story. *Economic and Political Weekly,* 28 (22): 1084–1087.

1132 • 1994 Prasad, K. and B. D. Singh. *Drought Disaster and Development: Profile, Performance and Potential.* New Delhi: Mittal Publication.

1133 ▪ 1995 Mohan, H. S. R. and H. Raghavan. An agro-climatic appraisal of droughts over Kerala state. *Geographical Review of India,* 57 (1): 40–47.

1134 • 1996 Sainath, P. *Everybody Loves a Good Drought: Stories from India's Poorest Districts.* New Delhi: Penguin Books.

1135 ▪ 1996 Saseendran, S. A. *et al.* Distribution of monsoon rainfall in India during El–Nino associated drought situation. *Annals of Arid Zone,* 35 (1): 9–16.

1136 ▪ 1996 Sreedevi, P. Climatic water balance and droughts of the Pageru river basin, Cuddapah District, (Andhra Pradesh). *Environmental Geology,* 42 (6): 681–689.

1137 ▪ 1997 Homji, V. M. M. Drought, aridity and desertification in Indian sub-continent. *Annals of Arid Zone,* 36 (1): 1–18.

1138 ▪ 1997 Martin, M. and R. Grover. Droughts in India were caused by El Nino. *Down to Earth,* 6 (5): 57.

1139 • 1997 Reddy, Y. G. Disparities in development: A study of drought prone regions in Andhra Pradesh. *Indian Association of Social Science Institutions, Quarterly,* 15 (4): 123–136.

1140 • 1997 Simhadri, S. River water politics in drought prone Telangana. *Economic and Political Weekly,* 32 (37): 2315–2317.

1141 ■ 1998 Khan, Y. *et al. Climate and Dry Land Ecology: Drought in Rajasthan.* New Delhi: National Centre for Disaster Management, Indian Institute of Public Administration.

1142 ■ 1998 Special Correspondent. Drought: A natural disaster. *Bhagirath,* 40: 28–32.

1143 ■ 1998 Tiwari, A. K. *et al.* Drought estimation through rainfall analysis for Bundelkhand region. *Indian Journal of Soil Conservation,* 30 (3): 280–283.

1144 ■ 1999 Singh, N. and N. A. Sontakke. Drought and desertification over Northwest India in the perspective of earth system climate dynamics. In A. S. Faroda *et al* (eds). *Management of Arid Ecosystem.* Jodhpur: Scientific Publisher. 5–20.

1145 • 2000 Agarwal, A. This drought is sponsored by Government of India. *Down to Earth,* 9 (2): 55–57.

1146 • 2000 Agarwal, A. Political drought. *Down to Earth,* 9 (9): 7.

1147 • 2000 Editorial. Drought: Outcome of long neglect. *Economic and Political Weekly,* 35 (18): 1496.

1148 • 2000 Ghosh, A. Droughts, traditional wisdom and the large dam mind set. *Mainstream,* 38 (April 29): 3–4.

1149 • 2000 Madhurkar, U. *et al.* How green is my village? *India Today,* (Sept. 18): 57–58.

1150 • 2000 Mehta, L. Drought diagnosis: Dry land blindness of planners. *Economic and Political Weekly,* 35 (27): 2439–2445.

1151 • 2000 Rangatiya, Y. The myth of drought. *Down to Earth,* 8 (24): 13–14.

1152 • 2000 Reddy, B. Correlates of poverty and development indicators in drought prone area of Andhra Pradesh. *Indian Association of Social Science Institutions, Quarterly,* 19 (2): 45–46.

1153 ■ 2000 Singh, T. Drought and El. Nino. *Down to Earth,* 8 (24): 24–26.

1154 ■ 2000 Sinha, K. and Jitendra. Droughts: Causes and remedies. *Kurukshetra,* 49 (2): 30–34.

1155 • 2000 Special Correspondent.Drought is in mind. *Down to Earth,* 9 (11): 47.

1156 ■ 2001 Asnani, G. C. El-Nino of 1997–98 and Indian monsoon. *Mausam,* 52 (1): 57–66.

1157 • 2001 Mahapatra, R. and R. Panda. The myth of Kalahandi. *Down to Earth,* 9 (21): 30–34.

1158 • 2001 Narasaiah, M. L. and K. Narayana. *Drought and Development.* New Delhi: Discovery Publishers.

1159 • 2002 Banik, D. *Democracy, Drought and Starvation in India: Testing Sen in Theory and Practice.* Oslo: University of Oslo.

1160 ■ 2002 Ramasastri, K.S. Hydrological aspects of droughts. In R. Nagarajan (ed). *National Workshop Proceedings on Recent Trends in Drought Assessment Monitoring and Management.* Mumbai: Centre of Studies in Resources, Engineering Indian Institute of Technology. 24–29.

1161 ■ 2002 Singh, R.D. and S.P. Singh. Rainfall analysis for investigation of drought at Jhansi in Bundelkhand region. *Indian Journal of Soil Conservation*, 30 (2): 117–121.

1162 • 2003 Bose, R. The end? *Down to Earth,* 12 (1): 38–41.

1163 ■ 2003 Gadgil, S. and P.N. Vinayakchandran. Drought of Indian summer monsoon: Role of clouds over the Indian ocean. *Current Science,* 85 (12): 1713–1719.

1164 ■ 2003 Guhathakurta, P. Drought in districts of India during the recent all India normal monsoon year and its probability of occurrence. *Mausam,* 54 (2): 542–545.

1165 ■ 2003 Hoerling, M. and A. Kumar. The perfect ocean for drought. *Science*, 299 (5607): 691–694.

1166 ■ 2003 Jayaraman, K.S. Monsoon rains start to ease India's drought. *Nature,* 423 (6941): 673.

1167 • 2004 Krishnakumar, R. A man made drought. *Frontline*, 21 (9): 44–45.

1168 ■ 2004 Nandargi, S. and O.N. Dhar. Co-existence of severe drought in India and extreme floods in Bangladesh during 1987 monsoon season. *The International Journal of Meteorology,* 29: 161–167.

1169 ■ 2004 Sikka, D.R. Monsoon drought: A natural hazard in India. In K.S. Valdiya (ed). *Coping with Natural Hazards: Indian Context.* Hyderabad: Orient Longman. 216–228.

1170 • 2005 Khan, S. *Drought in the Baran District of Rajasthan: A Study of Vulnerability and Response.* Delhi: Department of Geography, Delhi School of Economics, University of Delhi (Unpublished Master of Philosophy Dissertation).

1171 ■ 2006 Kumar, B. Sans ecologic. *Down to Earth,* 15 (1):33.

Cause: Earthquake

1172 ■ 1883 Oldham, T. *The Cachar Earthquake of 10 January 1869.* Memoirs of the Geological Survey of India, No.19 (1). Kolkata.

1173 ■ 1883 Rogers, R.E. Memorandum on the earthquake of the 31 December 1881, and the greatest sea waves resulting there from as shown on the diagrams of the tidal observatories in the Bay of Bengal. In *General Report on The Operations of the Survey of India,1881–82.* Kolkata: Government Printing Office. 70–73.

1174 ■ 1883 Walker, J.T. The earthquake of 31 December 1881. In *General Report on the Operations of The Survey of India,1881–82.* Kolkata: Survey of India. 60.

1175 ■ 1884 Oldham, R.D. Note on the earthquake of 31 December 1881. In *Records of the Geological Survey of India,* 17 (2): 47–100.

1176 ■ 1885 Jones, E.J. Note on the Kashmir earthquake of 30 May 1885. In *Records of the Geological Survey of India,* 18 (3): 153–155.

1177 ■ 1885 Jones, E.J. Report on the Kashmir earthquake of 30 May 1885. In *Records of the Geological Survey of India,* 18 (4): 221–227.

1178 ■ 1885 Middlemiss, C.S. Report on the Bengal earthquake of 14 July 1885. In *Records of the Geological Survey of India,* 18 (4): 200–221.

1179 ■ 1899 Oldham, R.D. *Report on the Great Assam Earthquake of 12 June 1897.* Memoirs of the Geological Survey of India, No. 29. Kolkata.

1180 ■ 1904 Ballore, D. and Monlessus, C.F.M. *The Seismic Phenomenon in British India, and their Connection with its Geology.* Memoirs of the Geological Survey of India, No. 35 (3). Kolkata.

1181 ■ 1905 Middlemiss, C.S. Miscellaneous notes on the Kangra Earthquake of the 4 April 1905. In *Records of the Geological Survey of India,* 32 (3).

1182 ■ 1905 Middlemiss, C.S. Preliminary account of the Kangra earthquake of 4 April 1905. In *Records of the Geological Survey of India,* 32 (4): 258–294.

1183 ■ 1908 Middlemiss, C.S. Two Kolkata earthquakes of 1906. In *Records of the Geological Survey of India,* 36 (3): 214–232.

1184 ■ 1910 Middlemiss, C.S. *The Kangra earthquake of 4 April 1905.* Memoirs of the Geological Survey of India, No. 38. Kolkata.

1185 ■ 1926 Oldham, R.D. *The Cutch (Kachchh) Earthquake of 16 June 1819 with a Revision of the Great Division Earthquake of 12 June 1897.* Memoirs of the Geological Survey of India, No. 46 (2). Kolkata.

1186 ■ 1929 Coulson, A.L. Second note on the Northwest Himalayan earthquake of the 1 February 1929. In *Records of the Geological Survey of India,* 63 (4): 434–443.

1187 ■ 1934 Fermor, L.L. Geological aspects of the North Bihar earthquake of 15 January 1934. *Current Science,* 2 (11): 442–445.

1188 ■ 1934 Fremor, L.L. Causes of the recent earthquake. *Indian Geographical Journal,* 9 (1).

1189 ■ 1934 Gee, E.R. *Dubri Earthquake of 3 July 1930.* Memoirs of the Geological Survey of India, No. 65 (1). Kolkata.

1190 ■ 1935 Anden, J.B. and A.M.N. Ghosh. Preliminary account of earthquake of 15 January 1934 in Bihar and Nepal. In *Records of the Geological Survey of India,* 68 (2): 177–239.

1191 ■ 1935 Fermor, L.L. *et al.* Symposium on the origin of North Bihar earthquake of 15 January 1934. In *Proceedings of the 22 Indian Science Congress, held at Kolkata. Indian Science Congress,* 452–461.

1192 ■ 1936 Vivekananda, M. Earthquakes and seismic activity in India. *Indian Geographical Journal,* 11 (1).

1193 ■ 1939 Mukherjee, S.M. Satpura earthquake 1938. *Proceedings of Indian Academy of Sciences, Earth and Planetary Sciences,* 13.

1194 ■ 1939 Roy, S.C. *The Bihar-Nepal Earthquake of 1934: Seismometric Study.* Memoirs of the Geological Survey of India, No. 73. Kolkata.

1195 ■ 1949 Auden, J.B. Note on the earthquake tremors at Shahkot, Sheikhupur district (Punjab) September 1943. In *Records of the Geological Survey of India,* 78 (1): 135–140.

1196 ■ 1950 Poddar, M.C. Assam earthquake of 15 August 1950. *Indian Minerals,* 4 (4): 167–176.

1197 ■ 1952 Jhingran, A.G. Note on an earthquake in the Andaman Islands, 26 June 1941. In *Records of the Geological Survey of India,* 82 (2): 300–307.

1198 ■ 1952 Poddar, M.C. *Preliminary Report on the Assam Earthquake, 15 August 1950.* Bulletin of the Geological Survey of India, Series B, No. 2. Kolkata.

1199 ■ 1953 Boileau, V.H. Assam earthquake of 15 August 1950: Results of the preliminary geological investigations. *Central Board of Geophysics,* India, 1: 50–55.

1200 ■ 1953 Krishnan, M.S. *The Structural and Tectonic History of India.* Memoirs of the Geological Survey of India, No. 81. Kolkata.

1201 ■ 1954 Tandon, A.N. Study of the Great Assam earthquake of 1950 and its aftershocks. *Indian Journal of Meteorology and Geophysics,* 5 (2): 95–137.

1202 ■ 1960 Srivastava, J.P. *A Report on the Delhi Earthquake 27 August 1960.* Kolkata: Geological Survey of India (Unpublished Report).

1203 ■ 1962 Arogyaswamy, R.N.P. Some aspects of tectonics of the Peninsula. In *Records of the Geological Survey of India,* 89 (2): 359–362.

1204 ■ 1964 Hazra, P.C. and D.K. Roy. Short note on seismic phenomena in India and their relation to tectonics. *Indian Minerals,* 16 (3): 241–246.

1205 ■ 1964 Hukku, B.M. *Report on the Geological Investigation of the Minor Earthquakes Around Sonipat.* Kolkata: Geological Survey of India (Unpublished Report).

1206 ■ 1967 Geological Survey of India. *Geological Report on the Koyna Earthquake of 11 December 1967. Satara District (Maharashtra).* Kolkata: Geological Survey of India (Unpublished Report).

1207 ■ 1968 Balasundaram, M. S. and M. C. Poddar. Koyna earthquake. *Journal of Indian Geophysical Union,* 5 (1): 21–27.

1208 ■ 1968 Chatterjee, G. C.Koyna earthquake. *Journal of Indian Geophysical Union,* 5 (1): 35–36.

1209 • 1968 Dutta, T. K. Recent earthquake series in Koyna region: A possible mechanism at source. *Bulletin of Indian Society of Earthquake Technology*, 5: 55–61.

1210 ■ 1968 Gupta, R. B. Koyna earthquake. *Journal of Geological Society of India*, 9 (1): 97–105.

1211 ■ 1968 Jacob, K. Scientific discussions on Koyna earthquake. *Indian Journal of Meteorology and Geophysics*, 19 (1): 5–12.

1212 ■ 1968 Krishnan, M. S. Koyna earthquake. *Journal of Indian Geophysical Union,* 5 (1): 9–10.

1213 ■ 1968 Mane, P. M. History of earth tremors and shocks in the Koyna hydroelectric project area. *Journal of Indian Geophysical Union,* 5 (1): 15–20.

1214 ■ 1968 Murthi, N. G. K. Koyna earthquake. *Journal of Indian Geophysical Union,* 5 (1): 11–14.

1215 ■ 1968 Nath, M. *et al.* Delhi earthquake of 27 August 1960. In *Records of the Geological Survey of India,* 95 (2): 267–382.

1216 • 1968 Ramiangar, A. S. *et al.* Geological paradox of the Koynanagar earthquake. *Indian Journal of Power and River Valley Development*, 18 (1): 45–46.

1217 ■ 1968 Wadia, D. N. and J. B. Auden. Koyna earthquake of December 1967. *Journal of Indian Geophysical Union*, 5 (1): 6–8.

1218 ■ 1969 Gupta, H. *et al.* Study of the Koyna earthquake of 10 December1967. *Bulletin of the Seismological Society of America*, 59 (3): 1149–1162.

1219 ■ 1969 Kailasam, L. N. Geophysical investigations in earthquake affected areas of Koyna, Satara district (Maharashtra). In G. C. Chatterjee (ed). *Contributions to Petrology, Geochemistry and Geophysics of Upper Mantle, India.* Kolkata: Memoirs of the Geological Survey of India. 117–122.

1220 ■ 1970 Chandrasekaran, A. R. and P. N. Agarwal. Strong motion data from 29 October 1968 Koyna earthquake. *Bulletin of Indian Society of Earthquake Technology,* 7 (3): 153–158.

1221 • 1970 Chatterjee, S. Probable causes of Koyna quake. *Geographical Review of India*, 32(2): 125–133.

1222 • 1970 Chatterjee, S. Probable causes of Koyna earthquake. *Science*, 291: 2101–2102.

1223 ■ 1970 Gubin, I. E. Koyna earthquake of 1967. *Physics of Solid Earth*, 6: 339–346.

1224 ■ 1970 Gupta, H.K. *Studies on Seismic Surface Depression: Crustal Structure of Himalaya Region and Global Seismicity*. Roorkee: University of Roorkee (Unpublished Ph.D thesis).

1225 ■ 1970 Ramakotaiah, G. *et al.* Geological conclusion of the Broach earthquake March 1970. *Bulletin of the Natural Gas Commission,* 7 (1): 48–50.

1226 ● 1970 Sengupta, S. *A Study of Seismic Characteristics in the Vicinity of Small Explosions.* Kharagpur: Indian Institute of Technology, Kharagpur (Unpublished Ph.D thesis).

1227 ■ 1971 Gupta, H.K. *et al.* Koyna earthquake ,10 December 1967: A multiple seismic event. *Bulletin of the Seismological Society of America,* 61 (1): 167–176.

1228 ■ 1971 Tsai, Y.B. and K. Aki. Koyna India earthquake of 10 December 1967. *EOS American Geophysical Union Transaction,* 52 (4): 277.

1229 ● 1972 Agrawal, P.N. 10 December 1967 Koyna earthquake and reservoir filling. *Bulletin* of the *Seismological Society of America,* 62 (1): 661–662.

1230 ■ 1972 Bhanghar, A.R. Focal mechanism of Indian earthquakes. *Bulletin of the Seismological Society of America,* 62 (2): 603–608.

1231 ● 1972 Gupta, H.K. *et al.* Common features of the reservoir-associated seismic activities. *Bulletin of the Seismological Society of America,* 62: 481.

1232 ■ 1973 Chaturvedi, A. *et al.* Tectonogenesis and seismicity of the Kumaon Himalaya *Himalayan Geology*, 3: 336–344.

1233 ■ 1973 Gupta, H.K. *et al.* Accuracy of determination of epicentre and origin time of small magnitude earthquakes in the Indian subcontinent. *Bulletin of the Seismological Society of America,* 63: 1901–1912.

1234 ■ 1973 Rastogi, B.K. *et al.* Earthquake mechanism and tectonics in the Assam-Burma region. *Tectonophysics,* 18 (3–4): 355–366.

1235 ■ 1974 Ben-Menahem, A. *et al.* The source of the Great Assam earthquake interplate wedge motion. *Physics of the Earth and Planetary Interiors,* 9 (4): 265–289.

1236 ■ 1974 Chaudhury, H.M. *et al.* Seismotectonic investigations of the Himalaya. *Himalayan Geology,* 4: 481–491.

1237 ● 1974 Gupta, H.K. and B.K. Rastogi. Investigations of the behaviour of reservoir-associated earthquakes. *Engineering Geology*, 8 (1–2): 29–38.

1238 ● 1974 Rastogi, B.K. Earthquake mechanism and plate tectonics in the Himalayan region. *Tectonophysics,* 21 (1–2): 47–56.

1239 ● 1975 Athavale, R.N. Induced seepage along a coastal parallel system of faults as a possible cause of the Koyna earthquakes. *Bulletin of the Seismological Society of America,* 65: 183–191.

1240 ■ 1976 Chaudhary, H. M. and Srivastava. Seismicity and focal mechanism of some recent earthquakes in Northeast India and neighbourhood. *Annali di Geofisica*, 29 (1–2): 41–57.

1241 ● 1976 Gupta, H. K. and B. K. Rastogi. *Dams and Earthquakes*. Amsterdam: Elsevier.

1242 ■ 1976 Gupta, H.K. and J. Combs. Continued seismic activity at the Koyna reservoir sites. *Engineering Geology*,10 (2–4): 307–313.

1243 ■ 1976 Kaila, K.L. and H. Narayan. Evolution of Himalaya: Based on seismotectonics and deep seismic sounding. In *Proceedings of Himalayan Geology Seminar, Structure, Tectonics, Seismicity and Evolution*. 1–30.

1244 ● 1976 Singh, S. *et al.* Filling of Ramganga reservoir, Kalagarh, Uttar Pradesh (India) and its possible influence on seismic activity. *Bulletin of the Seismological Society of America*, 66: 1727–1731.

1245 ■ 1976 Verma, R.K. *et al.* Seismicity of Himalaya and Tibetan region: Implications of the continental plate convergence. In *Proceedings Himalayan Geology Seminar, Section 1–B Structure, Tectonics, Seismicity and Evolution*. 193–206.

1246 ■ 1977 Chandra, U. Earthquakes of Peninsular India: A seismotectonics study. *Bulletin of the Seismological Society of America*, 67 (5): 1387–1413.

1247 ■ 1978 Chandra, U. Seismicity, earthquake mechanism and tectonics along the Himalayan mountain range and vicinity. *Physics of the Earth and Planetary Interiors*, 16 (2): 109–131.

1248 ■ 1979 Singh, D.D. *et al.* Spectral analysis of body waves for earthquakes and their source parameters in the Himalaya and nearby regions. *Physics of the Earth and Planetary Interiors,* 18 (2): 143–152.

1249 ● 1980 Gupta, H.K. *et al.* An investigation of earthquake in Koyna region for the period October 1973 through December 1976 to October 1980. *Bulletin of the Seismological Society of America*, 70 (5): 1833–1847.

1250 ■ 1980 Singh, D.D. and H.K. Gupta. Source dynamics of two Great Earthquakes of the Indian subcontinent: The Bihar-Nepal earthquake of 15 January 1934 and Quetta earthquake of 30 May 1935. *Bulletin of the Seismological Society of America*, 70 (3): 757–773.

1251 ■ 1981 Seeber, L. *et al.* Seismicity and continental subduction in the Himalayan arc. *American Geophysical Union,* Series 3: 215–242.

1252 ■ 1981 Srivastava, V.K. *et al.* Note on earthquake of 29 July 1980 around Dharchula Area, Pithoragarh District (Uttar Pradesh). *Indian Minerals*, 35 (3): 27–28.

1253 ● 1982 Bhatia, K.G. *Failure Phenomenon of Rockfill Dam Subject to Earthquake*. New Delhi: Indian Institute of Technology (Unpublished Ph.D thesis).

1254 ■ 1983 Agrawal, P.N. A study of 20 January 1982 earthquake near the Great Nicobar Island . *Bulletin of the Seismological Society of America,* 73 (4): 1139–1159.

1255 • 1984 Gupta, H.K. *et al.* Are reservoir induced earthquakes on M 5.0 at Koyna preceded by couple of earthquakes of M 4.0. *Bulletin of the Seismological Society of America,*74 (3–4): 863–873.

1256 • 1985 Gupta, H.K. The present status of reservoir induced seismicity investigations with special emphasis on Koyna earthquakes. *Tectonophysics,* 118 (3 and 4): 257–279.

1257 • 1985 Valdiya, K.S. Himalayan tragedy: Big dams, siesmicity, erosion and drying up of springs in the Himalayan region. *Central Himalayan Environmental Association Bulletin,* (1): 1–24.

1258 ■ 1985 Verma, R.K. *Gravity Field, Seismicity and Tectonics of the Indian Peninsula and the Himalayas.* New Delhi: Allied Publishers.

1259 • 1986 Bhattacharya, S.N. *et al.* Microearthquake observations around the dam in Northwest Himalayas. *Physics of the Earth and Planetary Interiors,* 44 (2): 169–178.

1260 ■ 1986 Gupta, G.D. *Himalayan Seismicity.* Bangalore: Memoirs of the Geological Society of India, No. 23.

1261 • 1986 Gupta, H.K. and K. Rajendran. Larger artificial water reservoirs in the vicinity of Himalayan foothill and reservoir induced seismicity. *Bulletin of the Seismological Society of America,* 76 (1): 205–215.

1262 • 1986 Patil, D.N. *et al.* Reservoir induced seismicity in vicinity of Lake Bhatsa, Maharashtra . *Physics of the Earth and Planetary Interiors.* 44 (2): 73–81.

1263 ■ 1986 Rastogi, B.K. *et al.* Precursory phenomenon in the microearthquake sequence near Osmansagar reservoir, Hyderabad . *Tectonophysics,* 138 (1): 17–24.

1264 ■ 1987 Banghar, A.R. Seismotectonics of the Andaman-Nicobar islands. *Tectonophysics,* 133 (1–2): 95–107.

1265 ■ 1987 Kayal, J.R. Microseismicity and source mechanism study: Shillong Plateau, (Northeast India). *Bulletin of the Seismological Society of America,* 77.

1266 ■ 1987 Wei, L. *et al.* Deformation and displacement fields of the Himalayan arc: Seismicity and large earthquake hazards in its eastern segment and vicinity. *Tectonophysics,* 138 (1): 93–107.

1267 • 1988 Chandrakanth, S.R. A mystery still. *Frontline,* 21 (September 3): 30–32.

1268 ■ 1988 Narula, P.L. Seismotectonic evaluation of Northwest Himalayas. In *Records of the Geological Survey of India,* 121 (8): 148–149.

1269 ■ 1988 Singh, H.N. *et al.* Investigations of the Idukki earthquake: Sequences of 7 and 8 June 1988. *Journal of Geological Society of India*, 34 (2): 133–146.

1270 ■ 1989 Chander, R. On applying the concept of rupture propagation to deduce the location of the 1905 Kangra earthquake epicentre. *Journal of Geological Society of India*, 33 (2): 150–158.

1271 ■ 1989 Misra, D.C. *et al.* Idukki earthquake and the associated tectonics from gravity study. *Journal of Geological Society of India,* 34 (2): 147–151.

1272 ■ 1989 Sarkar, I. Plate tectonics with resulting global physical landscape and seismicity: A geographical review of the meso-Cenozoic era. *Geographical Review of India,* 51 (3): 39–51.

1273 ■ 1990 Molnar, P. A review of the seismicity and the rate of active under thrusting. *Journal of Himalayan Geology*, 1 (2): 131–154.

1274 ■ 1990 Ramakrishan, R. *Seismotectonics of Southern Peninsula of India.* Dhanbad: Indian School of Mines (Unpublished Ph.D thesis).

1275 ■ 1991 Banghar, A.R. Mechanism solution of Nepal-Bihar earthquake of 20 August 1988. *Journal of Geological Society of India*, 37 (1): 25–30.

1276 ■ 1991 Biswas, S. *Seismicity, Earthquake Mechanism, Tectonics and Stress Distribution in Himalayas, Tibet, Burmese Arc and the Andaman Sea.* Kolkata: Jadavpur University (Unpublished Ph.D thesis).

1277 • 1991 Chadha, R.K. *Studies on Reservoir Induced Seismicity for Some Cases in Peninsular India.* Delhi: Indian Institute of Technology (Unpublished Ph.D thesis).

1278 • 1991 Jayaraman, K.S. Disaster prevention: Quake raises doubts about Indian dams. *Nature*, 354 (November 14): 96.

1279 • 1991 Viswanathan, A. Reservoir induced seismicity: A man made disaster. *Economic and Political Weekly*, 26 (52): 2979–2980.

1280 ■ 1992 Chander, R. An assessment of Oldham's model for the cause of the 1897 Shillong plate earthquake. *Journal of Himalayan Geology*, 3 (1): 119–120.

1281 ■ 1992 Gahalaut, V.K. and R. Chander. A rupture model for the Great Earthquake of 1897 (Northeast India). *Tectonophysics*, 204 (1–2): 163–174.

1282 • 1992 Gupta, H.K. Reservoir induced earthquake seismicity. *Current Science,* 62 (1–2): 183–198.

1283 ■ 1992 Gupta, H.K. *et al.* Uttarkashi earthquake of 20 October 1991. *Journal of Geological Society of India*, 39.

1284 • 1992 Gupta, H.K. and B.K. Rastogi. *Reservoir Induced Earthquakes.* Amsterdam: Elsevier.

1285 • 1992 Maikhuri, R.K. and V. Joshi. Causes of devastation of Uttarkashi (Garhwal Himalaya) earthquake and strategies for future. *Science and Culture,* 58 (5–8): 115–121.

1286 ■ 1992 Narula, P.L. and S.K. Shome. Macro seismic studies of recent earthquakes in Northwest Himalaya: A review. *Current Science*, 62 (1–2): 24–33.

1287 ■ 1992 Rastogi, B.K. Fault plane solutions of some earthquakes in India during 1980s. *Journal of Himalayan Geology*, 3 (2): 163–170.

1288 • 1992 Roy C, A. Building up a dangerous trend. *Down to Earth*, 1 (2): 27–30.

1289 ■ 1993 Chengappa, R. The Himalayas: Startling discoveries. *India Today,* (March 15): 90–101.

1290 ■ 1993 Chung, W.Y. Source parameters of two rift associated intraplate earthquakes in Peninsular India: The Bhadrachalam earthquake of 13 April 1969 and the Broach earthquake of 23 March 1970. *Tectonophysics*, 225 (3)219–230.

1291 ■ 1993 Chung, W.Y. and H. Gao. Source parameters of three rift associated intraplate earthquakes in Peninsular India. *EOS American Geophysical Union Transaction,* 74 (16): 205.

1292 • 1993 Gaur, V.K. (ed). *Earthquake Hazard and Large Dams in the Himalayas.* New Delhi: Indian National Trust for Art and Cultural Heritage.

1293 ■ 1993 Gupta, H.K. *et al.* An investigation into Latur earthquake of 29 September 1993 in Southern India. *Tectonophysics*, 287 (1–4): 299–318.

1294 ■ 1993 Khattri, K.N. Himalaya: Earthquakes and geophysical facts. *Himalayan Geology,* 14: 223–228.

1295 ■ 1993 Lakshmipathi, T. Rethinking seismic theory. *Frontline*, 26 (October 22): 128.

1296 ■ 1993 Prasad, C.V. Uttarkashi earthquake 1991: A geological analysis. In G.S. Rajwar (ed). *Garhwal Himalaya: Ecology and Environment.* New Delhi. Ashish Publishing House. 49–58.

1297 ■ 1993 Sarkar, I. *et al.* Evidence from small magnitude earthquakes on the active tectonics of Northwest Garhwal Himalaya. *Journal of Himalayan Geology,* 4 (2): 279–284.

1298 ■ 1993 Thakur, V.C. and S. Kumar. Seismotectonics of the 20 October 1991 Uttarkashi earthquake (North India). *Terra Nova,* 6 (2): 90–94.

1299 • 1994 Chandra Shekhar, A.R. The houses upon sand. *Down to Earth,* 2 (17): 11.

1300 • 1994 Rastogi, B.K. and R.K. Chadha. On dams and earthquakes. *Down to Earth*, 3 (6): 30.

1301 • 1994 Srivastava, R. Building on wrong premise. *Down to Earth,* 2 (24): 5–8.

1302 ■ 1994 Valdiya, K.S. Strong motion earthquake in Himalaya: Geological perspective *Current Science*, 67 (5): 313–323.

1303 • 1994 Veladapu, B, *Studies in Water Reservoir Induced Seismic Characteristics and Mechanism.* Pune: Pune University (Unpublished Ph.D thesis).

1304 ■ 1995 Bilham, R. Location and magnitude of the 1833 Nepal earthquake and its relation to the rupture zones of contiguous Great Himalayan earthquake. *Current Science*, 69: 101–128.

1305 ■ 1995 Gahalaut, V.K. *On Rupture Zones and Geodynamic Processes of Great Earthquakes along the Himalayan Convergent Plate Margin.* Roorkee: University of Roorkee (Unpublished Ph.D thesis).

1306 ■ 1995 Gupta, H.K. and G.D. Gupta (eds). *Uttarkashi Earthquake.* Bangalore: Memoirs of the Geological Society of India, No. 30.

1307 ■ 1995 Jain, A.K. and R. Chander. Geodynamic models for Uttarkashi earthquake of 20 October 1991. *Journal of Geological Society of India*, 45: 225–233.

1308 • 1995 Kalpana. *Studies Related to Occurrence of Earthquakes Near New Reservoirs.* Roorkee: University of Roorkee (Unpublished Ph.D thesis).

1309 • 1995 Mitha, A. House of cards. *Down to Earth*, 3 (20): 25–30.

1310 ■ 1995 Nandy, D.R. Neotectonism and seismic hazards in India. *Indian Journal of Geology*, 67 (1): 34–48.

1311 ■ 1995 Singh, D.D. Shear wave velocity structure beneath Northeast India and nearby regions and seismic source parameters of earthquakes in its vicinity. *Himalayan Geology,* 16: 285–298.

1312 ■ 1995 Singh, R.P. *et al.* The geodynamic context of Latur earthquake, 30 September 1993. *Physics of the Earth and Planetary Interiors*, 91 (4): 245–251.

1313 ■ 1995 Thakur, V.C. *et al.* A report on 30 September 1993 Latur earthquake of Maharashtra. *Journal of Himalayan Geology,* 6 (1): 31–50.

1314 • 1996 Chander, V.K. and R.K. Gahalaut. On the transience of Tehri reservoir load on Garhwal seismicity. *Current Science*, 70 (10): 873.

1315 ■ 1996 Joshi, D.D. Quaking mountains. *Down to Earth*, 5 (November 15): 14.

1316 ■ 1996 Kayal, J.R. Earthquake source process in Northeast India: A review. *Himalayan Geology*, 17: 53–70.

1317 ■ 1996 Kumar, S. Earthquake and tectonostratigraphy of the middle sector of the Himalaya between longitude 80° E and 90° E. In *Proceedings of the Symposium on Recent Advances in Geological Studies of Northwest Himalayas and Foredeep.* Special Publication No. 21. Kolkata: Geological Survey of India. 19–28.

1318 ■ 1996 Rajendran, C. P. and K. Rajendran. Application of paleoseismological methods in seismic hazard evaluation: An overview based on some examples of intraplate seismicity. *Himalayan Geology,* 17: 137–146.

1319 ■ 1996 Rajendran, C. P. *et al.* The 1993 Killari, Central India, earthquake: An example of fault reactivation in the Pre-Cambrian crust. *Geology,* 24 (7): 651–654.

1320 ■ 1996 Seeber, L. *et al.* The 1993 Killari earthquake in Central India: A new fault in Mesozoic flows? *Journal of Geophysical Research,* 101(B4): 8543–8560.

1321 ■ 1996 Shanker, R. Thermal and crustal structure of Indian Peninsula and its relation to recent earthquakes. *Himalayan Geology,* 17: 113–118.

1322 ■ 1997 Ramachandran, C. and M. Kesavamani. Fault segmentation and earthquake hazard zones in Deccan Trap region: Gravity evidence. *Journal of Geological Society of India,* 49(2): 23–32.

1323 • 1997 Rao, D. K. The making of a quake. *Frontline,* 30 (June 13): 32.

1324 ■ 1997 Sagina, R. *et al. Report on Seismotectonic Evaluation of Kangra Block: Micro Earthquake Study in Parts of Kangra and Chamba districts (Himachal Pradesh).* Kolkata: Geological Survey of India (Unpublished Report).

1325 ■ 1997 Shanker, D. and V. P. Singh. Seismic risk analysis for the occurrence of medium size earthquake in Kangra region,Himachal Pradesh. *Proceedings of Indian Natural Science Academy,* Part A, 63 (2): 71–76.

1326 ■ 1998 Bilham, R. Slip parameters for the Rann of Kachchh, (India), 16 June 1819 earthquake, quantified from contemporary accounts. In I.S. Stewart and C. Vitafinzi (eds). *Coastal Tectonics.* London: Geological Society. 295–318.

1327 ■ 1998 Joshi, A. Study of Uttarkashi earthquake in terms of rupture model and isoseismals. *Journal of Geophysics,* 19 (2): 133–140.

1328 ■ 1998 Kapur, N. *Analysis of Earthquake Related Upper Crustal Stresses and Ground Displacements in Himalayas.* Roorkee: University of Roorkee (Unpublished Ph.D thesis).

1329 ■ 1998 Khanal, K. N. *et al.* Source mechanisms and the depths of foci of two moderate magnitude earthquakes of the Central Himalaya, using waveform modelling. *Himalayan Geology,* 19 (1): 1–10.

1330 ■ 1998 Kumar, S. Intraplate seismicity and geotectonics near the focal area of the Latur earthquake (Maharashtra). *Journal of Geodynamics*, 25 (1–2): 109–128.

1331 ■ 1998 Ramesh, D.S. and C.H. Estabrook. Rupture histories of two stable continental region earthquakes of India. *Proceedings of Indian Academy of Sciences, Earth and Planetary Sciences*, 107 (3): 225–233.

1332 ■ 1998 Shankar, R. and P.L. Narula. Chamoli earthquake of 29 March 1999: A preliminary appraisal. *Indian Minerals,* 52: 141–150.

1333 ■ 1998 Sankaran, A.V. Remote triggering of earthquake: Some fresh views. *Current Science,* 75 (11): 1108–1109.

1334 ■ 1998 Satyabala, S.P. The Chamoli earthquake of 29 March 1999. *Journal of Geological Society of India,* 53 (6): 711–712.

1335 ■ 1998 Srivastava, H.N. and S.N. Bhattacharya. Application of principal component analysis to some earthquake related data in the Koyna region, (India). *Engineering Geology,* 50 (1–2): 141–151.

1336 ■ 1998 Thakur, V.C. *et al.* 22 May 1997 Jabalpur earthquake: Reactivation of East-West trending Jabalpur fault in the Narmada-Son lineament zone. *Himalayan Geology,* 19 (1): 119–126.

1337 ■ 1998 Yeats, R.S. and V.C. Thakur. Reassessment of earthquake hazard based on a fault-bend fold model of the Himalayan plate boundary fault. *Current Science*, 74 (3): 230–233.

1338 ■ 1999 Paul, A. *Source Parameters of Uttarkashi Earthquake of October 1991 by Wave-form Modelling.* Roorkee: University of Roorkee (Unpublished Ph.D thesis).

1339 ■ 1999 Rai, S.S. *et al.* What triggers Koyna earthquakes? Preliminary results from seismic tomography digital array. *Proceedings of Indian Academy of Sciences, Earth and Planetary Sciences,* 108 (1): 1–14.

1340 ■ 1999 Rajendran, C.P. and K. Rajendran. Geological investigations at Killari and Ter (Central India) and their implications for paleoseismicity in the shield region. *Tectonophysics,* 308 (1–2): 67–81.

1341 ■ 1999 Ravindran, K.V. and G. Philip. 29th March 1999 Chamoli earthquake: A preliminary report on earthquake induced landslides using Indian remote sensing-central data recorder data. *Science and Culture,* 65 (1): 21–25.

1342 ■ 1999 Shankar, R. Lineament tectonics: Its relation to seismic activity in India. *Shelter,* special issue, October: 54–56.

1343 ■ 1999 Verma, R.K. *et al.* An analysis of seismicity, maximum magnitude and acceleration for Kashmir, Punjab, Kumaon and Garhwal Himalaya region. *Himalayan Geology,* 20 (1): 123–130.

1344 • 2000 Bapat, A. Dams and earthquake. *Frontline*, 17 (1): 87–88.

1345 ■ 2000 Gupta, I. D. Using strong motion accelerograms for estimation of local magnitude of earthquakes in Himalayan region. *ISET Journal of Earthquake Technology*, 37 (1–3): 1–10.

1346 ■ 2000 Joshi, A. Modelling of rupture plains of peak ground acceleration and its application to the isosiesmal map of modified Mercalli intensity scale in Indian region. *Journal of Seismology*, 4 (2): 143–160.

1347 ■ 2000 Kanwar, S. S. *et al.* (eds). *Jabalpur Earthquake 22 May 1997: A Geo-Scientific Study*. Special Publication No. 51. Kolkata: Geological Survey of India.

1348 ■ 2000 Kayal, J. R. Seismotectonic study of the two recent stable continental region earthquakes in Central India. *Journal of Geological Society of India*, 55 (2): 123–138.

1349 ■ 2000 Nandy, D. R. Plate tectonics, earthquake and earthquake hazards. *Science and Culture*, 66 (1–2): 2–8.

1350 ■ 2000 Rajendran, K. and C. M. Harish. Mechanism of triggered seismicity of Koyna: Evaluation based on relocated earthquakes. *Current Science*, 79 (3): 358–363.

1351 ■ 2000 Rajendran, K. and C. P. Rajendran. Geological investigation in the 1819 Kutch Northwestern Indian seismic zone and implications for mechanics of stable continental crust earthquake. In *Proceedings of the International Symposium and School on Active Faulting*. Japan: 387–391.

1352 ■ 2001 Bandaru, V. L. *Geologial Evidences and Dating of a Paleoseismic Event in the Mesoseismal Area of the 1993 Latur Earthquake, Deccan Plateau (Maharashtra)*. Hyderabad: Osmania University (Unpublished Ph.D thesis).

1353 ■ 2001 Bendick, R. *et al.* The 26 January 2001 Republic Day earthquake (India). *Seismological Research Letters*, 72 (3): 328–335.

1354 ■ 2001 Bhatt, N. Y. Seismic wave amplification: One more example as evidenced at Ahmedabad on 26 January 2001. *Current Science*, 80 (9): 1091–1092.

1355 ■ 2001 Biswas, S. K. Structure and tectonics of Gujarat, with special reference to earthquakes in Kutch. In *Proceedings of the International Conference on Seismic Hazard with Particular Reference to Bhuj Earthquake of 26 January 2001*. New Delhi: Government of India Department of Science and Technology India Meteorological Department. 349–351.

1356 ■ 2001 Dasgupta, S. Targeting east-west normal faults in Peninsular India for seismic hazard studies. *Journal of Geological Society of India*, 57 (5): 467–468.

1357 ■ 2001 Ghevariya Z.G. and B.K. Sahu. Assessment of active fault and lineaments in Kutch and Saurashtra regions of Gujarat: A post 26 January earthquake study. In *Proceedings of the Role of Earth Sciences in Integrated Development and Related Societal Issues*. Special Publication No. 65 (2). Kolkata: Geological Survey of India. 96–103.

1358 ■ 2001 Gupta, H.K. and N.M. Rao *et al.* The deadliest intraplate earthquake. *Science,* 291 (5511): 2101–2102.

1359 ■ 2001 Gupta, H.K. *et al.* Bhuj earthquake of 26 January 2001. *Journal of Geological Society of India,* 57 (3): 275–278.

1360 ■ 2001 Kumar, S. *et al.* Earthquake recurrence and rupture dynamics of the Himalayan frontal thrust. *Science,* 294 (5550): 2328–2331.

1361 ■ 2001 Madhurkar, U. Gujarat earthquake: Terra infirma. *India Today,* (Dec. 31): 160.

1362 • 2001 Menon, P. Built-in danger. *Frontline,* 18 (March 3): 116–118.

1363 ■ 2001 Narayanpethkar, A.B. and A. Vasanthi. Possible relationship between temporal electrical resistivity variation and occurrence of the earthquake of 30 September 1993 in Latur region . *Current Science,* 81 (9): 1224–1230.

1364 ■ 2001 Pandey, Y. *et al.* Estimation of source parameters of Chamoli Earthquake, (India). *Proceedings of Indian Academy of Sciences, Earth and Planetary Sciences,* 110 (2): 171–177.

1365 ■ 2001 Patel, S.G. and B.G. Desai. The Republic day Kachchh earthquake of 2001: Trauma in oratosquilla strata. *Journal of Geological Society of India,* 58 (3): 215–216.

1366 • 2001 Photo file. Quake did not kill them: Killed by greed of retailer and government's convenience. *New Age,* 49 (2): 16.

1367 ■ 2001 Rajendran, K. and C.P. Rajendran. The 2001 Kutch earthquake: Coseismic surfaces and their significance. *Current Science,* 80 (11): 1397–1405.

1368 ■ 2001 Raju, M. Role of seismicity and rainfall as causative factors of extensive landslides in Mizoram. In *Proceedings of the National Seminar on Geotechniques and Geological Hazards in the Indian Context.* Kolkata: Geological Survey of India.

1369 ■ 2001 Raval, U. Earthquake over Kutch: A region of 'trident' space-time geodynamics. *Current Science,* 81 (7): 809–815.

1370 ■ 2001 Shankar, R. Seismotectonics of Kutch rift basin and its bearing on Himalayan seismicity. *ISET Journal of Earthquake Technology,* 38 (2–4): 59–65.

1371 ■ 2001 Thanavelu, C. Seismotectonic evaluation of Tamil Nadu. In Ganesh *et al* (eds). *Proceedings of the Workshop on Popular Lectures on Facts and Myths of Earthquake, Indian Context.* Tiruchirapalli: Bharathidasan University. 48–50.

1372 • 2001 Wieland, M. Discussion on failed and safe large dams in earthquake prone Himalayan region. *ISET Journal of Earthquake Technology*, 37 (1): 57.

1373 • 2002 Editorial. Tehri: Is it curtains? *Down to Earth,* 10 (18): 50–53.

1374 ▪ 2002 Thakur, V.C. and S.G Wesnousky. Seismotectonics of 26 January 2001 Bhuj earthquake affected region. *Current Science*, 82 (4): 396–399.

1375 ▪ 2002 Thakur, V.C. and S. Kumar. Seismotectionics of Chamoli earthquake 29 March 1999 and earthquake hazard assessment of Garhwal-Kumaon region (Northwest Himalaya). *Himalayan Geology,* 23 (1 and 2): 113–120.

1376 ▪ 2003 Bilham, R. *et al.* Flexure of the Indian plate and intra-plate earthquake. *Proceedings of Indian Academy of Sciences, Earth and Planetary Sciences,* 112 (3): 315–330.

1377 ▪ 2003 De, R. Fault plane solutions of the 26 January 2001 Bhuj earthquake sequence. In *Proceedings of Indian Academy of Sciences, Earth and Planetary Sciences*, 112 (3): 413–420.

1378 ▪ 2003 Ganesha, R.K. *Major Lineaments of Karnataka and their Significance with Respect to Seismicity: Remote Sensing Based Analysis.* Gulbarga: Gulbarga University (Unpublished Ph.D thesis).

1379 ▪ 2003 Ortiz, M. and R. Bilham. Source area and rupture parameters of the 31 December 1881 M 7.9 Car Nicobar earthquake estimated from tsunamis recorded in Bay of Bengal. *Journal of Geophysical Research,* 108 (B4): 2215.

1380 • 2003 Pandey, A. and R.K. Chadha. Surface loading and triggered earthquake in the Koyna-Warna region. *Physics of the Earth and Planetary Interiors,* 139 (3–4): 207–223.

1381 ▪ 2003 Rajendran, C.P. *et al.* The 13 September 2002 North Andaman earthquake: An analysis in the context of regional seismicity. *Current Science,* 84: 919–924.

1382 ▪ 2003 Rajendran, C.P. *et al.* The odds of a seismic source near Dwarka (Northwest Gujarat): An evaluation based praxis. *Current Science,* 84: 695–701.

1383 ▪ 2003 Rajendran, K. and C.P. Rajendran. Seismogenesis in the stable continental regions and implications for hazard assessment: Two recent examples from India. *Current Science*, 86 (7): 896–903.

1384 ▪ 2003 Singh, S.K. *Seismic Imaging and Tectonics of Koyna Seismic Zone.* Hyderabad: Osmania University (Unpublished Ph.D thesis).

1385 ▪ 2003 Yadav, D.K. *Seismotectonics of Arunachal Himalayas.* Dibrugarh: Dibrugarh University (Unpublished Ph.D thesis).

1386 ▪ 2004 Bilham, R. Earthquakes in India and the Himalaya: Tectonics, geodesy and history. *Annals of Geophysics*: 839–858.

1387 ■ 2004 Gahalaut, V.K. and R. Burgmann. Constraints on the source parameters of the 26 January 2001 Bhuj (India) earthquake from satellite images. *Bulletin of the Seismological Society of America*, 94 (6): 2407–2413.

1388 ■ 2004 Gahalaut, V.K. *et al.* On the mechanism and source parameters of the deep crustal Jabalpur earthquake of 21 May 1997: Constraints from aftershocks and changes in static stress. *Geophysical Journal International*, 159 (2): 345–351.

1389 ■ 2004 Hough, S.E. *et al.* Revisiting the 1897 Shillong and 1905 Kangra earthquakes in Northern India: Site response, Moho reflections and a triggered earthquake. *Current Science*, 87: 1632–1638.

1390 ■ 2004 Khattri, K.N. A perspective of seismic hazard and risk in Northern India. In K.S. Valdiya (ed). *Coping with Natural Hazards: Indian Context*. Hyderabad: Orient Longman. 48–58.

1391 ■ 2004 Misra, K.S. *et al.* Seismogeology of the Kutch and adjoining region with special reference to 26 January 2001 earthquake in the vicinity of Bachau (Gujarat). *Journal of Geological Society of India*, 64 (2): 153–164.

1392 ■ 2004 Rajendran, C.P. *et al.* Interpreting the style of faulting and paleoseismicity associated with the 1897 Shillong (Northeast India) earthquake: Implications for regional tectonism. *Tectonics*, 23.

1393 ■ 2004 Rajendran, K. *et al.* Mild tremors near Thalassei (November 2003): Evidence for a seismic source in North Kerala. *Journal of Geological Society of India*, 64 (2): 233–236.

1394 ■ 2004 Singh, S.K. *et al.* A source study of the Bhuj earthquake of 26 January 2001 (M 7.6). *Bulletin of the Seismological Society of America*, 94 (4): 1195–1206.

1395 ■ 2004 Thakkar, M.G. and B. Goyal. On the relation between magnitude and liquefaction dimension at the epicentral zone of 2001 Bhuj earthquake. *Current Science*, 87 (6): 811–817.

1396 ■ 2005 Rajendran, C.P. and K. Rajendran. The status of central seismic gap: A perspective based on the spatial and temporal aspects of the large Himalayan earthquakes. *Tectonophysics*, 395 (1–2): 19–39.

1397 ■ 2005 Rajendran, C.P. *et al.* Towards developing a better appreciation of the seismicity and tectonics of Andaman and Nicobar Islands subduction zone. *Current Science*, 88 (12): 1979–1980.

1398 ■ 2005 Rao, N.P. and A.H. Chary. What caused the Great Sumatran earthquakes of 26 December 2004 and 28 March 2005? *Current Science*, 89 (3): 449–452.

1399 ■ 2005 Santosh, M. The Gondwana quake. *Gondwana Research-International Geoscience Journal*, 8 (2): 105–106.

1400 ■ 2005 Sarkar, I. Role of small earthquakes in occurrence of moderate earthquakes in the Garhwal Himalaya. *Himalayan Geology,* 26 (1): 357–366.

1401 ■ 2005 Sukhija, B. S. and H. K. Gupta. Paleoseismicity and recurrence of great earthquakes in Shillong Plateau, (India). *Himalayan Geology,* 26 (1): 187–192.

1402 ■ 2005 Wilson, M. Modelling the Sumatra-Andaman earthquake reveals a complex, non-uniform rupture. *Journal of Geological Society of India,* 66 (4): 499–502.

1403 • 2006 Akhtar, A. S. Making of a disaster. *Economic and Political Weekly,* 41(6): 479–480.

1404 • 2006 Editorial. Everybody's fault. *Nature.* 440 (March 2) 1.

1405 ■ 2006 Gahalaut, V. K. 2005 Kashmir earthquake: Not a Kashmir Himalaya seismic gap event. *Current Science,* 90 (4): 507–508.

1406 ■ 2006 Gupta, R. P. and I. Sarkar. Earthquake evidence for active extension of some major lineaments in Garhwal Himalaya. *Himalayan Geology,* 27 (1): 123–128.

1407 • 2006 Jayant, T. V. Shaky foundation. *Down to Earth,* 14 (22): 1–15.

1408 ■ 2006 Mahajan, A. K. Seismicity, seismotectonics and seismic hazard of 1905 Kangra earthquake affected regions (Northwest Himalaya). In K. K. Sharma *et al* (eds). *Environmental Geohazards: Science and Society.* New Delhi: Research India Press. 64–86.

1409 ■ 2006 Narula, P. L. Seismotectonics and seismic hazard evaluation in Indian perspective. In K. K. Sharma *et al* (eds). *Environmental Geohazards: Science and Society.* New Delhi: Research India Press, 31–52.

1410 ■ 2006 Saikia, C. K. Modelling of the 21 May 1997 Jabalpur earthquake in Central India: Source parameters and regional path calibration. *Bulletin of the Seismological Society of America,* 96: 1396–1421.

Cause: Flood

1411 ■ 1941 Mahalanobis, P. C. Rainstorm and River Floods in Orissa. Bhubaneswar: *Irrigation and Power Department,* Government of Orissa.

1412 ■ 1954 Parthasarthy, K. Weather in relation to floods in Bihar, Bengal and Assam during July and August. *Indian Journal of Meteorology and Geophysics,* 5 (4).

1413 • 1959 Chakrabarty, A. K. Floods in Himalayan rivers of Ganga plain: Causes and control. *Geographical Review of India,* 21 (2–4): 33–42.

1414 ■ 1962 Dhar, O. N. A study of rainfall and floods in the Yamuna catchment. *Indian Journal of Meteorology and Geophysics,* 13 (3).

1415 ■ 1965 Dhar, O.N. and J. Narayanan. A brief study of rainfall and flood producing rainstorm in the Beas catchment (upto Pong). *Indian Journal of Meteorology and Geophysics*, 16 (1): 81–92.

1416 ■ 1966 Dhar, O.N. and T.G. Changraney. A study of meteorological situations associated with major floods in Assam during the monsoon months. *Indian Journal of Meteorology and Geophysics*, 17: 111–118.

1417 ● 1969 Sen, S. *A Geographical Study of Drainage Condition and Flood Problem in the Lower Reach of Damodar River and Probable Effect of Damodar Valley Corporation Dam*. Kolkata: University of Calcutta (Unpublished Ph.D thesis).

1418 ● 1971 Special Correspondent. West Bengal: The gathering crisis. *Economic and Political Weekly*, 6 (13): 1245–1246.

1419 ■ 1974 Pal, S.K. and K. Bagchi. Recurrence of floods in Brahmaputra and Kosi basins: A study in climatic geomorphology. *Geographical Review of India*, 36: 242–248.

1420 ■ 1975 Chansarkar, R.A. Sudden floods in Himalayan rivers: Causes and effects. In *Proceedings of the Seminar on Landslides and Toe Erosion Problems with Special Reference to Himalayan Region*. Kolkata: Indian Society of Engineering Geology. 84–79.

1421 ● 1975 Das, A.N. The Bihar floods: Unnatural deluge. *Economic and Political Weekly*, 10 (27): 1420–1421.

1422 ■ 1975 Pal, S.K. and K. Bagchi. Geomorphic significance of flood: A comparative study of Brahmaputra and Kosi valleys. *Geographical Review of India,* 37 (3): 242–248.

1423 ■ 1976 Panchadhyayi, S. A diagnostic study of flood problems in Subarnekha basin. *Indian Journal of Power and River Valley Development,* 26 (10): 319–323.

1424 ■ 1976 Rieger, H.C. Floods and droughts: The Himalaya and the Ganges Plain as an ecological system. In *Proceedings of the International Symposium on Mountain Environment and Development.* Katmandu: International Centre for Integrated Mountain Development. 13–29.

1425 ● 1976 Sinha, A. Bihar floods: Familiar patterns. *Economic and Political Weekly*, 11 (39): 1560–1561.

1426 ● 1976 Special Correspondent. What really caused the floods. *Economic and Political Weekly*, 11 (49): 1987–1988.

1427 ● 1978 Bahuguna, S.L. Himalayan trauma: Forest, faults, floods: Chipko seeks a new policy. In *Proceedings of the National Workshop on Integrated Development of the Ganga Brahmaputra Barak Basin.* Delhi: Gandhi Peace Foundation.

1428 ■ 1978 Kumar, T.S.K. *Geomorphic Study of Sahibi River Basin with Special Reference to Floods*. Delhi: Department of Geography, University of Delhi (Unpublished Master of Philosophy Dissertation).

1429 ■ 1978 Panchadhyayi, S. and H.R. Betal. Study of rainstorms and estimation of maximum flood volume in Subernrekha basin. *Geographical Review of India*, 40 (1): 82–87.

1430 • 1978 Roy, A. West Bengal: A man made calamity. *Economic and Political Weekly*, 13 (47): 1728.

1431 ■ 1979 Mukherji, S. *Floods in West Bengal*. New Delhi: Centre for Studies in Regional Development, Jawaharlal Nehru University (Unpublished Master of Philosophy Dissertation).

1432 ■ 1979 Panda, G.K. *Drainage and Floods in the Mahanadi Delta: A Study in Applied Geomorphology*. Bhubaneswar: Department of Geography, Utkal University (Unpublished Master of Philosophy Dissertation).

1433 ■ 1979 Prasad, C. and G.S. Rawat. Bhagirathi flash-floods: A geomorphological appraisal. *Himalayan Geology*, 9 (2): 734–43.

1434 ■ 1979 Rudra, K. Causes of floods: Some investigation in fluvial geomorphology. *Indian Journal of Power and River Valley Development*, 29 (2): 43–46.

1435 ■ 1979 Special Correspondent. Morvi disaster: An unprecedented natural calamity. *Bhagirath*, 26 (October-December): 151–153.

1436 ■ 1980 Ehrich, C. The next flood knocking at the door: A foreigner's view on the ecological crises of India. In T.V. Singh and J. Kaur (eds). *Studies in Himalayan Ecology and Development Strategies*. New Delhi: English Book Store, 234–242.

1437 ■ 1981 Bhalme, H.N. and D.A. Mooley. Cyclic fluctuations in the flood-drought area and relationship with the double (Hale) sunspot cyclone. *Journal of Applied Meteorology* 20 (9): 1041–1048.

1438 ■ 1981 Sarma, A.A.L.N. and D.C. Rao. Some climatic studies of the incidence of floods at selected stations of the river Mahanadi basin. *Indian Journal of Power and River Valley Development*, 31 (3–4): 42–45.

1439 ■ 1982 Qasim, S.Z. Environmental aspects of flood disaster. *Bhagirath*, 29 (October): 152–154.

1440 ■ 1983 Gaytri, V. *Ecological Perspectives on Floods in the Yamuna in its Middle Reach*. Delhi: Department of Geography, Delhi School of Economics, University of Delhi (Unpublished Master of Philosophy Dissertation).

1441 ■ 1985 Banerjee, B. Genesis of floods in West Bengal and their impact on society. *Deccan Geographer*, 23 (1): 40–54.

1442 ■ 1985 Panchadhyayi, S. *Fluvial Morphology and Flood Characteristics of the Subarnarekha Basin*. Kolkata: University of Calcutta (Unpublished Ph.D thesis).

1443 ■ 1986 Roy, S. *Meteorological Aspects of Severe Floods in India*. Vishakhapatnam: Andhra University, Waltair (Unpublished Ph.D thesis).

1444 ■ 1987 Banerjee, A. River system and flood problem in Jalpaiguri district (West Bengal). *Geographical Review of India*, 49 (1): 28–32.

1445 ■ 1987 Dhar, O.N. and B.N. Mandal. Appraisal of the heavy rainfall and the highest floods in Bihar. *Indian Journal of Power and River Valley Development*, 37 (11 and 12): 302–308.

1446 ■ 1987 Government of Bihar. *Background of Floods in Bihar*, July–August 1987. Patna.

1447 ■ 1987 Kamaraju, M.V. *et al.* Water balance approach for assessment of floods in Haryana: A case study. *Indian Journal of Power and River Valley Development*, 27 (5): 119–123.

1448 ■ 1989 Panda, G.K. *The Drainage and Floods in the Orissa Coastal Plain: A Study in Applied Geomorphology*. Bhubaneswar: Department of Geography, Utkal University (Unpublished Ph.D thesis).

1449 ■ 1989 Subaramayya, I. and G.M. Rao. Hydrometeorological aspect of the 1986 floods in Godavari river. *Indian Journal of Power and River Valley Development,* 39 (4): 202–209.

1450 ■ 1990 Singh, N. *et al.* Rainfall index for hydrological floods and droughts over India. *Mausam,* 41 (3): 469–474.

1451 • 1991 Agrawal, A. and A. Chak (eds). *Floods, Flood Plains and Environmental Myths*. New Delhi: Centre for Science and Environment.

1452 ■ 1992 Banerjee, A. River system and flood problems, Jalpaiguri district of West Bengal. *Transactions, Institute of Indian Geographers*, 14 (1): 73–94.

1453 ■ 1992 Chatterji, A.K. *Monsoons, Cyclones and Floods in India*. Kolkata: Birla Economic Research Foundation.

1454 ■ 1992 Dhar, O.N. and S.S. Nandargi. Study of rainfall and floods in the 1987–1991, monsoon seasons over the contiguous Indian region. *The International Journal of Meteorology,* 17 (174): 330–336.

1455 ■ 1992 Prasad, K. A diagnostic study of flood producing rainstorm of September 1988 over Northwest India with the aid of a fine mesh numerical analysis system. *Mausam*, 43 (1): 71–76.

1456 • 1992 Raja, D. Flood and politics in Tamil Nadu. *New Age,* 40 (50): 12.

1457 ■ 1992 Ramchandran, K. *et al.* Rainfall abnormality in India. *Geographical Review of India,* 54 (2): 21–34.

1458 • 1993 Bora, J. *Drainage in a Flood Prone City: Guwahati*. New Delhi: School of Planning and Architecture (Unpublished Postgraduate Dissertation).

1459 • 1993 Hofer, T. Deforestation, changing discharge and increasing floods: Myth or reality? In B. Messerli *et al* (eds). *Himalayan Environment Pressure-Problems-Processes 12 Years of Research*. Berne: Institute of Geography, University of Berne.

1460 • 1993 Hofer, T. Himalayan deforestation, changing river discharge and increasing floods: Myth or reality. *Mountain Research and Development,* 13 (3): 213–233.

1461 ■ 1994 Gupta, G. L. *The Problems of Floods in Eastern Rajasthan: A Climatic Morphological Study.* Jaipur: Rajasthan University (Unpublished Ph.D thesis).

1462 • 1994 Hofer, T. Deforestation, changing discharge and increasing floods: Myth or reality? In K.R. Dikshit *et al* (eds). *India: Geomorphological Diversity.* New Delhi: Rawat Publications, 220–255.

1463 • 1994 Hofer, T. Demystifying the deluge. *Down to Earth,* 2 (2): 37.

1464 ■ 1994 Kale, V. S. *et al.* Geomorphic and hydrological aspects of monsoon floods on the Narmada and Tapi rivers in Central India. *Geomorphology,* 10 (1–4): 157–168.

1465 • 1994 Saha, P. Before the flood. *Down to Earth,* 3 (5): 33–36.

1466 • 1994 Seuffert, O. Hot rains-high floods-heavy erosion: Rainfall, runoff and erosion patterns in India and the role of natural and anthropogenic factors and their interaction. In K.R. Dikshit *et al* (eds). *India: Geomorphological Diversity.* Jaipur: Rawat Publications. 33–56.

1467 ■ 1996 Nandy, D.R. and D.G.S. Chattopadhayay. Geological factors related to river behaviour including flood with special reference to North Bengal. In R.B. Singh (ed). *Disasters, Environment and Development.* New Delhi: Oxford and IBH Publishing Co. Pvt. Ltd. 197–206.

1468 ■ 1996 Rangasami, A. 1993, one of the wettest years in Brahmaputra basin during the decade 1983–93. *Bhagirath,* 43: 14–19.

1469 ■ 1997 Bhuya, T.R. *et al.* Geoenvironmental hazard study of flood and erosion prone Chapar-Dhubri area of Lower Brahmaputra valley (Assam). In *Proceedings of the Symposium on the Earth Sciences in Environmental Assessment and Management.* Special Publication No. 48 (2). Kolkata: Geological Survey of India. 79–84.

1470 • 1997 D' Souza, R. *Damming the Mahanadi: The Final Solution (1928–46).* New Delhi: Center for Historical Studies, Jawaharlal Nehru University (Unpublished Master of Philosophy Dissertation).

1471 • 1997 Dogra, B. More floods with flood control. *Economic and Political Weekly,* 32 (18): 933–934.

1472 • 1997 Mishra, D.K. The Bihar flood story. *Economic and Political Weekly,* 32 (35): 2206–2217.

1473 ■ 1997 Nayak, T.R. *et al.* Flood hazards in Faridabad district of Haryana caused by river Yamuna during September 1995. In *Proceedings of the Symposium on the Earth Sciences in Environmental Assessment and Management.* Special Publication No. 48 (2). Kolkata: Geological Survey of India. 75–78.

1474 • 1998 Banerjee, R. No dykes, no deluge. *Down to Earth,* 7 (14): 58–59.

1475 ■ 1998 Dhar, O.N. and S.S. Nandargi. Floods in Indian rivers and their meteorological aspects. In V.S. Kale (ed). *Flood Studies in India.* Bangalore: Memoirs of the Geological Society of India, No. 41. 1–26.

1476 • 1998 Editorial. Preventable damage. *Economic and Political Weekly,* 33: 2265.

1477 • 1998 Hofer, T. Do land use changes in the Himalayas affect downstream flooding? Traditional understanding and new evidences. In V.S. Kale (ed). *Flood Studies in India.* Bangalore: Memoirs of the Geological Society of India, No. 41. 119–141.

1478 ■ 1998 Kale, V.S. Monsoon floods in India: A hydro-geomorphic perspective. In V.S. Kale (ed). *Flood Studies in India.* Bangalore: Memoirs of the Geological Society of India, No. 41. 229–256.

1479 ■ 1998 Starkel, L. *et al.* Floods in Sikkim Himalayas: Their causes, course and effects. In V.S. Kale (ed). *Flood Studies in India.* Bangalore: Memoirs of the Geological Society of India, No. 41. 101–118.

1480 • 1999 Kumar, A. *Urbanisation and Flooding in Thiruvanthapuram.* New Delhi: School of Planning and Architecture (Unpublished Ph.D thesis).

1481 ■ 1999 Sonule, B.B. and V.B. Changole. Terrain components of flood-hazards in the Shahanur watershed: A geographical analysis. *Deccan Geographer,* 37 (2): 35–41.

1482 • 1999 Special Correspondent. Dam travails. *Down to Earth,* 8 (2): 14.

1483 • 1999 Vania, R. Engineered chaos. *Down to Earth,* 8 (9): 27–31.

1484 ■ 2000 Dhar, O.N. and S.S. Nandargi. A study of floods in the Brahmaputra basin in India. *International Journal of Climatology,* 20 (7): 771–781.

1485 ■ 2000 Jain, V. *et al.* Neotectonics versus floods hazard in the North Bihar plains: An integrated study aided by Indian remote sensing data. In *Proceedings of the Annual Convention of Indian Institute of Remote Sensing.* Kanpur: Indian Institute of Technology.

1486 • 2000 Kumar, V. Whose baby are the floods?. *Down to Earth,* 9: 28.

1487 ■ 2000 Prasad, E. Tryst with rain. *Down to Earth,* 9 (9): 32.

1488 • 2000 Special Correspondent. Nature's wrath. *Down to Earth,* 9 (August 31): 22.

1489 • 2000 Special Correspondent. Sponsored blight. *Down to Earth,* 9 (8): 28–29.

1490 ■ 2001 Khole, M. and U.S. De. Flood and drought in association with cold and warm El-Nino southern oscillation events and related circulation features. *Mausam,* 52 (1): 13–24.

1491 • 2001 Putatunda, N. West Bengal floods: Avoidable tragedy. *Economic and Political Weekly,* 36 (2): 101–103.

1492 • 2001 Saifuddin and Iqbaluddin. Flood beyond the floodplain in parts of Yamuna basin are corollary of the area development: A remote sensing view. *Photonirvachak, Journal of the Indian Society of Remote Sensing,* 29 (3): 183–186.

1493 • 2002 Editorial. A flood without rain. *Down to Earth,* 11 (7): 42.

1494 ■ 2002 Mukhopadhayay, S.C. Geomorphology and natural hazards in the lower Brahmaputra basin with special reference to floods. *Indian Journal of Geomorphology*, 7 (1–2): 73–80.

1495 • 2002 News Editorial. Damning report. *Down to Earth,* 10 (19): 8.

1496 ■ 2002 Singh. Y. Floods: An ecological hazard of the semi-natural drainage ecosystem of the Inter-State Chandigarh Region (India). *Transactions, Institute of Indian Geographers*, 23 (2): 85–97.

1497 ■ 2003 Dhar, O.N. *et al.* Hydrometeorological aspects of floods in India. *Natural Hazard Journal,* 28 (1): 1–33.

1498 ■ 2003 Nandargi, S. *et al.* Floods in India during the drought year of 2002. *International Journal of Meteorology*, 28 (281): 249–255.

1499 ■ 2004 Basu, G.C. and R.N. Goldar. A study of catastrophic flood in September 2000 over Gangetic West Bengal by rational method of rainfall run off. *Mausam,* 55 (1): 197–204.

1500 ■ 2004 Vishwas, S.S. and P.S. Hire. Effectiveness of monsoon floods on the Tapi river (India):Role of channel geometry and hydrologic regime. *Geomorphology,* 57 (3–4): 275–291.

1501 • 2005 Arunachalam, B. Drainage problems of Brihan Mumbai. *Economic and Political Weekly,* 40 (36): 3909–3911.

1502 • 2005 Bunsha, D. When dams swelled. *Frontline,* 22 (17): 22–23.

1503 • 2005 Chandrashekhar, P. Why Mumbai choked?. *Frontline,* 22 (17): 17–19.

1504 • 2005 Goswami, R. *Vulnerability to Flood Hazards in Majuli (Assam)*. Delhi: Department of Geography, Delhi School of Economics, University of Delhi (Unpublished Master of Philosophy Dissertation).

1505 • 2005 Jamwal, N. Disaster reigned in. *Down to Earth,* 14 (9): 14.

1506 • 2005 Jamwal, N. There's nothing august about Mumbai. *Down to Earth,* 14 (9): 24.

1507 • 2005 Jamwal, N. Not providential Gujarat flood: Blame poorly planned roads and cities. *Down to Earth,* 14 (6): 7–8.

1508 • 2005 Ramachandraiah, C. Hyderabad's floods: Nature's revenge. *Economic and Political Weekly*, 40 (38): 4115–4116.

1509 ■ 2005 Special Correspondent. Rain, rain go away. *Down to Earth,* 14 (6): 6.

1510 • 2006 Amarnath. Arrested development. *Down to Earth,* 14 (16): 42.

1511 ■ 2006 Miral, M.S. and K. Kumar. Glacial lake outburst flood: A glacial hazard in higher Himalayas. In K.K. Sharma *et al* (eds). *Environmental Geohazards: Science and Society*. New Delhi: Research India Press. 229–234.

1512 • 2006 Viswanathan, S. An avoidable tragedy. *Frontline*, 22 (27): 123–124.

Cause: Heavy Rain

1513 ■ 1970 Starkel, L. Cause and effect of a heavy rainfall in Darjeeling and in the Sikkim Himalayas. *Journal of Mumbai Natural History Society*, 67: 1–6.

1514 ■ 1974 Abbi, S.D.S. *et al.* Rainfall and floods in India during 1973 southwest monsoon period. *Indian Journal of Meteorology and Geophysics*, 25 (3–4): 363–370.

1515 ■ 1986 Dhar, O.N. *et al.* Meteorology of heavy rainfall over the Garhwal-Kumaon region of the Himalayas: A brief appraisal. In *Proceedings of the Workshop on Flood Damage Assessment in the Himalayan Region*. New Delhi: Central Board of Irrigation and Power. 82–91.

1516 ■ 1986 Reddy, R.S. *et al.* An investigation of heaviest rainfall over Coastal Andhra Pradesh of India during October. *Monthly Weather Review*, 114 (4): 777–779.

1517 ■ 1989 Dhar, O.N. and S.S. Nandargi. Distribution of rainfall over different sections of the Narmada basin associated with the highest floods in recent times. *Indian Journal of Power and River Valley Development*, 39 (7): 210–213.

1518 ■ 1992 Dhar, O.N. and S.S. Nandargi. A study of rainfall and floods in the 1987–1991 monsoon seasons over the contiguous Indian region. *Indian Journal of Meteorology, Hydrology, and Geophysics*, 17 (174): 330–336.

1519 ■ 1993 Rao, C.P. and P.V.R. Rao. Influence of cyclonic disturbances from sea on rainfall over Eastern Ghats in Andhra Pradesh. *Mausam*, 44 (2): 153–162.

1520 • 1996 Special Correspondent. Monsoons: Avoidable havoc. *Economic and Political Weekly*, 31 (27): 1716–1717.

1521 ■ 1999 Kulkarni, A.K. *et al.* Analysis of a heavy rainspell of the monsoon 1995 over Punjab and Haryana region. *Indian Journal of Power and River Valley Development*, 49 (3–4): 54–57.

1522 ■ 2005 Madan, O.P. and U.C. Mohanty. Offshore trough and very heavy rainfall events along the West Coast of India during Arabian Sea Monsoon Experiment 2002. *Mausam*, 56 (1): 37–48.

1523 ■ 2006 Sharma, K.K. Understanding cloudburst mechanism in Himalayas: An approach in modeling an environmental hazard. In K.K. Sharma *et al* (eds). *Environmental Geohazards: Science and Society*. New Delhi: Research India Press. 210–214.

Cause: Landslide

1524 ■ 1894 Holland, T.H. Report on the Gohna Landslip, Garhwal. In *Records of the Geological Survey of India,* 27 (2): 55–65.

1525 ■ 1951 Auden, J.B. *Geological Note on Slips and Subsidence in Shimla.* Kolkata: Geological Survey of India (Unpublished Report).

1526 ■ 1960 Raju, K.C. *A Geological Report on the Investigation of Slips and Subsidence in Shimla.* Kolkata: Geological Survey of India (Unpublished Report).

1527 ■ 1962 Bhaskaran, A.R. Landslides, their origin and control in the Kosi catchment. *Journal of Soil and Water Conservation in India*, 10: 79–83.

1528 ■ 1964 Ashraf, Z. *A Reconnaissance Note on the Nashri Landslide, Srinagar-Jammu National Highway.* Kolkata: Geological Survey of India (Unpublished Report).

1529 ■ 1965 Sen Sharma, S.B. *Geotechnical Note on the Landslide Problem and Road Alignments in Part of Kameng Subansiri and Siang Frontier Divisions under the Project Vartak.* Kolkata: Geological Survey of India (Unpublished Report).

1530 ■ 1966 Ashraf, Z. and R.B. Nag. *A Report on the Detailed Geological Investigations of the Nashri Landslide on the Jammu-Srinagar National Highway (Jammu and Kashmir).* Kolkata: Geological Survey of India (Unpublished Report).

1531 ■ 1966 Nautiyal, S.P. *et al. Landslides and Hillside Stability in the Western Himalayas.* Kolkata: Bulletin of the Geological Survey of India, Series B, No. 15 (2).

1532 ■ 1966 Sondhi, V.P. *et al. Landslides and Hillside Stability in the Eastern Himalayas.* Kolkata: Bulletin of the Geological Survey of India, Series B, No. 15 (1).

1533 ■ 1967 Ghosh, D.K. *Geological Report on the Landslides along the National Highway between Jeori and Khal in the Satluj Valley and on the Reconnaissance of the Proposed Road Alignment between Khal and Chango in the Spiti Valley of Himachal Pradesh.* Kolkata: Geological Survey of India (Unpublished Report).

1534 ■ 1968 Ashraf, Z. *Inspection Note on the Malori Landslide on the Batote-Bhadarwah Road (Jammu and Kashmir State).* Kolkata: Geological Survey of India (Unpublished Report).

1535 ■ 1968 Hukku, B. M. *et al. Second Progress Report on the Geological Investigations of the Stability of the Hill Slopes around Nainital (Uttar Pradesh).* Kolkata: Geological Survey of India (Unpublished Report).

1536 ■ 1968 Pawde, M.B. A geological note on landslides at Port Blair (Andaman). In *Records of the Geological Survey of India,* 95 (2): 483–486.

1537 ■ 1968 Verma, R. S. *Geological Report on the Landslide along the Jammu-Srinagar National Highway (Jammu and Kashmir).* Kolkata: Geological Survey of India (Unpublished Report).

1538 ■ 1969 Chaturvedi, S. N. and A. K. Tikku. *A Report on the Slides and Ground Subsidence in the Srinagar Valley (Jammu and Kashmir).* Kolkata: Geological Survey of India (Unpublished Report).

1539 ■ 1971 Chadda, O.P. and V.S. Krisnaswamy. *Report of Central Team for Study and Evaluation of the Landslides and Subsidence in Simla during August 1971.* Kolkata: Geological Survey of India (Unpublished Report).

1540 ■ 1971 Chatterji, G.C. and P. Ramachandran. Landslides and floods in North Bengal and adjoining areas. *Indian Journal of Power and River Valley Development,* 21 (2): 45–47.

1541 ■ 1972 Kalvoda, J. Geomorphological studies in the Himalaya, with special reference to the landslide and allied phenomena. In K.S. Jhingran *et al* (eds). *Himalayan Geology.* Delhi: Wadia Institute of Himalayan Geology. Volume 2, 301–316.

1542 ■ 1972 Nand, N. L. and C. Prasad. Alaknanda tragedy: A geomorphological appraisal. *National Geographical Journal of India,*18 (3–4): 206–212.

1543 ■ 1972 Shome, S.K. *A Note on the Examination of the Dhak Nala Slide, Chamoli District*. Kolkata: Geological Survey of India (Unpublished Report).

1544 ■ 1974 Shome, S.K. *A Geological Note on the Reconnaissance of the Landslide Affected Areas of Chamoli District (Uttar Pradesh).* Kolkata: Geological Survey of India (Unpublished Report).

1545 ■ 1975 Chansarkar, R.A. Geologic and geomorphic factors in landslide investigations. In *Proceedings of the Seminar on Landslides and Toe Erosion Problems with Special Reference to Himalayan Region.* Kolkata: Indian Society of Engineering Geology. 54–66.

1546 ■ 1975 Hukku, B.M. and P.L. Narula. A geological assessment of slopes at Khuni Nallah landslide and Nashri landslide on National Highway (Northwest Himalayas). In *Proceedings of the Seminar on Landslides and Toe Erosion Problems with Special Reference to Himalayan Region.* Kolkata: Indian Society of Engineering Geology. 79–165.

1547 ■ 1975 Raghuraman, S. Hydrology and its contribution to landslides, with special reference to problems in Sikkim. *In Proceedings of the Seminar on*

Landslides and Toe Erosion Problems with Special Reference to Himalayan Region. Kolkata: Indian Society of Engineering Geology. 41–68.

1548 ■ 1976 Chaudhury, A.K. *A Report of the Investigation of a Number of Slides and Bridge Studies under Project Pushpak (Mizoram).* Kolkata: Geological Survey of India (Unpublished Report).

1549 ■ 1977 Chaudhury, A.K. *A Note on the Stability of Hill Slopes along Various Road Stretches under Project Pushpak.* Kolkata: Geological Survey of India (Unpublished Report).

1550 ■ 1977 Chaudhury, A.K. *A Short Note on the Investigation of Slides around Nonskhen and Other Places, Pynursla Development Block, (Meghalaya).* Kolkata: Geological Survey of India (Unpublished Report).

1551 ■ 1979 Chaudhury, A.K. *A Note on the Stability of the Hill Slope along Shillong-Cherapunji Road, Meghalaya.* Kolkata: Geological Survey of India (Unpublished Report).

1552 ■ 1979 Majumdar, N. *A Short Note on the Landslide Investigation along Border Roads in Lohit District of Arunachal Pradesh, Project Vartak.* Kolkata: Geological Survey of India (Unpublished Report).

1553 ■ 1979 Ray, S. *A Short Note on the Problem of Slope Stability along the Halflong Hill Section of Northeastern Railways.* Kolkata: Geological Survey of India.

1554 ■ 1980 Chaudhury, A.K. *A Note on the Investigation of a Slide in Phek District (Nagaland).* Kolkata: Geological Survey of India (Unpublished Report).

1555 ■ 1980 Natrajan, T.K. *et al.* A major landslide in Sikkim: Analysis, correction and efficacy of protective measures. In *Proceedings of the International Symposium on Landslides.* New Delhi: Sarita Pradhan. 397–402.

1556 ■ 1980 Sargunan, A. and M.S. Jegathesan. An investigation of landslides in and around Ooty. In *Proceedings of the International Symposium on Landslides.* New Delhi: Sarita Pradhan. 215–219.

1557 ■ 1980 Varadarajan, A. *et al.* Effect of water flow and earthquake on stability of natural slopes. In *Proceedings of the International Symposium on Landslides.* New Delhi: Sarita Pradhan. 303–307.

1558 ■ 1984 Haigh, M.J. Deforestation and disaster in Northern India. *Land Use Policy*, 1 (3): 187–198.

1559 • 1984 Haigh, M.J. Prediction and morphology of road induced landslides in the Himalaya: A case study from Central Himalaya. In S.C. Joshi (ed). *Rural Development in the Himalaya: Problems and Prospects.* Nainital: Gyanodaya Prakashan. 176–204.

1560 ■ 1985 Basu, S.R. Landslide and ecological imbalance: A case study in Darjeeling Himalayas. In *Proceedings of the Sixth Indian Geographical Congress.* Kharagpur: Department of Architecture and Regional Planning, Indian Institute of Technology.

1561 ■ 1986 Kamble, D.T. *Geotechnical Report on the Bridge Site Landslides and Sinking Areas on the Roads (Project Vartak) in Arunachal Pradesh.* Kolkata: Geological Survey of India (Unpublished Report).

1562 ■ 1987 Mallick, D.K. and S.K. Das. *A Report on the Investigation of Landslides Occurring between Km 37 and Km 108 along National Highway 39 in Nagaland and Manipur States.* Kolkata: Geological Survey of India (Unpublished Report).

1563 ■ 1987 Sinha, A.K. *A Report on the Slide and Sinking Zones between Km 4 and Km 8 on Churachandpur Tipainukn Road under Project Sewak.* Kolkata: Geological Survey of India (Unpublished Report).

1564 ■ 1987 Srivastava, V.B. and G.S. Jaggi. *Geotechnical Report on Landslides on Langlii-Tuipang road, Mizoram.* Kolkata: Geological Survey of India (Unpublished Report).

1565 ■ 1988 Haigh, M.J. Environmental correlations of landslide frequency along new highways in the Himalaya: Preliminary results. *Catena,* 15 (6): 539–553.

1566 • 1988 Mullik, B.B. Role of human inteference in the landslide hazard of the Himalaya and North-East region of India. In *Proceedings of the Workshop on Geohazards and Related Societal Issues.* Lucknow: Indian Society of Engineering Geology.

1567 ■ 1988 Rao, J.P. Some geological and geotechnical factors influencing landslides in Himalayas. In C. Bonnard (ed). *Proceedings of the Fifth International Symposium on Landslides.* Rotterdam: Balkema. 293–300.

1568 • 1989 Haigh, M.J. Environmental indicators of landslide activity along the Kilbury Road, Nainital, Kumaon (Lesser Himalaya). *Mountain Research and Development,* 8 (2): 25–33.

1569 • 1990 Chandra, A. *Landslide: An Expression of Man-Nature Interaction in the Western Himalayas.* Chandigarh: Punjab University (Unpublished Ph.D thesis).

1570 ■ 1990 Joshi, M. and P.D. Pant. Causes and preventional measures for rockfall and landslides on Naina Peak, Nainital (Uttar Pradesh). *Mountain Research and Development,* 10 (4): 343–346.

1571 ■ 1990 Pant, P.D. *Geological and Geomorphological Investigations of Loharkhit-Karmi Area, with Special Emphasis on Landslide Study, District Almora (Uttar Pradesh).* Nainital: Kumaon University (Unpublished Ph.D thesis).

1572 ■ 1992 Pande, R.K. *et al.* Landslide mechanism in Central Himalaya: Causes and controls. In O. Singh *et al* (eds). *Frontiers in Environmental Geography.* New Delhi: Concept Publishing Company. 113–123.

1573 ■ 1993 Basu, S.R. and L. Ghosh. A comprehensive study of landslides and floods in the link basins of the Darjeeling Himalayas. *Indian Journal of Power and River Valley Development,* 43 (10): 196–203.

1574 ■ 1993 Haigh, M.J. Landslide ecology: Mussoorie-Tehri road. In G.S. Rajwar (ed). *Garhwal Himalaya: Ecology and Environment.* New Delhi. Ashish Publishing House. 59–74.

1575 ■ 1993 Haigh, M.J. Environmental influences on landslide activity, Almora Bypass, Kumaon (Lesser Himalayas). *Natural Hazard Journal,* 8 (2): 153–170.

1576 ■ 1994 Joshi, K.C. and P. Pande. Post-earthquake landslides hazard evaluation, Uttarkashi district (Uttar Pradesh). In *Records of the Geological Survey of India,* 127 (8): 284–288.

1577 ■ 1994 Mehrotra, G.S. *et al.* A geological appraisal of landslide hazard in the East Sikkim Himalaya. *Journal of Himalayan Geology,* 5 (1): 35–45.

1578 ■ 1994 Pande, P. Landslide activity and Bhagirathi river blockade, Uttarkashi district (Uttar Pradesh). In *Records of the Geological Survey of India,* 127 (8): 287–288.

1579 ■ 1994 Rawat, R.K. *Landslides Investigation in Sikkim Region (Northeastern Himalayas)* Dhanbad: Indian School of Mines (Unpublished Ph.D thesis)

1580 ■ 1994 Thakur, D.K. *A Study of Landslides along National Highway 22, Pawari-Peo Road and Panji road in Kalpa Tehsil (District Kinnaur).* Shimla: University of Himachal Pradesh (Unpublished Master of Philosophy Dissertation).

1581 ■ 1995 Panikkar, S.V. *Landslides Around Dehradun and Mussorie: A Geomorphic Appraisal.* Mumbai: Indian Institute of Technology (Unpublished Ph.D thesis).

1582 ■ 1996 Anbalagan, R. and P. Gupta. A geotechnical appraisal of Chilla landslide (Outer Garhwal region). In K. Senneset (ed). *Proceedings of Seventh International Symposium on Landslides.* Rotterdam: Balkema. 949–952.

1583 • 1996 Arya, A.S. *et al.* Geoenvironmental appraisal of Tehri dam reservoir with special reference to landslide hazards. *Himalayan Geology,* 17: 157–160.

1584 ■ 1996 Ashraf, Z. Geological studies of the Nashri landslide on the Jammu-Srinagar Highway. In *Proceedings of the National Seminar on Prevention of Soil Erosion and Stabilisation of Hill Slopes along Hill Road.*

1585 ■ 1996 Gupta, S.K. Geological approach for the study of landslide hazard of a part of Bhagirathi basin, Garhwal Himalaya. *Journal of Engineering Geology,* 25:1–4.

1586 ■ 1996 Gupta, V. *Structure and Gemorphology of the Upper Satluj Valley, District Kinnaur, with Special Reference to Landslides.* Srinagar: H.N.B Garhwal University (Unpublished Ph.D thesis).

1587 ■ 1996 Mehrotra, G.S. *et al.* Terrain analysis and spatial assessment of landslide hazards in parts of Sikkim Himalaya. *Journal of Geological Society of India,* 47 (4): 491–498.

1588 ■ 1996 Mukhopadhyay, S. C. Functional effects of precipitation and human intervention in generating landslides in East and Central Himalayan terrain, with special emphasis on Sikkim-Darjeeling areas. In *Proceedings of the Seminar on Earthquakes and Landslides: Natural Disaster and Mitigation.* Kolkata: West Bengal Academy of Science and Technology. 29–42.

1589 ■ 1996 Mullick, B. B. and C. A. Ghosh. Landslides in the Himalaya: Causes, types and management. In K. Senneset (ed). *Proceedings of Seventh International Symposium on Landslides.* Rotterdam: Balkema. 381–386.

1590 ■ 1996 Panikkar, S. V. and V. Subramanyam. A geomorphic evaluation of the landslides around Dehradun and Mussoorie, Uttar Pradesh (India). *Geomorphology*, 15 (2): 169–181.

1591 ■ 1996 Sharda, Y. P. and Sinha, A. K. Detailed analysis of a slide along Batote-Kishtwar road (Jammu Kashmir). In *Proceedings of the Symposium on Recent Advances in Geological Studies of Northwest Himalayas and Foredeep*, Lucknow, 21–23 Febuary, 1995. Special Publication No. 21. Kolkata: Geological Survey of India. 65–68.

1592 ■ 1997 Gupta, S.K. Morphometric approach to the study of landslides in Central Himalayan region of Uttarkashi . *In Proceedings of the Symposium on the Earth Sciences in Environmental Assessment and Management.* Special Publication No. 48 (2). Kolkata: Geological Survey of India. 181–190.

1593 ■ 1997 Narula, P. L. *et al.* Landslide incidences: A reflection of contemporary crustal adjustment and associated geomorphic anomalies. In *Proceedings of the Symposium on the Earth Sciences in Environmental Assessment and Management.* Special Publication No. 48 (2). Kolkata: Geological Survey of India. 43–50.

1594 • 1998 Sah, M. P. and R. K. Mazari. Anthropogenically accelerated mass movement, Kulu valley (Himachal Pradesh). *Geomorphology,* 26 (1–3): 123–138.

1595 • 1998 Singh, R. B. and S. Ghai. Genesis and impacts of a Landslide: The case study of Kharadi-Khanera slope section (Uttar Pradesh Himalayas). In R. B. Singh (ed). *Sustainable Development of Mountain Environment in India and Canada.* New Delhi: Oxford and IBH Publishing Co. Pvt. Ltd. 298–308.

1596 ■ 1998 Special Correspondent. Malpa disaster: Ominous portent. *Economic and Political Weekly*, 33: 2204.

1597 ■ 1999 Raju, N. and S. A. Chore. *A Report on Preliminary Landslide Appraisal Studies in Parts of Assam and Meghalaya.* Kolkata: Geological Survey of India (Unpublished Report).

1598 ■ 2000 Bagri, D. S. *Geological Appraisal of the Area around Bhatwari District, Uttarkashi (Uttar Pradesh) With Special Reference to Landslides and*

Other Problems. Srinagar: H. N. B. Garhwal University (Unpublished Ph.D thesis).

1599 ■ 2000 Saraf, A. K. Indian remote sensing satellite-panchromatic depicts Chamoli earthquake induced landslides in Garhwal Himalayas (India). *International Journal of Remote Sensing*, 21 (12): 2345–2352.

1600 ■ 2000 Satendra, D. R. The turmoil beneath. *Down to earth,* 8 (24): 57.

1601 ■ 2000 Starkel, L. *et al* (eds). *Rains, Landslides and Floods in the Darjeeling Himalaya*. New Delhi: Indian National Science Academy.

1602 • 2001 Barnard, P. L. *et al.* Natural human induced landslides in the Garhwal Himalaya of Northern India. *Geomorphology*, 40 (1–2): 21–35.

1603 ■ 2001 Bhambry, G. C. and S. Pradip. Geotechnical evaluation of Shoulding-Khad landslide on National Highway 22, Kinnaur district (Himachal Pradesh). In *Proceedings of the National Seminar on Geotechniques and Geological Hazards in the Indian Context*. Kolkata: Geological Survey of India.

1604 ■ 2001 Muthu, P. R. *Amboori Landslides, Thiruvanthapuram District (Kerala) on 9 November 2001*. Kolkata: Geological Survey of India (Unpublished Report).

1605 ■ 2002 Muthu, R. P. and C. Muraleedharan. *Report on the Geotechnical Investigations Carried out for Amboori Landslide of 9 November 2001*, Thiruvananthapuram District (Kerala). Kolkata: Geological Survey of India (Unpublished Report).

1606 ■ 2003 Gangta, P. *A Study of Landslide Hazards along Shimla-Rohru Highway in Himachal Pradesh (India)*. Shimla: Department of Geography, University of Himachal Pradesh (Unpublished Master of Philosophy Dissertation).

1607 ■ 2003 Kanungo, D. P. and S. Sarkar. Landslides and terrain parameters in Darjeeling Himalaya. *Himalayan Geology*, 24 (1–2): 55–62.

1608 • 2005 Special Correspondent.No plastic bags no landslides. *Down to Earth*, 13 (16): 32.

1609 ■ 2005 Unthu, R. P. and C. Muraleedharan. Causes and mechanism of Amboori landslide of 9 November 2001, Thiruvananthapuram district (Kerala). *Journal of Geological Society of India*, 66 (2) 203–208.

1610 ▲2006 Kumar, A. and Sangeeta. Landslides: An ecological disaster. In K. K. Sharma *et al* (eds). *Environmental Geohazards: Science and Society*. New Delhi: Research India Press. 147–161.

1611 ■ 2006 Sarda, V. K. and A. K. Gupta. Geotechnical investigations related to landslides. In K. K. Sharma *et al* (eds). *Environmental Geohazards: Science and Society*. New Delhi: Research India Press. 138–146.

Cause: Storm

1612 ■ 1966 Raman, P.K. and B.M. Chinabra. An empirical relationship between average rainfall and major rainstorms of Bihar and Madhya Pradesh. *Indian Journal of Meteorology and Geophysics*, 17: 97–102.

1613 ■ 1967 Janardhan, S. Storm induced sea level changes at Sagar island situated in North Bay of Bengal. *Indian Journal of Meteorology and Geophysics*, 18 (3): 205–212.

1614 ■ 1972 Dhar, O.N. and G.C. Ghose. On some hydro-meteorological aspects of space-time distribution of depressions and storms over major river basins. *Vayumandal*, 2 (2).

1615 ■ 1981 Dhar, O.N. *et al.* The rainstorm which caused Morvi dam disaster in August 1979. *Hydrological Sciences Bulletin*, 26: 71–81.

1616 ■ 1982 Sharma, K.D. and N.S. Vangani. Some rainfall features of July 1979 storm over Luni basin. *Annals of Arid Zone*, 21 (1): 23–32.

1617 ■ 1986 Rao, P.G. and P.R. Pakhecha. Analysis of 1982 heavy rainstorm over Lower Mahanadi. *Indian Journal of Power and River Valley Development*, 36 (5 and 6): 165–167.

1618 ■ 1988 Dhar, O.N. and S.S. Nandargi. Rainstorms which contributed to the greatest areal rain depths in Andhra Pradesh. *Indian Journal of Power and River Valley Development*, 38 (9): 261–268.

1619 ■ 1989 Sen, A.K. Norwesters: Local severe storms and atmospheric electricity. *Science and Culture*, 55 (8): 306–313.

1620 ■ 1991 Kulkarni, A.K. *et al.* Analysis of severe rainstorms of 23–25 July 1989 over northern half of Indian Peninsula. *Transactions, Institute of Indian Geographers*, 13 (1): 29–41.

1621 ■ 2002 Kandalgaonkar, S.S. *et al.* Thunderstorm activity and sea surface temperature over the island stations and along the East and West Coast of India. *Mausam,* 53 (2): 245–248.

1622 ■ 2005 Manohar, G.K. and A.P. Kesarkar. Climatology of thunderstorm activity over the Indian region: Latitudinal and seasonal variation. *Mausam,* 56 (3): 581–592.

Cause: Tsunami

1623 ■ 2005 Aitchison, J.C. The Great Indian Ocean tsunami disaster. *Gondwana Research-International Geoscience Journal*, 8 (2): 107–108.

1624 ■ 2005 Geological Society of India. *Great Tsunami of 26 December 2004*. Bangalore: Memoirs No. 64.

1625 ■ 2005 Hopkin, M. Triple slip of tectonic plates caused seafloor surge. *Nature*, 433 (7021): 3–3.

1626 ■ 2005 Narayan, J. P. *et al.* Effects of Medu and coastal topography on the damage pattern during the recent Indian Ocean tsunami along the coast of Tamil Nadu. *Science of Tsunami Hazards,* 23 (2): 19.

1627 ■ 2005 Reic, L. Geist. Rapid tsunami models and earthquake sources parameters: Far field and local applications. *ISET Journal of Earthquake Technology*, 42 (4): 127.

1628 ■ 2005 Subramanian, K. S. The 26 December 2004 tsunami. *Journal of Geological Society of India*, 65 (4): 515.

1629 ■ 2005 Waltham, T. The Asian tsunami disaster December 2004. *Geology Today*, 21 (1): 22–26.

1630 ■ 2006 Murthy, T. S *et al.* The role of oscillations of the first and second class in tsunami around Andaman and Nicobar Islands. In S. M. Ramaswamy *et al* (eds). *Geomatics in Tsunami.* Delhi: New India Publishing Agency. 11–16.

1631 ■ 2006 Subramaniyan, B. R. *et al. Twenty Sixth December 2004 Tsunami: Causes, Effects, Remedial Measures Pre and Post Tsunami Disaster.* New Delhi: New Academic Publications.

Cause: Assorted

1632 • 1984 Raju, P. *et al.* Human induced geomorphic hazards in Vishakapatnam coastal strip (Andhra Pradesh). *Transactions, Institute of Indian Geographers*, 6 (1): 79–84.

1633 • 1989 Baliga, B. B. Disaster natural and man made. *Science and Culture*, 55 (9): 321–325.

1634 • 1991 Chaudhuri, S. Towards disaster. *Economic and Political Weekly*, 26 (45): 2265.

1635 ■ 1992 South Asian Association for Regional Cooperation. *Regional Study on the Causes and Consequences of Natural Disaster and the Protection and Preservation of the Environment.* Kathmandu: SAARC Secretariat.

1636 • 1992 Special Correspondent. Don't blame it on nature. *Down to Earth*, 1 (2): 3–4.

1637 • 1994 Khurana, M. L. I will not say Delhi is an environmental disaster. *Down to Earth*, 3 (13): 50–51.

1638 • 1996 Bora, P. P. A man-made horror. *India Today*, (November 29): 54–59.

1639 • 1996 Singh, R. B (ed). *Disasters, Environment and Development.* New Delhi: Oxford and IBH Publishing Co. Pvt. Ltd.

1640 • 1998 Special Correspondent. Disasters waiting to happen. *Economic and Political Weekly,* 33 (37): 1419–1420.

1641 • 2000 Baweja, H. and U. Madhuarkar. Man made disaster. *India Today*, (May 8): 40–42.

1642 ■ 2000 Martin, M. Environmental disasters. In S. Parasuraman and P.V. Unnikrishnan (eds). *India Disasters Report: Towards a Policy Initiative.* New Delhi: Oxford University Press. 59–65.

1643 • 2000 Pandey, B.W. *Hazard Risk Assessment and Sustainable Land Development in Upper Beas Basin.* Delhi: Department of Geography, Delhi School of Economics, University of Delhi (Unpublished Ph.D thesis).

1644 • 2000 Singh, R.B. and S. Ghai. Human induced hazards and development in a part of Upper Yamuna Watershed. In A.M. Pathan and S.S. Thigale (eds). *Contributions to Environmental Geoscience.* New Delhi: Aravali Books. 151–160.

1645 ■ 2003 Soni, V.K. and P.S. Kannan. Temporal variation and the effect of volcanic eruption on the atmospheric turbidity over India. *Mausam*, 54 (4): 881–890.

1646 ■ 2004 Das, P.K. Meteorological dimensions of natural hazards. In K.S. Valdiya (ed). *Coping with Natural Hazard: Indian Context.* Hyderabad: Orient Longman. 37–47.

1647 ■ 2004 De, U.S. and M. Kahole. Natural hazards associated with meteorological extreme events. *Natural Hazard Journal*, 31 (2): 487–497.

1648 • 2004 Gulia, K.S. *Genesis of Disaster: Ramifications and Ameliorations.* New Delhi: Kalpaz.

1649 ■ 2004 Valdiya, K.S. Geodynamic perspective of natural hazards in India. In K.S. Valdiya (ed). *Coping with Natural Hazards: Indian Context.* Hyderabad: Orient Longman. 7–36.

1650 • 2005 David, A.W.P. A few authorities more. *Down to Earth*, 14 (7): 51.

1651 ■ 2006 Gupta, V.J. *Natural Disasters and Environment.* New Delhi: S.S. Publications

1652 • 2006 Kumria, P. Deforestation and disasters. In K.K. Sharma *et al* (eds). *Environmental Geohazards: Science and Society.* New Delhi: Research India Press. 304–328.

1653 • 2006 Saikia, A. Disaster unpreparedness in Assam. In P.S. Ranade (ed). *Natural Disasters: Management and Preparedness.* Hyderabad: Institute of Charted Finanacial Analysts of India University Press. 156–162.

Impact

Impact is the effect or pervasive influence of any event or programme, autonomous or designed. Under its banner fall references that research into a single or cascading effect of the disaster. A force could impinge on the physical features like crustal deformation, shift in river regime or change in a slope. By ruining transport and infrastructure it could well damage the life lines of a region. Its impact on economy are far reaching. By disrupting the lives and livelihoods, disasters affect many a human facet. From dislocation to displacement, a disaster is about the physical and mental health of the people. Be it the old, children or women the research that focuses on the toll of a disaster is clubbed under the category 'impact'. There are 797 references under the group of 'impact' and these form one-fifth of the total of 4004 (Tables 6.16, 6.17 and 6.18).

Table 6.16 Reference for each Category

Natural Disaster	References	Number
Cyclone	1654–1735	82
Drought	1736–1891	156
Earthquake	1892–2134	243
Floods	2135–2272	138
Heat and Cold Wave	2273	1
Heavy Rain	2274–2282	9
Landslide	2283–2326	44
Storm	2327–2329	3
Tsunami	2330–2402	73
Assorted	2403–2450	48

Table 6.17 Research References on Impact

'Natural' disaster	Book	Report	Thesis	Article[1]	Chapter[2]	Feature[3]	Total
Cyclone	4	10	3	24	5	36	82
Drought	9	8	7	52	14	66	156
Earthquake	2	8	8	137	23	65	243
Floods	5	5	9	36	20	63	138
Heat and Cold Wave	0	0	0	0	0	1	1
Heavy Rain	0	0	0	0	2	7	9
Landslide	0	2	3	14	14	11	44
Storm	0	0	0	2	0	1	3
Tsunami	4	2	1	27	15	24	73
General Hazard	5	3	2	14	15	9	48
Total	29	38	33	306	108	283	797

[1] *in a journal;* [2] *in a book/report/proceedings;* [3] *in a magazine.*

Table 6.18 Dominance of Human over Natural in Research References on Impact

Phase	Natural	Human	Ecological	Total
1830–1947	4	11	0	15
1948–1989	38	113	0	151
1990–2006	140	485	6	631
Total	182	609	6	797

Impact: Cyclone

1654 • 1971 Frank, N.L. and S.A. Hussain. The deadliest tropical cyclone in history. *Bulletin of the American Meteorological Society*, 52 (6): 438–445.

1655 ■ 1981 Dube, S.K. *et al.* The response of different wind stress forces on the surges along the east coast of India. *Mausam*, 32 (3): 315–320.

1656 • 1988 School of Planning and Architecture. *Cyclone and Building Behaviour: A Study of Divi Seema.* New Delhi: Budua' s Press.

1657 • 1988 Vinayak, M. Agony in West Bengal. *Frontline*, 21 (December 24): 35–36.

1658 • 1989 Naidu, B.R. *Economic Consequences of 1977 Cyclone in Andhra Pradesh.* Tirupati: Sri Venkateswara University.

1659 • 1990 Government of Andhra Pradesh. *Memorandum Presented to the Government of India on Cyclone Damages.* Hyderabad.

1660 • 1990 Rao, K.V. In the face of challenge. *Frontline*, 23 (May 26): 120–121.

1661 • 1990 Rao, M.M. Guntur woes. *Frontline*, 23 (May 26): 13–14.

1662 • 1990 Rao, M.V. A battered coast. *Frontline*, 23 (May 26): 4–10.

1663 • 1990 Special Correspondent. Devastated districts. *Frontline*, 23 (May 26): 118–119.

1664 • 1990 Special Correspondent. After the cyclone. *Frontline*, 23 (May 26).

1665 • 1990 Viswanathan, A. Killer wind. *India Today*, (May 8): 46.

1666 • 1991 Subramaniyam, T.S. The storm trail in devastated Tamil Nadu. *Frontline*, 24 (December 20): 121–122.

1667 ■ 1994 Satyanarayan, A.M. *Some Studies on Storm Surge and Sea Level Variations along the East Coast of India.* Cochin: Cochin University of Science and Technology (Unpublished Postgraduate Dissertation).

1668 • 1995 Ponnuswamy, S. and C.T. Sriramula. Impact of cyclones on transport in Chennai: A case study. In G.K. Mishra and G.C. Mathur (eds). *Natural Disaster Reduction.* New Delhi: Reliance Publishing House, 180–193.

1669 • 1996 Bavadam, L. Ships in distress: Monsoon arrives with cyclone in Mumbai. *Frontline*, 29: 119.

1670 ● 1996 Sridhar, V. Lost at sea. *Frontline*, 29: 91–92.

1671 ● 1997 Editorial. Cyclone havoc. *Down to Earth*, 6 (11): 13.

1672 ■ 1997 Thiyagesan, K. and R. Ragarajan. Effect of a cyclone on waterbird populations at the Pichavaram mangroves, Southern India. *Wader Study Group Bulletin*, 80: 47–51.

1673 ● 1998 Chaudhury, K. On the tornado trail. *Frontline*, 31(April 8): 115–116.

1674 ● 1998 Dhaulta, J.P. The cyclonic disaster of Gujarat. *Nursing Journal of India*, 89 (7): 145.

1675 ● 1998 Editorial. Tornado havoc. *Down to Earth*, 6 (23): 13.

1676 ● 1998 Madhurkar, U. The storm subsides. *India Today*, (June 29): 30–32.

1677 ● 1998 Padmanabhan, R. Cyclone devastation. *Frontline*, 31 (July 3): 112–113.

1678 ■ 1998 Rao, M.S. Impact of cyclonic storms on delta and coastal environments of the east coast of India. In B. Sinha (ed). *Ecosystem Degradation in India.* New Delhi: Ashish Publishing House. 36–45.

1679 ● 1998 Vidyasagar, B. *Impact of Cyclone on Housing in Coastal Andhra Pradesh.* New Delhi: Department of Housing, School of Planning and Architecture (Unpublished Postgraduate Dissertation).

1680 ● 1999 Banerjee, R. Bent and broken. *India Today*, (December 13): 32–33.

1681 ● 1999 Banerjee, R. Silence of the grave. *India Today*, (November 29): 28–30.

1682 ● 1999 Chattopadhyay, S. Killer cyclone. *Frontline*, 16 (24): 22–25.

1683 ● 1999 Chattopadhyay, S.S. A ravaged region. *Frontline*, 16 (25): 63–65.

1684 ● 1999 Chattopadhyay, S.S. The trail of misery. *Frontline*, 16 (24): 27.

1685 ● 1999 Editorial. Orissa: Compounding the disaster. *Economic and Political Weekly*, 34 (46 and 47): 3248.

1686 ● 1999 Mandal, G.S. Cyclones: The problem size. *Shelter,* special issue, October: 6.

1687 ● 1999 Purie, A. Editorial. *India Today*, (November 15): 1.

1688 ● 1999 Raghavan, S. and A.K. Sharma. Tropical cyclone: Impacts in India and neighbourhood. In P. Roger (Jr) and P. Roger (Sr) (eds). *Storms.* London: Routledge, 339–365.

1689 ● 1999 Ramachandran, R. Scientific failures. *Frontline*, 16 (24): 28–31.

1690 ● 1999 Sanyal, S. A crippled economy. *Frontline*, 16 (25): 73–76.

1691 ● 1999 Sehgal, S.C. *et al.* Outbreak of leptospirosis after the cyclone in Orissa. *The National Medical Journal of India*, 15(1): 22–23.

1692 ● 1999 Sen, A and B. Paul. Land's end. *India Today*, (November 29): 24–26.

1693 ● 1999 Shanmugasundaram, J. *et al.* Behaviour of buildings to withstand cyclonic winds: An experience of Gujarat and Andhra Pradesh. *Shelter,* special issue, October: 61–64.

1694 ● 1999 Sharma, V.K. *Cyclone and its Impact: Orissa Super Cyclone 1999.* New Delhi: National Centre for Disaster Management, Indian Institute of Public Administration.

1695 • 2000 Agency for International Development Bureau for Humanitarian Response. *India: Cyclone.* Washington D.C: Office of the United States of America Foreign Disaster Assistance.

1696 • 2000 Behera, A. *et al* (eds). *Voicing Silence Experience of Women with Disaster in Orissa.* Bhubaneswar: Orissa State Disaster Management Authority.

1697 ■ 2000 Bhatt, M.R. Coastal disaster in the villages of Dhandhuka. In R.B. Singh (ed). *Disaster Management.* Jaipur: Rawat Publications. 202–224.

1698 • 2000 Chattopadhyay, S.S. Left in the lurch. *Frontline*, 17 (9): 43–45.

1699 • 2000 Dhara, S. Deaths foretold. *Down to Earth*, 8 (23): 39–42.

1700 • 2000 Editorial. Tidal wave kills 26. *Down to Earth*, 9 (7): 13.

1701 • 2000 Kanchan, A.J. The 1999 Super Cyclone in Orissa. In S. Parasuraman and P.V. Unnikrishnan (eds). *India Disasters Report: Towards a Policy Initiative.* New Delhi: Oxford University Press. 199–200.

1702 • 2000 Kaur, U. Historical perspective of cyclonic storms and their impact in Coastal Andhra Pradesh. *Annals of National Association of Geographers, India,* 20 (2): 66–84.

1703 • 2000 Naidu, T.B. *My Experiences with the Hurricane.* Vijayawada: VGS Publication

1704 ■ 2000 Roy, P. *Orissa Cyclone: Rapid Assessment Survey of Needs. New Delhi*: Indian Council of Social Science Research (Project Report).

1705 • 2000 Shanmugasundaram, J. *et al.* Cyclone damage to buildings and structures: A case study. *Journal of Wind Engineering and Industrial Aerodynamics*, 84 (3): 369–380.

1706 • 2000 Special Correspondent. Yet another tragedy. *Down to Earth*, 9 (11): 10.

1707 • 2000 Suri, S. *Orissa Disaster: Agony of the Living.* New Delhi: Author Press.

1708 • 2000 United Nations International Central Fund. *Orissa Super Cyclone: Emergency Information Pack.* New Delhi.

1709 • 2001 Anantakrishnan, P. Coastal calamity: The Orissa cyclone and its aftermath. *India Today*, (January 29): 63.

1710 ■ 2001 Chinthalu, G.R. *et al.* Response of the Bay of Bengal to Gopalpur and Paradip Super Cyclones during 15–31 October 1999. *Current Science*, 81 (3): 283–291.

1711 ■ 2001 Deo, A.A. *et al.* Oceanic response to cyclone moving in different directions over Indian seas using IRG model. *Mausam*, 52 (2): 163–174.

1712 • 2001 Hare, O.G. Hurricane 07b in the Godavari delta : Vulnerability, mitigation and the spatial impact. *Geographical Journal*, 167: 23.

1713 ■ 2001 Kundu S.N. *et al.* Change analysis using Indian remote sensing satellite-P4-ocean colour monitor data after the Orissa Super Cyclone. *International Journal of Remote Sensing*, 22 (7): 1383–1389.

1714 • 2001 Mahapatra, R. Tottering on tragedies: An analysis. *Down to Earth*, 10: 30–41.

1715 ■ 2001 Nayak, S. R. *et al.* Application of Indian remote sensing satellite P4 ocean colour monitor data to study the impact of cyclone on coastal environment of Orissa. *Current Science*, 80 (9): 1208–1213.

1716 • 2001 Special Correspondent. Swept apart (Orissa cyclone). *Down to Earth*, 10: 35.

1717 • 2001 Special Correspondent. Erasama's agony continues. *New Age*, 49 (44): 6.

1718 • 2002 Banerjee, R. Road to nowhere. *India Today*, (May 27): 52–53.

1719 • 2002 Dash, B. Whither winds of change? *Down to Earth*, 11(10): 40.

1720 • 2002 Revenue Department. *White Paper on 1999 Super Cyclone.* Bhubaneswar: Government of Orissa.

1721 ■ 2002 Singh, O. P. *et al.* Impact of Southern oscillation on the frequency of monsoon depressions in the Bay of Bengal. *Natural Hazard Journal*, 25 (2): 101–115.

1722 • 2002 Sur, D. *et al.* Super Cyclone in Orissa: An assessment of psychological status of survivors. *Journal of Traumatic Stress*, 15(4): 313–319.

1723 • 2002 Swain, M. *Impact of Super Cyclone on Life and Livelihood of Women: An Analytical Study of the Two Coastal Districts of Orissa.* Bhubaneswar: Nabakrishna Choudhury Centre for Development Studies.

1724 ■ 2002 Trivedi, D. K. and J. Sanjay. *Numerical Simulation of a Super Cyclonic Storm Orissa 1999: Impact of Initial Conditions.* Pune: Indian Institute of Tropical Meteorology.

1725 • 2003 Dhara, S. *Tropical Cyclones in the East Coast of India and their Impact on Fishing and Tourism in West Bengal Coast.* Burdwan: University of Burdwan (Unpublished Ph.D thesis).

1726 • 2003 Raghavan, S. and S. Rajesh. Trends in tropical cyclone impact: A study in Andhra Pradesh. *Bulletin of the American Meteorological Society*, 84 (5): 635–644.

1727 ■ 2003 Smith, A. *et al.* Effect of May 2003 tropical cyclone on physical and biological processes in the Bay of Bengal. *International Journal of Remote Sensing*, 24.

1728 • 2004 Das, P. Orissa: The agony of the living. *Frontline*, 21 (23): 46–48.

1729 • 2004 Kumar, K. V. K. *Cyclone Devastation: Its Implications.* Delhi: Serial Publications.

1730 ■ 2004 Rao, A. D. *et al.* Impact of tropical cyclones on coastal upswelling processes. *Natural Hazard Journal*, 31 (2): 415–435.

1731 • 2005 Acharyya, A. *et al.* Mid–1997 tidal surge of Bay of Bengal at Digha on the East Coast of India: Its damage evaluation and mitigation. *Journal of Nepal Geological Society*, 21: 75–86.

1732 • 2005 Rao, M. M. Cyclone havoc. *Frontline*, 22 (21): 44–46.

1733 • 2005 Samal, K. C. Sudden impact of natural disaster: Experience from 1999 Super Cyclone in Orissa. *Man and Development*, 27 (3).

1734 • 2005 Shultz, J. *et al.* Epidemiology of tropical cyclone: The dynamics of disaster, disease and development. *Epidemiologic Reviews*, 27 (1): 21–35.

1735 ■ 2006 Jena, B. *et al.* Tropical rainfall measuring mission derived sea surface temperature in the wake of a cyclonic storm over the central Bay of Bengal. *International Journal of Remote Sensing*, 27 (14): 2065–3072.

Impact: Drought

1736 • 1871 Henvey, F. *A Narrative of the Drought and Famine which Prevailed in the Northwestern Provinces during 1868–1869 and the Beginning of 1870*. Allahabad: Government Press.

1737 • 1963 Mallick, A. K. and T. S. Govindaswamy. The drought problem in India in relation to agriculture. *Annals of Arid Zone*, 2 (1): 106–113.

1738 • 1969 Jodha, N. S. Drought and scarcity in Rajasthan desert: Some basic issues. *Economic and Political Weekly*, 4 (16): 699–703.

1739 • 1969 Swaminathan, M. C. *et al.* Food and nutrition situation in drought affected areas of Bihar. *Indian Journal of Nutrition and Dietetics*, 6: 209–217.

1740 • 1974 Krishnamachari, K. A. V. R. *et al.* Food and nutritional situation in the drought affected areas of Maharashtra: A survey in recommendations. *Indian Journal of Nutrition and Dietetics*, 11: 20.

1741 • 1974 Rao, K. N. V. Impact of drought on social system in a Telangana village. *Eastern Anthropologist*, 27 (4): 299–314.

1742 • 1975 Borkar, V. V. and M. V. Nadkarni. *Impact of Drought on Rural Life*. Mumbai: Popular Prakashan.

1743 • 1975 Subramaniam, V. *Parched Earth: The Maharashtra Drought 1970–73*. Mumbai: Orient Longman.

1744 • 1976 Mundle, S. Bihar: The bonded of Palamau. *Economic and Political Weekly*, 11 (18): 653–656.

1745 • 1979 Arunachalam, B. Realities of agricultural drought in semi-arid Maharashtra. In N. B. K Reddy (ed). *Proceedings of the All India Symposium on Drought Prone Areas of India*. Tirupati: Rayalseema Geographical Society. 147–152.

1746 ■ 1979 Bagchi, K. Agriculture in drought prone areas. *Geographical Review of India*, 41(4): 293–296.

1747 • 1979 Chohan, T. S. *A Geographical Study of the Impact of Drought Hazard and Relief Measures in the Malani Region, Rajasthan*. Jodhpur: University of Jodhpur (Unpublished Ph.D thesis).

1748 • 1979 Mustaja, F.R. Droughts in Maharashtra: Experience and strategy. In N.B.K Reddy (ed). *Proceedings of the All India Symposium on Drought Prone Areas of India.* Tirupati: Rayalseema Geographical Society. 153–156 .

1749 • 1979 Singh, T. Drought at Palamau. In N.B.K Reddy (ed). *Proceedings of the All India Symposium on Drought Prone Areas of India.* Tirupati: Rayalseema Geographical Society. 243–245.

1750 • 1980 Agroeconomic Research Centre. *Impact of Drought on Consumption Standard in the Villages of Allahabad District (Uttar Pradesh).* Allahabad.

1751 • 1980 Jha, G.P. *Agriculture in Drought Prone Region: A Geographical Perspective.* New Delhi: Inter-India Publication.

1752 • 1980 Special Correspondent. Bihar: In a drought hit village. *Economic and Political Weekly*, 15 (35): 1518–1519.

1753 • 1981 Balakrishna, S. *et al. Socio-Economic Conditions of Drought Prone Areas in Madhya Pradesh.* Hyderabad: National Institute of Rural Development.

1754 • 1981 Mooley, D.A. and B. Parthasarathy. Large scale droughts over India and their impact on economy during the period 1871–1978. *Annals of National Association of Geographers, India*, 1 (2): 17–26.

1755 • 1981 Singh I.J. and R.N. Pandey. *Socio-Economic Impact of Drought on Farming Community in Haryana.* Hissar: Haryana Agricultural University.

1756 • 1982 Fukuoka, Y. Drought and agroclimatology in South India. In K. Fujiwara (ed). *Geographical Field Research in South India, 1980.* Hiroshima: Department of Geography, University of Hiroshima. 27–39.

1757 • 1982 Gupta, A.K. Drought, deficit, indebtedness: Deprivational and developmental alternatives before small farmers. In *Proceedings of the Symposium on Coping with Scarcity in International Agricultural Development.* Lexington.

1758 • 1982 Oughton, E. The Maharashtra drought of 1970–73: An analysis of scarcity. *Oxford Bulletin of Economics and Statistics*, 44 (3): 169–197.

1759 • 1983 Mishra, S. *et al.* Agricultural drought over Gangetic West Bengal in relation to *aman* rice. *Geographical Review of India*, 45 (2):78–80.

1760 • 1983 Moses, B.C. and M.S.S. Pandian. Tamil Nadu drought situation: View from a village *Economic and Political Weekly*, 18 (23): 996.

1761 • 1984 Biswas, B.C. and P. Nayar. Quantification of drought and crop potential. *Mausam*, 35 (3): 281–286.

1762 • 1984 Rao, G.G.S.N. and Y.S. Ramakrishna. Incidence of drought and its impact on food production in Rajasthan. *Annals of Arid Zone*, 23 (4): 327–334.

1763 • 1985 Nadkarni, M.V. *Socio-Economic Conditions in Drought Prone Areas: A Bench Mark Study of Drought Districts in Andhra Pradesh, Karnataka and Tamil Nadu (Volume 1 and 2).* New Delhi: Concept Publishing Company.

1764 • 1985 Special Correspondent. Maharashtra: The silent drought. *Economic and Political Weekly*, 20 (2): 96–98.

1765 • 1986 Bandhopadhyay, J. and V. Shiva. Drought, development and desertification. *Economic and Political Weekly*, 21 (1): 1448–1450.

1766 • 1986 Caldewell, J.C. *et al.* Periodic high risk as a cause of fertility decline in a changing rural environment: Survival strategies in the 1980–83 South Indian drought. *Economic Development and Cultural Change*, 34 (4): 677–701.

1767 • 1986 Gupta, A.K. Drought and deprivation: Socio-ecology of stress, survival and surrender. In *Proceedings of the Seminar on Control of Drought, Desertification and Famine.* New Delhi: India International Centre.

1768 • 1986 Jaiyaram, A. Karnataka drought. *Frontline,* (February 22): 61–65.

1769 • 1986 Kuchhadiya, D.B. *An Economic Analysis of Risk on Drought Prone Farms in Amreli District of Gujarat State.* Vishakhapatnam: Andhra University, Waltair (Unpublished Ph.D thesis).

1770 • 1986 Lakshmaiah, N.G. *Determinants of Fertility and Family Planning among the Sugalis in Drought Prone Areas.* Anantapur: Sri Krishna Devaraya University (Unpublished Ph.D thesis).

1771 • 1986 Llaiah, K. Andhra Pradesh drought: Worsening situation. *Economic and Political Weekly*, 21 (24): 909–912.

1772 • 1986 Rao, G.G.S.N. *et al.* Impact of drought on productivity of pearl millet in Rajasthan. *Mausam,* 37 (3): 197–202.

1773 • 1986 Reddy, Y.V.R. and M. Sudha. Impact of Drought Prone Area Programme on rural economy and employment in arid and semi-arid regions. *Indian Journal of Regional Science*, 18 (2): 87–95.

1774 • 1986 Sitaramaiah, P. *Cropping Pattern and Farming Techniques in a Drought Prone Area, Anantpur District.* Anantpur: Sri Krishna Devaraya University (Unpublished Ph.D thesis).

1775 • 1986 Special Correspondent. Scorching sun, scarred souls. *Frontline*, (Feb. 22): 40.

1776 • 1986 Special Correspondent. Gujarat: A thirsty land. *Frontline*, (April 19): 30–35.

1777 • 1987 Badhwar, I. and M. Rehman. Devastating drought. *India Today*, (September 15): 28–40.

1778 • 1987 Ghosh, A. Drought: The Rajasthan scenario. *Economic and Political Weekly*, 22 (41): 1427–1429.

1779 • 1987 Kumar, P.H. Thirsty. *Frontline*, (May 2).

1780 ● 1987 Ramanathan, C. Tamil Nadu: Worst in many years. *Frontline*, 20 (May 2): 31–33.

1781 ● 1987 Rao, D.K. and M.V. Rao. A difficult district, Mahbubnagar (Andhra Pradesh). *Economic and Political Weekly*, 22 (28): 43–46.

1782 ● 1987 Sinha, S.K. The 1982–83 drought in India: Magnitude and impact. In M. Glantz *et al.* (eds). *The Societal Impacts Associated with the 1982–83 Worldwide Climate Anomalies.* Boulder, Colorado: National Centre for Atmospheric Research. 37–42.

1783 ● 1987 Velayudhar, T.K. Monsoons and the economy. *Economic and Political Weekly*, 22 (49): 2108–2111.

1784 ● 1987 Vijapurkar, M. Gujarat drought spreads again. *Frontline*, 20 (May 16): 30–31.

1785 ● 1988 Banerjee, S. Identification of drought and its impact on agriculture in Uttar Pradesh. *Geographical Review of India*, 50 (3): 50–56.

1786 ● 1988 Nanda, S.K. Drought and its fall out. *Mainstream*, 26 (January 2): 11–12.

1787 ● 1988 Special Correspondent. Kalahandi: A stark picture. *Economic and Political Weekly*, 23 (12): 886–888.

1788 ■ 1989 Balgopal, K. Drought and terrorist and disruptive activities in Adilabad. *Economic and Political Weekly*, 24 (47): 2587–2591.

1789 ● 1989 Banerjee, B. Problems of drought: Indian perspective. *Science and Culture*, 55 (9): 343–352.

1790 ● 1989 Menon, P. Drought in Orissa: The children of hunger. *Frontline*, 22 (April 29): 99–101.

1791 ● 1989 Shankari, U. What is happening to the cows and bulls of Sundrapalle? *Economic and Political Weekly*, 24 (25): 1164–1170.

1792 ● 1990 Chatterji, S. (Translated by S. Sinha). *The Drought and Other Stories* (Second Edition). New Delhi: Sahitya Academy Publication.

1793 ● 1990 Kumar, B.G. Consumption disparities, food surpluses and effective demand failures: Reflections on macro economics of drought vulnerability. *Economic and Political Weekly*, 25 (10): 499–508.

1794 ● 1990 Rajendra, M. *A Study of Drought Impact Assessment Mechanisms and Relief Operations.* New Delhi: Indian Council of Social Science, (Project Report).

1795 ● 1990 Sahai, K. Sorrow of the Thar. *Frontline*, 23 (May 26): 96–101.

1796 ● 1991 Bagchi, K. *Drought Prone India: Problems and Perspectives* (Volume I and II). New Delhi: Agricole Publishing Academy.

1797 ● 1991 Bidinger, P.D. *et al.* Consequences of mid–1980s drought: Longitudinal evidences from Mahbubnagar. *Economic and Political Weekly*, 25 (39): A105–A114.

1798 • 1991 Saradamoni, K. In good season and bad season. *Economic and Political Weekly*, 26 (44): 2513–2515.

1799 • 1991 Sastri, A.S.R.S.A. Drought climatology in subhumid climates and its impact on agriculture: A case study for Raipur district. *Bhagirath,* 38 (July–September): 113–122.

1800 • 1992 Alam, M.J. Drought leads people to starvation in Bihar. *New Age*, 40 (3): 16.

1801 • 1992 Dyson, T. *et al.* Bihar famine 1966–67 and Maharashtra drought 1970–73: The demographic consequences. *Economic and Political Weekly*, 27 (26): 1325–1332.

1802 • 1992 Muralidharan, S. Drought: Looming food crises. *Frontline*, 25 (June 5): 94–97.

1803 • 1992 Ramachandran, K. Impact of drought on foodgrain production in India. *Indian Geographical Journal*, 67 (2): 29–36.

1804 • 1992 Special Correspondent. From deluge to drought. *Down to Earth*, 1 (7): 11.

1805 • 1992 Special Correspondent. Drought Maharashtra. *Frontline*, 25 (June 5): 96.

1806 • 1992 Special Correspondent. Drought: Withering away. *Frontline*, 25 (June 5): 97.

1807 • 1992 Sridhar, V. Drought: A problem of procurement. *Frontline*, 25 (June 5): 100.

1808 • 1992 Thiruvengadachari, S. Satellite surveillance of drought conditions. *Kurukshetra*, 41(3): 33.

1809 • 1993 Chadha, S.K. Devastating droughts of India. In S.K. Chadha (ed). *Understanding Environment.* Delhi: Amar Prakashan, 83–87.

1810 • 1993 Dash, J. Impact of drought on agricultural production, forest utilisation and mitigation. In P.K. Nayak and A. Mahajan (eds). *Kalahandi-Human Encounter with Drought.* New Delhi: Reliance Publishing House. 52–90.

1811 • 1993 Mitra, A. and K. Kumar. Death by starvation. *Down to Earth*, 2 (2): 32–36.

1812 • 1993 Narayana, N. *Socio-Cultural Development and its Impact on Cropping Pattern in a Drought Prone District: A Study in Rayalseema Region of Andhra Pradesh.* New Delhi: Indian Council of Social Science, (Project Report).

1813 • 1993 Nayak, P.K. and A. Mahajan (eds). *Kalahandi: Human Encounter with Drought.* New Delhi: Reliance Publishing House.

1814 • 1993 Ramakrishnan, V.A perennial problem: Drought in Madhya Pradesh. *Frontline*, 26 (June 18): 28–29.

1815 • 1993 Ray, T. Valley of death: Hunger and thirst stocks Orissa. *Frontline*, 26 (June 4): 35–40.

1816 • 1993 Venkasteshwaralu, J. Effect of drought on *kharif* foodgrain production: A retrospect and prospect. *Annals of Arid Zone*, 32 (1): 1–6.

1817 • 1994 Bandyopadhyay, S. *Study of Drought in Different Parts of West Bengal and Its Impact on Agriculture.* Kolkata: University of Calcutta (Unpublished Ph.D thesis).

1818 • 1994 Samal, K. C. Drought and its toll in Kalahandi. *Mainstream*, 32 (19 February): 24–26.

1819 • 1995 Padmanabhan, R. Growing agony: Drought stalks Maharashtra. *Frontline*, 28 (June 16): 74–79.

1820 ▲1995 Sharma, S. *Nature and Society: A Study of Social Dimensions of Drought.* Jaipur: Rajasthan University (Unpublished Ph.D thesis).

1821 ■ 1995 Singh, T. *Drought Disaster and Agricultural Development in India.* New Delhi: People's Publishing House.

1822 • 1996 Krishnaswamy, C. Madanapalle's tragedy. *Frontline*, 29 (July 12): 116.

1823 • 1997 Dandge, R. G. *Growth and Instability in Agricultural Production in the Drought prone Region of Maharashtra: A Regional Analysis.* Kolhapur: Shivaji University (Unpublished Ph.D thesis).

1824 • 1998 Editorial. India. *Down to Earth*, 7 (8): 12.

1825 • 1998 Malik, J. N. *et al.* Depth of despair. *India Today*, (August 31): 26–29.

1826 • 1999 Editorial. Rajasthan affected. *Down to Earth*, 8 (15): 11.

1827 • 1999 Hassan, I. U. Dry valley. *Down to Earth*, 8 (10): 14–15.

1828 • 2000 Banerjee, R. Orissa: Sizzler of a state. *India Today*, (May 8): 48.

1829 • 2000 Bavadam, L. Retrograde development. *Frontline*, 17 (11): 73–76.

1830 ■ 2000 Bavadan, L. Gujarat's thirst. *Economic and Political Weekly*, 35 (22): 120–122.

1831 • 2000 Baweja, H. Rajasthan: Desert of woes. *India Today*, (May 8): 44–48.

1832 • 2000 Dasgupta, S. A. Mirage chasers. *India Today*, (May 8): 54–55.

1833 ■ 2000 Dash, J. Dried up water sources. *Down to Earth*, 8 (20): 16.

1834 • 2000 Editorial. Nutritional status of pre-school children in the drought affected Kalahandi district of Orissa. *Indian Journal of Medical Research*, 111: 90–94.

1835 • 2000 Editorial. Starving cattle. *Down to Earth*, 8 (23): 11.

1836 ■ 2000 Joshi, S. Drought land. *Economic and Political Weekly*, 35 (49): 4304–4306.

1837 • 2000 Madhurkar, U. Pain model. *India Today*, (May 15): 54.

1838 • 2000 Madhurkar, U. A lifeless landscape. *India Today*, (May 8): 49–53.

1839 • 2000 Menon, P. Hungry in Anantpur. *Frontline*, 17 (13): 43–45.

1840 • 2000 Mishra, S. Curse of Vindhyas. *India Today*, (May 8): 52–53.

1841 • 2000 Prasad, E. When 'calamity' strikes. *Down to Earth*, 9 (1): 13–15.

1842 • 2000 Rajesh. Rajasthan: Drought, debt and poverty. *Economic and Political Weekly*, 35 (25): 2101–2104.

1843 • 2000 Rajlakshmi, T.K. Gujarat's thirst. *Frontline*, 17 (10): 120–122.

1844 • 2000 Ramachandraiah, C. Dying for water. *Down to Earth*, 8 (24): 56.

1845 • 2000 Self-Employed Women's Association. *The Gum Collectors: Struggling to Survive in the Dry Areas of Banaskantha.* Research Report. Ahmedabad.

1846 ■ 2000 Singh, S.P. and A. Dwivedi. Impact of drought on fish production: A case study of some ponds in Menewa block of Etawah district (Uttar Pradesh). In R.B. Singh (ed). *Disaster Management.* Jaipur: Rawat Publications. 144–151.

1847 ■ 2000 Singh, T. Drought: A challenging natural hazard in subhumid climatic region (case study of Palamau). In R.B. Singh (ed). *Disaster Management.* Jaipur: Rawat Publications. 18–131.

1848 • 2000 Sridhar, V. Food security in peril. *Frontline*, 17 (11): 66–72.

1849 • 2000 Tiwari, M. Riots for water. *Down to Earth*, 8 (16): 7–8.

1850 • 2001 Bunsha, D. Marathwada distress. *Frontline*, 18 (21): 45–47.

1851 • 2001 Choudhary, A.R. Drought, state and people. *Seminar*, 505: 102–104.

1852 • 2001 Menon, P. Karnataka's agony. *Frontline*, 18 (17): 42–43.

1853 • 2001 Mishra, N. Drought and deaths. *Frontline*, 18 (April 27): 71–73.

1854 • 2001 Rajalakshmi, T.K. and V.V. Krishnan. Rajasthan's thirst. *Frontline*, 18 (6): 129–133.

1855 • 2001 Rajappan, M. Water riots and mass migration in Madhya Pradesh. *New Age*, 49 (22): 5–6.

1856 ■ 2001 Saha, A. Time locked: An analysis. *Down to Earth*, 10 (6): 33–35.

1857 • 2001 Singh, M.M. and K.K. Kumar. The devastating drought. *Bhagirath*, 48 (4) (October–December): 159–166.

1858 • 2001 Special Correspondent. Parched lands: Drought in Orissa. *Down to Earth*, 10 (10): 34.

1859 • 2001 Special Correspondent. Testing time. *Down to Earth*, 10 (2): 40–42.

1860 • 2001 Special Correspondent. Orissa grapples with drought and starvation. *New Age*, 49 (7): 14.

1861 • 2002 Bharwada, C. and V. Mahajan. Drinking water crisis in Kutch: A natural phenomenon. *Economic and Political Weekly*, 37 (50): 4859–4866.

1862 • 2002 Das, H.P. *et al.* Evaluation of agricultural drought for dryland crop. *Mausam*, 53 (3): 375–380.

1863 • 2002 Department of Agriculture and Cooperation. *Drought, Status Report of Gujarat, Maharashtra and Rajasthan.* New Delhi: Government of India.

1864 • 2002 Economic and Political Weekly Research Foundation. Dire impact of drought on savings and investment. *Economic and Political Weekly*, 37 (36): 3391–3397.

1865 • 2002 *Frontline* Team. The ground realities. *Frontline*, 19 (17): 9.

1866 • 2002 *Frontline* Team. The southern swath. *Frontline*, 19 (4): 125.

1867 • 2002 Jadhav, F. Effect of drought on drinking water supply and solutions: Some case studies from Maharashtra. In R. Nagarajan (ed). *National Workshop Proceedings on Recent Trends in Drought Assessment Monitoring and Management.* Mumbai: Centre of Studies in Resources, Engineering Indian Institute of Technology, 98–108.

1868 • 2002 Mishra, N. Hunger deaths in Baran. *Frontline*, 19 (2): 40–41.

1869 • 2002 Rao, V. M. and R. S. Deshpande. Food security in drought prone areas: A study in Karnataka. *Economic and Political Weekly*, 37 (35): 3677–3681.

1870 • 2002 Singh, K. *Situation Report: Rajasthan Drought.* New Delhi: United Nations Development Programme.

1871 • 2003 Aiyar, S. Thirsty India. *India Today*, (January 9): 28–45.

1872 • 2003 Menon, P. In search of livelihood. *Frontline*, (June 6): 55–56.

1873 • 2003 Rajalakshmi, T. K. Baran's call. *Frontline*, (April 26): 108–109.

1874 • 2003 Rajalakshmi, T. K. From bad to worse in Rajasthan. *Frontline*, (April 26): 107.

1875 • 2003 Satyasundar, B. Passage to hunger. *Down to Earth*, 12 (12): 22–23.

1876 • 2003 Subramanian, T. S. Despair and death. *Frontline*, 20 (3): 46.

1877 • 2003 Subramanian, T. S. Drought in the delta. *Frontline*, 20 (3): 44–47.

1878 ■ 2004 Dandge, R. G. *Indian Agriculture in Drought prone Areas: A Case Study of Maharashtra.* New Delhi: Classical Publishing.

1879 • 2004 Department of Agriculture and Cooperation. *Drought 2002.* New Delhi: Government of India.

1880 • 2004 Krishnakumar, R. Kerela's plight. *Frontline*, 21 (6): 41–45.

1881 • 2004 Krishnakumar, R. The cascading effect. *Frontline*, 21 (9): 42.

1882 • 2004 Krishnakumar, R. Tribal misery. *Frontline*, 21 (9):43.

1883 • 2004 Krishnakumar, R. Driven to death. *Frontline*, (April 24): 40–45.

1884 • 2004 Menon, P. Karnataka woes. *Frontline*, 21 (6): 40.

1885 • 2004 Singh, R. B. Nature and extent of disguised employment in drought affected area: A case study of Jamunia Kamal village of Sailana block of Ratlam district (Madhya Pradesh). *Indian Journal of Regional Science*, 36 (1): 127.

1886 • 2004 Subramanian, T. S. Drought and despair. *Frontline*, 21 (6): 35–40.

1887 • 2005 Bavadam, L. Vidharbha's despair. *Frontline*, 22 (12): 47–49.

1888 • 2005 Kumar, P. *et al.* Drought intensity and yields of rice and wheat in dry tracts of Bihar. *Mausam*, 56 (4): 915–918.

1889 • 2005 Kumar, R.H. Diet and nutritional situation of the population in the severely drought affected areas of Gujarat. *Journal of Human Ecology*, 18(4): 319–326.

1890 • 2005 Nageswara, E.B. *et al.* Satellite based assessment of agricultural drought in Karnataka state. *Photonirvachak, Journal of the Indian Society of Remote Sensing*, 33 (3): 429–434.

1891 • 2006 Selvaraj, K.N. and C. Ramaswamy. Drought, agriculture risk and rural income: Case of a water limiting rice production environment (Tamil Nadu). *Economic and Political Weekly*, 41 (26): 2739–2746.

Impact: Earthquake

1892 • 1869 Oldham, T. Exhibition of photographs of the result of the Cachar earthquake of 10 January 1869. *Proceedings of the Asiatic Society of Bengal*, 38: 113–115.

1893 ■ 1884 Walker, J.T. Earthquake disturbances of the tides on the coast of India. *Nature*, 28: 358–360.

1894 • 1885 Editorial. The late earthquake in Kashmir. *The Lancet*, 126 (3236): 455.

1895 • 1897 Editorial. The earthquake in India. *The Lancet*, 150 (3856): 219.

1896 • 1897 Latouche, T.H.D. Kolkata earthquake of 12 June 1897.*Nature,* 56: 273–274.

1897 • 1905 Editorial. The earthquake in the Punjab: A narrative by a medical survivor. *The Lancet*, 166 (4278): 632.

1898 ■ 1934 Burrard, S. Ground levels in Bihar in relation to earthquake of 15 January 1934. *Nature*, 133 (3363):582–583.

1899 • 1935 Buxton, N. The Indian earthquake. *The Lancet* 225 (5832): 1358.

1900 ■ 1938 Banerji, S.K. Earthquake problems of India. *Science and Culture,* 3 (12): 637–642.

1901 ■ 1950 Ghosh, S.K. Crack on the ground due to earthquake of 1950. *Geographical Review of India,* 12 (4).

1902 • 1951 Central Water and Power Commission. *Report of the Advisory Committee for the Reconnaissance and Survey of the Earthquake and Flood Affected Areas in Northeast Assam.* New Delhi: Government of India.

1903 ■ 1953 Banerji, S.K. Origin and nature of the distrubance produced by the Assam earthquake of 15 August 1950 and its aftershocks. *Central Board of Geophysics, India,* 1: 11–15.

1904 • 1953 Dutt, G.N. Damage caused by earthquake of the 15 August 1950 in Assam. *Central Board of Geophysics, India*, 1: 72–75.

1905 ■ 1953 Garo, E.R. Earthquake of 15 August and its effect on the topography and the regimes of the rivers in Northeast Assam and damage to roads. *Central Board of Geophysics, India,* 1: 65–71.

1906 ■ 1955 Tandon, A.N. Direction of faulting in the Great Assam earthquake of 15 August, 1950. *Indian Journal of Meteorology and Geophysics*, 6 (1): 61–64.

1907 ■ 1955 Ward, F. Aftermath of the Great Assam earthquake of 1950. *Geographical Journal*, 121 (3): 290–303.

1908 • 1967 Chopra, A.K. Reservoir-dam interaction during earthquakes. *Bulletin of the Seismological Society of America,* 57: 675–687.

1909 • 1967 Special Correspondent. By quake and fire. *Economic and Political Weekly*, 2 (46): 2193.

1910 • 1968 Ambraseys, N. and S.K. Sharma. Large earthquake forces on gravity dam. *Nature*, 219: 1354–1356.

1911 ■ 1968 Phadke, A.V. Significance of fissures and cracks developed in earthquake affected area around Koynanagar. *Current Science*, 37 (5): 126–128.

1912 • 1969 Agarwal, P.N. Structural response results during the 27 June 1966 earthquake in Nepal-India border region. *Bulletin of the Seismological Society of America*, 59 (2): 771–775.

1913 • 1970 Agarwal, P.N. Structural response results from the 29 October 1968 earthquake in the Koyna-Pophali region. *Bulletin of the Seismological Society of America*, 60 (1): 79–88.

1914 ■ 1970 Basu, S. and A. V. Setlur. Transient response of dam like structures subjected to earthquake. *Bulletin of Indian Society of Earthquake Technology*, 7 (1): 25–40.

1915 • 1970 Bulsari, B. S.and M. S. Thakkar. Response of structures in Broach earthquake. *Bulletin of Indian Society of Earthquake Technology*, 7 (4): 197–206.

1916 ■ 1970 Rao, B.S.R. and T.K.S. Prakasa Rao. Crustal structure and the Godavari valley earthquake. *Pure and Applied Geophysics*, 82 (5): 119–123.

1917 ■ 1971 Agarwal, S.R. *et al.* An attempt to determine ground acceleration during Koyna earthquake of 10 December 1967. *Bulletin of Indian Society of Earthquake Technology*, 8 (1): 34–38.

1918 ■ 1971 Agrawal, P.N. Observations of earth surface tilting in Koyna-Pophali region, Maharashtra. *Bulletin of the Seismological Society of America*, 61: 975–982.

1919 • 1972 Agrawal, P.N. Structural response results during 29 July 1970 earthquake in Burma-India border region. *Bulletin of the Seismological Society of America*, 62: 101–114.

1920 • 1972 Chopra, A.K. and P. Chakraborti. Analysis of earthquake performance of Koyna dam. *Bulletin of Indian Society of Earthquake Technology*, 9 (2): 49–60.

1921 • 1973 Chopra, A.K. and P. Chakraborti. The Koyna earthquake and the damage to Koyna Dam. *Bulletin of the Seismological Society of America*, 63: 381–397.

1922 ■ 1987 Molnar, P. The distribution of intensity associated with the 1905 Kangra earthquake and bounces on the extent of rupture zone. *Journal of Geological Society of India*, 29 (2): 221–229.

1923 • 1988 Ahmed, F. A devastating dawn. *India Today*, (September 15): 42–47.

1924 • 1988 Lindley, D. Indian earthquake. *Nature*, 334: 640.

1925 • 1988 Upadhaya, R. Bihar: Badly shaken. *Frontline*, (September 3): 26–31.

1926 • 1988 Upadhyay, B.K.Bihar: Aftermath of earthquake. *Mainstream*, 26 (Sept 17): 7.

1927 ■ 1989 Chander, R. Interpretation of the observed ground level changes due to the 1905 Kangra earthquake. *Tectonophysics*, 149 (3–4): 289–298.

1928 ■ 1989 Chander, R. Southern limits of major earthquake ruptures along the Himalayas 75° E to 90° E latitude. *Tectonophysics*, 170 (1–2): 115–174.

1929 ■ 1989 Das, S. and N.K. Jha. *Reconnaissance Report on the 21 August 1988 Earthquake in the Nepal-India Border Region.* Tokyo: Science and Culture, National Government Publication, Japanese Ministry of Education.

1930 ■ 1989 Sato, T. and J. Kiyona. Estimation of seismic intensity of ground motion during the 21 August 1988 earthquake in the Nepal-India border region. *Journal of Natural Disaster Science*, 11 (2): 21–36.

1931 • 1991 Ahuja, R. Himalayan trembles: Trail of destruction in the North. *Frontline*, 24 (October 26): 126–129.

1932 • 1991 Chandra Mohan, C.K. Jolted to misery: Quake survivors face hardships. *Frontline*, (November 9): 62–75.

1933 • 1991 Dhar, A. The damage factor. *Frontline*, (October 26): 128.

1934 • 1991 K.K., M.S. The chronology. *Frontline*, 24 (November 9): 10.

1935 • 1991 Pathak, R. Tremor of destruction. *India Today*, (November 15): 22–24.

1936 • 1991 Special Correspondent. The earthquake: Low intensity, high damage. *India Today*, (November 15): 25–28.

1937 • 1991 Valdiya, K.S. The Uttarkashi earthquake of 20 October: Implications and lessons. *Current Science*, 61 (12): 801–803.

1938 ▲1992 Arya, A.S. Possible effects of a major earthquake in Kangra region of Himachal Pradesh. *Current Science*, 62 (1–2): 251–256.

1939 • 1992 Baweja, H. The woes of winter. *India Today*, (February 15): 142–144.

1940 • 1992 Dath, T.R.C. and L. Baker. *A Study of the Conditions of a Variety of the Earthquake Hit Villages in Garhwal (20 October 1991)*. New Delhi: Department of Science and Technology, Government of India.

1941 ■ 1992 Molnar, P. The distribution of intensity associated with the 1897 Assam earthquake and bounces on the extent of the rupture zone. *Journal of Geological Society of India*, 39 (1): 13–27.

1942 • 1993 Fernandes, C. Face of the tragedy. *Mainstream*, 31 (48): 5–6.

1943 • 1993 Gupta, H.K. The deadly Latur earthquake. *Science*, 262: 1666–1667.

1944 • 1993 Kumar, S.N. and D.K. Rao. Images of grief. *Frontline*, (October 22): 12.

1945 • 1993 Menon, P. Life after devastation. *Frontline*, (November 5): 4–9.

1946 • 1993 Menon, P. Lost forever: The trail of death. *Frontline*, (November 5): 10–11.

1947 • 1993 Menon, P. Orphaned overnight. *Frontline*, (November 5): 20–21.

1948 • 1993 Mohan, M.M. Lives in ruins. *Frontline*, (October 22): 15.

1949 • 1993 Radhakrishna, B.P. A tragedy too deep for words: Latur earthquake of 30 September 1993. *Journal of Geological Society of India*, 42 (5): 523–526.

1950 • 1993 Special Correspondent. Earthquakes. *India Today*, (July 31): 43–71.

1951 • 1993 Special Correspondent. India's latest earthquake. Nature, 365: 476.

1952 ■ 1994 Gupta, R.P. *et al.* Indian remote sensing satellite detection of surface effects of the Uttarkashi Himalaya earthquake of 20 October 1991. *International Journal of Remote Sensing*, 15 (11): 2153– 2156.

1953 • 1994 Kagami, H. *Reconnaissance Report on the 30 September 1993 Earthquake in the State of Maharashtra*. Mumbai: Government of Maharashtra.

1954 ■ 1994 Khattri, K.N. *et al.* Inversion of strong motion waveforms of source slip function of 1991 Uttarkashi earthquake, Himalaya. *Journal of Himalayan Geology*, 5 (2): 163–192.

1955 • 1994 Kulkarni, N. *Non Engineered Houses in Earthquake Prone Areas*. Ahmedabad: Center for Environmental Planning and Technology (Unpublished Ph.D thesis).

1956 • 1994 Kumar, S. Health of Indian earthquake survivors. *The Lancet*, 344 (8920): 466.

1957 • 1994 Maybin, E. *et al.* Forty seconds that shook their world: The 1993 earthquake in India. *Focus on Gender*, 2 (1): 34–38.

1958 • 1994 Murthy, C.V.R. The 29 September 1993 M 6.4 Killari (Maharashtra) earthquake in Central India. *Shelter, Special Report*, 28 (1).

1959 • 1994 Nigam, N.C. Earthquake induced loads on structures. *Indian Concrete Journal*, 68 (11).

1960 • 1994 Rai, S. Mental aftershocks. *India Today*, (April 30): 82–84.

1961 • 1994 Rao, D.K. and S.N. Kumar. When the earth shook. *Frontline*, 27 (October 22): 4–11.

1962 • 1994 Singh, N.K. The terror of tremors. *India Today*, (November 15): 125–129.

1963 • 1994 Sinha, R. and A. Goyal. Damage to buildings in Latur earthquake. *Current Science*, 67: 380–385.

1964 ■ 1995 Bhat, G.K. and A. Revi. Water crisis in earthquake affected areas of Marathwada. *Economic and Political Weekly*, 30 (29): 1826–1830.

1965 ■ 1995 Chung, W.Y. and H. Gao. Source parameters of the Anjar earthquake of 21 July 1956 India, and its seismotectonic implications for Kutch rift basin. *Tectonophysics*, 242 (3–4): 281–292.

1966 ■ 1995 Geological Survey of India. *Collection of Papers Presented at the Workshop on 30 September 1993 Killari Earthquake (Maharashtra).* Special Publication No. 27. Kolkata.

1967 • 1995 Hagde, D. and N. Acharya. Double victims. *Down to Earth*, 4 (3): 47.

1968 ■ 1995 Majumdar, R.K. Intraplate deformation in the North Indian plate and Latur earthquake: A discussion. *Indian Journal of Geology*, 67 (1): 77–88.

1969 • 1995 Parasuraman, S. The impact of 1993 Latur-Osmanabad (Maharashtra) earthquake on lives, livelihoods and property. *Disasters: The Journal of Disaster Studies, Policy and Management*, 19(2): 156–169.

1970 • 1995 Parasuraman, S. *Impact of Earthquake on Lives, Livelihoods and Property in Maharashtra.* Mumbai: Tata Institute of Social Sciences.

1971 ■ 1995 Yu, G. *et al.* Strong ground motion from the Uttarkashi (Himalaya) India earthquake: Comparison of observations with synthetics using the composite source model. *Bulletin of the Seismological Society of America*, 85 (1): 31–50.

1972 • 1996 Kailash. A demographic profile of the Marathwada earthquake victims. *Population Geography*, 18 (1–2): 47–60.

1973 • 1996 Mishra, G. Earthquake in Latur and Osmanabad districts of Maharashtra: Destruction caused and relief rendered. *Himalayan Geology*, 17: 217–224.

1974 • 1996 Sharan, P. and G. Chaudhary. Preliminary report of psychiatric disorders in survivors of earthquake. *American Journal of Psychiatry*, 153 (4): 556–558.

1975 • 1996 Sinvhal, A. and P. Bose. Performance of hospitals and clinics in the Latur (Osmanabad) earthquake. *Himalayan Geology*, 17: 235–239.

1976 ■ 1997 Joshi, D. D. *et al.* Jaisalmer earthquake of 1991: Its effects and tectonic implications. *Journal of Geological Society of India*, 49(4): 433–436.

1977 • 1997 Sinha, R. and N. Adarsh. Earthquake damage scenario for Mumbai. In *Proceedings of the Workshop on Earthquake Disaster Preparedness*, Indian Society of Earthquake Technology. Roorkee: Indian Institute of Technology.

1978 ■ 1998 Gupta, R. P. *et al.* Cover discrimination of areas susceptible to earthquake induced liquefaction from Landsat data. *International Journal of Remote Sensing*, 19 (4): 569–572.

1979 ■ 1998 Rajendra, C. P. *et al.* Surface deformation related to the 1819 Kachchh earthquake: Evidence for recurrent activity. *Current Science*, 75 (6): 623–626.

1980 • 1999 Arya, A. S. Damage scenarios of probable earthquakes of M6.5 in Uttar Pradesh Himalaya. *Himalayan Geology*, 20 (1): 93–104.

1981 • 1999 Chakravarty, S. Killer quakes. *India Today*, (April 12): 34–35.

1982 • 1999 Patwal, S. *Chamoli Earthquake and its Aftermath*. New Delhi: Advanced Professional Programme in Public Administration, Indian Institute of Public Administration (Unpublished Master of Philosophy Dissertation).

1983 • 1999 Ramakrishnan, V. Shattered by an earthquake. *Frontline*, 16 (April 23): 128.

1984 ■ 1999 Rao, S. *et al.* Tectonic stress field in the epicentral zone of the Latur earthquake of 1993. *Proceedings of Indian Academy of Sciences, Earth and Planetary Sciences*, 108 (2): 93–98.

1985 • 1999 Ravindran, K.V. and G. Philip. Chamoli earthquake: A preliminary report on earthquake induced landslides using Indian remote sensing satellite-1C/ID data. *Current Science*, 77 (1): 21–25.

1986 • 1999 Siavoshina, M. *Behaviour of Reinforced Earthern Embankments during Earthquakes*. Roorkee: University of Roorkee (Unpublished Ph.D thesis).

1987 • 1999 Special Correspondent. Darkness at midnight. *Down to Earth*, 8 (April 30): 5.

1988 • 1999 Vta-finzi, C. Destruction and deliberation. *Nature*, 397 (January 21): 223–225.

1989 • 2000 Acharaya, N. Double victims of Latur earthquake. *Indian Journal of Social Work*, 61(4): 558–564.

1990 • 2000 Aubrey, J. Gender issues in the Marathwada earthquake. In S. Parasuraman and P.V. Unnikrishnan (eds). *India Disasters Report: Towards a Policy Initiative*. New Delhi: Oxford University Press. 214–216.

1991 • 2000 Chopra, R. Earthquake: Reconfiguring lives and landscapes. In S. Parasuraman and P.V. Unnikrishnan (eds). *India Disasters Report: Towards a Policy Initiative.* New Delhi: Oxford University Press. 201–206.

1992 • 2000 Gandevia, N.R. Killari: Psychological health of a village ravaged by an earthquake. *Indian Journal of Social Work*, 61(4): 652–663.

1993 • 2000 Jain, S.K. *Seismic Base Isolation of Multistoreyed Buildings.* Roorkee: University of Roorkee (Unpublished Ph.D thesis).

1994 • 2000 Parasuraman, S. and N. Acharaya. Analysis in forms of vulnerability in a disaster. *Indian Journal of Social Work*, 61(4): 581–597.

1995 ■ 2000 Rao, B.R. Historical seismicity and deformation rates in the Indian Peninsular shield. *Journal of Seismology*, 4 (3): 247–258.

1996 ■ 2000 Sarkar, I. and A.K. Saraf. Some observations of Chamoli earthquake-induced damage using ground and satellite data. *Current Science*, 78 (1): 91–96.

1997 • 2001 Bavadam, L. The medical crisis. *Frontline*, 18 (4): 14–15.

1998 • 2001 Bezbaruah, S. Death: The way out, preventing collapse. *India Today*, (February 12): 62.

1999 ■ 2001 Bhoi, S. and A.K. Sahoo. Significant changes in ocean parameters after the Gujarat earthquake. *Current Science*, 80 (11): 1376–1377.

2000 ■ 2001 Bilham, R. Intraplate earthquakes and flexure of the Indian plate. In *Proceedings of the International Conference on Seismic Hazard with Particular Reference to Bhuj Earthquake of 26 January 2001.* New Delhi: Government of India Department of Science and Technology India Meteorological Department. 352–353.

2001 ■ 2001 Bilham, R. Slow tilt reversal of the Lesser Himalaya between 1862–1992 at 78° E. and bounds to the southeast rupture of the 1905 Kangra earthquake. *Geophysical Journal International*, 144 (3): 1–23.

2002 • 2001 Building Material Technology Promotion Council. *Rapid Assessment of Damage to Buildings Affected by Kachchh Earthquake of 26 January 2001.* New Delhi: Housing and Urban Development Corporation.

2003 • 2001 Chakravarti, S. Hell on the earth. *India Today*, (February 12): 20–36.

2004 • 2001 Earthquake Engineering Research Institute. *Annotated Images from the Bhuj India Earthquake of 26 January 2001.* Oakland.

2005 • 2001 *Frontline* Team. Gujarat's black Friday. *Frontline*, 18 (2): 128–134.

2006 • 2001 Goyal, A. *et al.* Performance of reinforced concrete buildings in Ahmedabad during Bhuj earthquake 26 January 2001. In H.R. Wason, *et al* (eds). *Proceedings of the Workshop on Recent Earthquakes of Chamoli and Bhuj.* Roorkee.

2007 ■ 2001 Jayasundar, R. Anomalous very low frequency electric field perturbations associated with Chamoli earthquake. *Current Science*, 80 (9): 1416–1421.

2008 ■ 2001 Kumar, S. Lives after earthquake. *The Lancet*, 357 (9254): 447.

2009 ● 2001 Kumar, S. and H. Ashraf. Authorities overstretched as earthquake death toll increases. *The Lancet*, 357 (9253): 369.

2010 ● 2001 Kumar, S. and R. Kumar. Bhuj earthquake: Performance of masonry buildings. In H.R. Wason, *et al* (eds). *Proceedings of the Workshop on Recent Earthquakes of Chamoli and Bhuj*. Roorkee. 81–89.

2011 ● 2001 Lahiri, A. *et al.* Economic consequences of Gujarat earthquake. *Economic and Polit-ical Weekly*, 36 (26): 1319–1332.

2012 ● 2001 Madhurkar, U. and S. Raval. Starting over. *India Today*, (May 7): 57–61.

2013 ● 2001 Madhurkar, U. and S. Bezbaruah. Double disaster. *India Today*, (Feb19): 20.

2014 ● 2001 Mahapatra, R. Republic quake. *Down to Earth*, 9 (19): 42–48.

2015 ● 2001 Manjrekar, S.K. Earthquake damaged structures in Gujarat: Selection of repair material and case studies. *Indian Concrete Journal*, 75 (12).

2016 ■ 2001 Nakata, T. *et al.* Extensive surface deformation around Budharmora associated with the 26 January 2001 Republic Day (Bhuj) earthquake of India. *Active Fault Research*, 20: 127–136.

2017 ● 2001 Prakash, V. *et al.* Damage observed in temples, prayer halls and community centers due to the Kutch earthquake of 26 January 2001. In H.R. Wason, *et al* (eds). *Proceedings of the Workshop on Recent Earthquakes of Chamoli and Bhuj*. Roorkee. 407–421.

2018 ● 2001 Prakash, V. *et al.* Effects of Kutch earthquake of 26 January 2001 on Gujarat Electricity Board substations. In H.R. Wason, *et al* (eds). *Proceedings of the Workshop on Recent Earthquakes of Chamoli and Bhuj*. Roorkee. 443–449.

2019 ● 2001 Prakash, V. *et al.* Effects of the Kutch earthquake of 26 January 2001 on lifeline structures. In H.R. Wason, *et al* (eds). *Proceedings of the Workshop on Recent Earthquakes of Chamoli and Bhuj*. Roorkee.

2020 ● 2001 Prakash, V. *et al.* Effects of the Kutch earthquake of 26 January 2001 on municipal overhead tanks, ground reservoirs and pumping units. In H.R. Wason, *et al* (eds). *Proceedings of the Workshop on Recent Earthquakes of Chamoli and Bhuj* Roorkee. 433–441.

2021 ● 2001 Prakash, V. *et al.* Effects of the Kutch earthquake of 26 January 2001 on oil industry. In H.R. Wason, *et al* (eds). *Proceedings of the Workshop on Recent Earthquakes of Chamoli and Bhuj*. Roorkee. 159–165.

2022 ● 2001 Prakash, V. *et al.* Oil industry and the Kutch earthquake of 26 January 2001. In H.R. Wason, *et al* (eds). *Proceedings of the Workshop on Recent Earthquakes of Chamoli and Bhuj*. Roorkee. 175–178.

2023 • 2001 Prasannarajan, S. and H. Chawala. Earthquake: Ground zero dateline fearscape. *India Today*, (February 12): 54–58.

2024 ■ 2001 Rastogi, B.K. *et al.* The deadliest stable continental region earthquake occurred near Bhuj on 26 January 2001. *Journal of Seismology*, 5 (4): 609– 615.

2025 ■ 2001 Ray, P.K.C. Detection and analysis of surface changes due to the recent earthquake in Gujarat using Indian remote sensing satellite and linear imaging and self scanning sensor-3 data. In H.R. Wason, *et al* (eds). *Proceedings of the Workshop on Recent Earthquakes of Chamoli and Bhuj*. Roorkee. 225–236.

2026 • 2001 Reddy, L.R. *The Pain and Horror: Gujarat Earthquake*. New Delhi: Ashish Publishing House.

2027 ■ 2001 Saraf, A.K. The Kutch earthquake of 26 January 2001: Satellite data reveals earthquake induced ground changes and appearance of water bodies. In H.R. Wason, *et al* (eds). *Proceedings of the Workshop on Recent Earthquakes of Chamoli and Bhuj*. Roorkee. 207–216.

2028 • 2001 Saram, R. and V.S. Aiyar. Earthquake: Economic impact, what goes down. *India Today*, (February 12): 69–71.

2029 • 2001 Shah, J. Disastrous building. *Frontline*, 18 (March 17): 114–115.

2030 • 2001 Sharma, R. Pneumonia, cholera and dysentery feared after earthquake. *The Lancet*, 357 (9254): 317.

2031 • 2001 Sharma, V.K. Gujarat earthquake: Some emerging issues. *Disaster Prevention and Management*, 10 (5): 349–355.

2032 • 2001 Shylendra, H. Are earthquake deaths over estimated? Assessment for Kutch villagers. *Economic and Political Weekly*, 36 (9): 722–723.

2033 • 2001 Singh, M. Gujarat earthquake and related issues. *Mainstream*, 39 (14): 14–15.

2034 ■ 2001 Singh, R.P. and A.K. Sahoo. Ground deformation of the Gujarat earthquake of 26 January 2001. *Journal of Geological Society of India*, 58 (3): 209–214.

2035 ■ 2001 Sinvhal, A. *et al.* Damage seismotectonics and isoseismals for the Kutch earthquake of 26 January 2001. In H.R. Wason, *et al* (eds). *Proceedings of the Workshop on Recent Earthquakes of Chamoli and Bhuj*. Roorkee. 61–70.

2036 ■ 2001 Sinvhal, A. *et al.* Isoseismals for the Kutch earthquake of 26 January 2001. In H.R. Wason, *et al* (eds). *Proceedings of the Workshop on Recent Earthquakes of Chamoli and Bhuj*. Roorkee. 179–182.

2037 • 2001 Sinvhal, A. *et al.* Damage observed to educational buildings due to Kutch earthquake of 26 January 2001. In H.R. Wason, *et al* (eds). *Proceedings of the Workshop on Recent Earthquakes of Chamoli and Bhuj*. Roorkee. 381–388.

2038 • 2001 Sinvhal, A. *et al.* Damage observed to hospitals and medical facilities due to Kutch earthquake of 26 January 2001. In H.R. Wason, *et al* (eds). *Proceedings of Workshop on Recent Earthquakes of Chamoli and Bhuj.* Roorkee. 389–397.

2039 • 2001 Sinvhal, A. *et al.* Damage observed to railway bridge at Dholawa and railway station at Bhachau in Kutch earthquake of 26 January 2001. In H.R. Wason, *et al* (eds). *Proceedings of the Workshop on Recent Earthquakes of Chamoli and Bhuj.* Roorkee. 399–406.

2040 • 2001 Sinvhal, A. *et al.* Damage observed to Surajbari bridge due to Kutch earthquake of 26 January 2001. In H.R. Wason, *et al* (eds). *Proceedings of the Workshop on Recent Earthquakes of Chamoli and Bhuj.* Roorkee. 423–431.

2041 • 2001 Sinvhal, A. *et al.* Destruction of multistoreyed buildings in Kutch earthquake of 26 January 2001. In H.R. Wason, *et al* (eds). *Proceedings of the Workshop on Recent Earthquakes of Chamoli and Bhuj.* Roorkee. 451–460.

2042 • 2001 Sinvhal, A. *et al.* Multistoryed buildings and the Kutch earthquake of 26 January 2001. In H.R. Wason, *et al* (eds). *Proceedings of the Workshop on Recent Earthquakes of Chamoli and Bhuj.* Roorkee. 183–187.

2043 • 2001 Sinvhal, A. *et al.* Earthquake scenario and tourism in Uttaranchal. In *Proceedings of the All India Seminar on Infrastructure Development in Uttaranchal-Problems and Prospects.* Roorkee: Indian Institute of Technology.

2044 • 2001 Special Correspondent. More shocks. *Down to Earth,* 10 (3): 10.

2045 • 2001 Special Correspondent. Earthquake tragedy in India spares nuclear power station. *Nature,* 409 (Februray 1): 552–553.

2046 ■ 2001 Subbaraman, J.V. Surface changes after the 2001 Gujarat earthquake. *Current Science,* 81 (8): 870–871.

2047 • 2001 Subramaniam, T.S. Shaken South. *Frontline,* 18 (October 26): 48.

2048 • 2001 Swami, P. Battling on the several fronts. *Frontline,* 18 (February 17): 19.

2049 • 2001 Swami, P. *et al.* The aftershocks. *Frontline,* 18 (4): 4–8.

2050 • 2001 United States Congress. *The Earthquake in India.* Washnigtion D. C. House Committee on International Relations.

2051 • 2001 Venkatesan, V. and S. Praveen.The killer earthquake. *Frontline,* 18 February: 4.

2052 • 2001 World Bank and Asian Development Bank. *A Preliminary Report on the Earthquake Damage in Gujarat.* New Delhi: World Bank and Asian Development Bank.

2053 • 2001 Yusuf, Y. *et al.* Damage assessment after 2001 Gujarat earthquake using LANDSAT-7 satellite image. *Photonirvachak, Journal of the Indian Society of Remote Sensing,* 29 (1–2): 17–22.

2054 ■ 2002 Aandahl, G. and K.O. Brien. *Damage, Seismotectonics and Isoseismals for the Kutch Earthquake of 26 January 2001*. New York: Ellis Horwood Limited.

2055 • 2002 Berz, Z. *et al.* Catastrophe portrait: The Gujarat earthquake in India. In G. Berz, *et al.* (eds). *Natural Catastrophes: Annual Review 2001*. USA: DIANE.

2056 • 2002 Editorial. Gujarat carnage and the health services: A public health crisis. *Indian Journal of Medical Research*, 10 (3)

2057 ■ 2002 Jade, S. and M. Mukul. Estimates of coseismic displacement and post-seismic deformation using global positioning system: Geodesy for the Bhuj earthquake of (January 26) 2001. *Current Science*, 82 (6): 748–752.

2058 • 2002 Kedare, J. and H.S. Dhavale. Effect of Marathwada earthquake on widows and married women. *Indian Journal of Social Work*, 63 (2): 183–193.

2059 • 2002 Kumar, R.P. Suffering and healing among the survivors of Bhuj earthquake. *Psychological Studies*, 47 (1–3): 106–112.

2060 ■ 2002 Moharan P.C. *et al.* Some observations on the surface deformities and their spatial variability in the earthquake affected areas in Kachchh district (India). *Indian Journal of Geomorphology*, 7 (1 & 2): 65–71.

2061 ■ 2002 Mukhopadhyay, S. Simultaneous inversion of the aftershock data of the 1993 Killari earthquake in Peninsular India and its seismotectonic implications. *Proceedings of Indian Academy of Sciences, Earth and Planetary Sciences*, 111 (1): 1–15.

2062 • 2002 Parikh, A. Skimming through socio-economic losses from an earthquake: A case study from Kachchh. *Indian Journal of Social Work*, 63 (2): 223–235.

2063 ■ 2002 Radhakrishna, M.G. and T.D. Sanu. Shallow seismicity, stress distribution and crustal deformation pattern in the Andaman, West Sunda Arc and Andaman Sea (North-Eastern Indian Ocean). *Journal of Seismology*, 6 (1): 25–41.

2064 ■ 2002 Rajendran, C.P. and K. Rajendran. Historical constraints on previous seismic activity and morphologic changes near the source zone of the 1819 Rann of Kutch earthquakes: Further light on the penultimate event. *Seismological Research Letters*, 73 (4): 470–479.

2065 ■ 2002 Sah, M.P. and S.K. Bartarya.The impact of 29 March 1999 Chamoli earthquake on slope stability and spring discharge in Chamoli-Rudraprayag districts, Garhwal Himalaya. *Himalayan Geology*, 23 (1): 121–133.

2066 • 2002 Samardjieva, E. and J. Badal. Estimation of the expected number of casualities caused by strong earthquakes. *Bulletin of the Seismological Society of America*, 92 (6): 2310–2322.

2067 ■ 2002 Saraf, A.K. *et al.* Satellite data reveals 26 January Kutch 2001 earthquake induced ground changes and appearance of water bodies. *International Journal of Remote Sensing*, 23 (9): 1749–1756.

2068 • 2002 Shah, C.B. Non destructive testing of earthquake-affected structures in Gujarat: Case study. *Indian Concrete Journal*, 76 (6).

2069 • 2002 Sharma, R. Gujarat earthquake causes major mental health problems. *British Medical Journal*, 324 (7332):259.

2070 ■ 2002 Singh, R.P. *et al.* Changes observed in the land and ocean after Gujarat earthquake of 26 January 2001, using Indian remote sensing data. *International Journal of Remote Sensing*, 23 (16): 3123–3128.

2071 ■ 2002 Tiwari, R.R. Environmental risk factors: A study in Kutch region. *Indian Journal of Medical Sciences*, 56 (3): 130–132.

2072 • 2002 Wilkinson, S.I. Putting Gujarat in perspective. *Economic and Political Weekly*, 37 (17): 1633.

2073 ■ 2003 Antolik, M. and D.S. Dreger. Rupture process of the 26 January 2001 M 7.6 Bhuj, India, earthquake from teleseismic broadband data. *Bulletin of the Seismological Society of America*, 93: 1235–1248.

2074 • 2003 Chopra, N. *Disasters in Greater Mumbai: A Geographical Inquiry.* Delhi: Department of Geography, Delhi School of Economics, University of Delhi (Unpublished Master of Philosophy Dissertation).

2075 ■ 2003 Gahalaut, V.K. *et al.* Rupture mechanism of the 1993 Killari earthquake, India: Constraints from aftershocks and static stress change. *Tectonophysics*, 369 (1–2): 71–78.

2076 • 2003 Gupta, M.K. An eyewitness account of the Bhuj earthquake. *Proceedings of Indian Academy of Sciences, Earth and Planetary Sciences*, 112 (3): 347–352.

2077 • 2003 Hough, S.E. *et al.* A media based assessment of damage and ground motions from the 26 January 2001, M 7.6 Bhuj earthquake. *Proceedings of Indian Academy of Sciences, Earth and Planetary Sciences*, 112 (3): 357–384.

2078 ■ 2003 Jade, S. *et al.* Pre-seismic, coseismic and post-seismic displacement associated with the Bhuj earthquake derived from recent and historic geodetic data. *Proceedings of Indian Academy of Sciences, Earth and Planetary Sciences*, 112 (3): 331–346.

2079 • 2003 Jagadish, K.S. *et al.* Behaviour of masonary structures during the Bhuj earthquake of January 2001. *Proceedings of Indian Academy of Sciences, Earth and Planetary Sciences*, 112 (3): 431–440.

2080 ■ 2003 Kayal, J.R. *et al.* Aftershock of the March 1999 Chamoli earthquake and seismotectonic structure of the Garhwal Himalaya. *Bulletin of the Seismological Society of America*, 93 (1): 109–117.

2081 ■ 2003 Midzi, V. *et al.* Local site effects in Latur, India: An empirical study based on the aftershocks of the Killari earthquake of 29 September 1993. *Engineering Geology,* 68 (3–4): 251–258.

2082 • 2003 Naqvi, T. *Vulnerability Study of Building Structures for Seismic Forces.* New Delhi: Indian Institute of Technology (Unpublished Ph.D thesis).

2083 • 2003 Rai, D. C. Performance of elevated tanks in M 7.7 Bhuj earthquake of January 2001. *Proceedings of Indian Academy of Sciences, Earth and Planetary Sciences*, 112 (3): 421–430.

2084 ■ 2003 Rajendran, C. P. and K. Rajendran. The surface deformation and earthquake history associated with the 1819 Kachchh earthquake. *Journal of Geological Society of India*, 54: 87–142.

2085 • 2003 Roshani, D. *Impact and Response to Bhuj Earthquake in Rural and Urban Areas of the District of Kachchh.* Delhi: Department of Geography, Delhi School of Economics, University of Delhi (Unpublished Master of Philosophy Dissertation).

2086 ■ 2003 Sarkar, I. and R. Chander. A simulation of earthquake induced undrained pore pressure changes with bearing on some soil liquefaction observations following the 2001 Bhuj earthquake. *Proceedings of Indian Academy of Sciences, Earth and Planetary Sciences*, 112 (3): 471–478.

2087 ■ 2003 Sinvhal, A. Isoseismals for the Kutch earthquake of 26 January 2001. *Proceedings of Indian Academy of Sciences, Earth and Planetary Sciences*, 112 (3): 353–356.

2088 ■ 2004 Bodin, P. and S. Horton. Source parameters and tectonic implications of aftershock of the M 7.6 Bhuj earthquake of 26 January 2001. *Bulletin of the Seismological Society of America*, 94 (3): 818–827.

2089 ■ 2004 Bodin, P. *et al.* Ground-motion scaling in the Kachchh Basin, India, deduced from the aftershock of the 2001 M 7.6 Bhuj earthquake. *Bulletin of the Seismological Society of America*, 94 (5): 1658–1669.

2090 ■ 2004 Desai, B. G. and S. J. Patel. Kutch earthquake of 2001: Seismically triggered deformed structures in the tertiary sediments of Mandui coast in the the Gulf of Kachchh (Gujarat). *Journal of Geological Society of India*, 64 (5): 637–646.

2091 • 2004 Dunbar, P. K. *et al.* Earthquake loss estimation for India based on macro-economic indicators. In T. Beer and A. Ismail-Zadeh (eds). *Risk Science and Sustainibility: Science for Reduction of Risk and Sustainable Development of Society.* Dordrecht: Kluwer Academic Publishers. 163–180.

2092 ■ 2004 Iyengar, R.N. and S.T.G. Raghukanth. Attenuation of strong ground motion in Peninsular India. *Seismological Research Letters*, 75 (4): 530–540.

2093 ■ 2004 Kayal, J.R. *et al.* Aftershock and seismotectonic implications of the 13 September 2002 earthquake (M 6.5) in the Andaman Sea basin. *Bulletin of the Seismological Society of America*, 94 (1): 326–333.

2094 ■ 2004 Pandey, P. and A.K. Pandey. Soft sediment deformation in the microseismal region of 1999 Chamoli earthquake of Garhwal Himalaya and their significance. *Himalayan Geology*, 25 (1–2): 79–88.

2095 • 2004 Rastogi, B.K. Damage due to the M 7.7 Kutch (India) earthquake of 2001. *Tectonophysics*, 390 (1 & 4): 85–103.

2096 • 2005 Alam, M. Disaster children: Initial assessment and management. *Indian Journal for Practising Doctors*, 2 (4).

2097 ■ 2005 Ammon, C.J. *et al.* Rupture process of the 2004 Sumatra-Andaman earthquake. *Science*, 308 (5725): 1133–1138.

2098 ■ 2005 Bilham, R. and N. Ambraseys. Apparent Himalayan slip deficit from the summation of seismic moment for Himalayan earthquakes, 1500–2000. *Current Science*, 88 (10): 1658–1663.

2099 ■ 2005 Champati Ray, P.K. *et al.* A quick appraisal of ground deformation in Indian region due to the 8 October 2005 earthquake, Muzaffarabad. *Photonirvachak, Journal of the Indian Society of Remote Sensing*, 33 (4): 466–473.

2100 ▲ 2005 Devi, R. Bhuj earthquake: Rural-urban differentiation. In A. Kapur, *et al. Disasters in India: Studies of Grim Reality.* New Delhi: Rawat Publication. 161–200.

2101 ■ 2005 Earnest, A. *et al.* Near-field observations on the coseismic deformation associated with the 26 December 2004 Andaman-Sumatra earthquake. *Current Science*, 88 (7): 1237–1244.

2102 • 2005 Editorial. A valley blighted. *Economic and Political Weekly*, 40 (43): 4567–4568.

2103 • 2005 Husain, A. Sacred innocence. *India Today*, (November 7): 22.

2104 • 2005 Khan, J.A. Disaster epidemiology and surveillance. *Indian Journal for Practising Doctors*, 2 (4).

2105 • 2005 Kumar, R. *et al.* Earthquake induced structures in Pinjor formation of Nadah area (Himalaya). *Journal of Geological Society of India*, 65 (3): 346–352.

2106 • 2005 Maheshwari, B.K. *et al.* Structural damages on the coast of Tamil Nadu due to tsunami caused by 26 December 2004 Sumatra earthquake. *ISET Journal of Earthquake Technology*, 42 (2–3): 63.

2107 • 2005 Malhotra, A.K. *et al.* Mental health problems following disaster: Management in primary care settings. *Indian Journal for Practising Doctors*, 2 (4).

2108 ■ 2005 Malik, J.N. and V. Joshi. Landscape change in Andaman and Nicobar Islands due to M 9.3 tsunamigenic Sumatra earthquake of 26 December 2004. *Science and Culture*, 88 (9):1384–1386.

2109 • 2005 Moszynski, P. Kashmir crisis is worse than the Asian tsunami, says World Health Organization. *British Medical Journal*, 326: 331–926.

2110 ■ 2005 Park, J. *et al.* Earth's free oscillations excited by the (December 26) 2004 Sumatra-Andaman earthquake. *Science*, 308 (5725): 1139–1144.

2111 • 2005 Pawar, A. T. *et al.* Rapid assessment survey of earthquake affected Bhuj block of Kachchh district . *Indian Journal of Medical Sciences*, 59 (11): 488–494.

2112 • 2005 Raghvebdra S. *et al.* Analysis of earthern dams affected by the 2001 Bhuj earthquake. *Engineering Geology*, 80 (3–4): 282–291.

2113 ■ 2005 Rajput, S. *et al.* Rupture parameters of the 1999 Chamoli earthquake in the Garhwal Himalayas: Constraints from aftershocks and change in failure issues. *Tectonophysics*, 404 (1–2): 23–32.

2114 • 2005 Reddy, B. M. Himalayan quake. *Frontline*, 22 (22): 3–8.

2115 • 2005 Reddy, B. M. Remains of the day.*Frontline*, 22 (22): 11–13.

2116 • 2005 Roy, N. *et al.* Surgical and psycho-social outcomes in the rural injured: A follow up study of the 2001 earthquake victims. *Applied Geography*, 36 (8): 927–934.

2117 • 2005 Saatcioglu, M. *et al.* Effects of the 26 December 2004 Sumatra earthquake and tsunami on physical infrastructure. *ISET Journal of Earthquake Technology*. 42 (4): 79–94.

2118 ■ 2005 Saraf, A. K. and S. Choudhury. National oceanic and atmospheric administration/ advanced very high resolution radiometer detects thermal anomaly associated with the 26 January 2001 Bhuj earthquake, Gujarat (India). *International Journal of Remote Sensing*, 26 (6): 1065–1073.

2119 • 2005 Special Correspondent. Paradise lost: Kashmir quake. *India Today*, (October 24): 48–57.

2120 • 2005 Special Correspondent. After the earth moved. *Science*, 307: 670.

2121 • 2005 Swami, P. The cold reality. *Frontline*, 22 (22): 14–16.

2122 • 2005 Swami, P. Grief without borders. *Frontline*, 22 (22): 17–19.

2123 • 2005 Swami, P. Terror amidst death. *Frontline*, 22 (22): 20–21.

2124 ■ 2005 Wallace, K. *et al.* Surface deformation in the region of the 1905 Kangra M 7.8 earthquake in the period of 1846–2001. *Geophysical Research Letters*, 32 (15): L15307.

2125 • 2005 Ways, M. Human losses expected in Himalayan earthquakes. *Natural Hazard Journal*, 34 (3): 305–314.

2126 • 2006 Gohil, J.R. Vanished smile, stray dogs, thalassemia and preterm transport: My experience of Gujarat earthquake. *Indian Journal of Pediatrics*, 73 (7): 633.

2127 • 2006 Jayant, T.V. Himalayan hazard. *Down to Earth*, 15 (1): 35.

2128 • 2006 Kapoor, A. *2001 Bhuj Earthquake: A Geographical Study of Disaster and Response.* Delhi: Department of Geography, Delhi School of Economics, University of Delhi (Unpublished Ph.D thesis).

2129 ■ 2006 Kumar, V.K. *et al.* Mapping damage in the Jammu and Kashmir caused by 8 October 2005 M 7.3 earthquake from the Cartosat-1 and Resourcesat-1 imagery category original. *International Journal of Remote Sensing*, 27 (20): 4449–4459.

2130 • 2006 Mitra, M.N. Sound of silence. *Down to Earth*, 15 (8): 38.

2131 • 2006 Rai, D.C. and C.V.R. Murthy. Effects of the 2005 Muzaffarabad (Kashmir) earthquake on built environment. *Current Science*, 90 (8): 1066–1070.

2132 ■ 2006 Rajendran, C.P. The style of crustal deformation and seismic history associated with the 2004 Indian Ocean earthquake: A perspective from the Andaman-Nicobar Islands. *Bulletin of the Seismological Society of America*, 96.

2133 ■ 2006 Rajendran, C.P. *et al.* The deformational characteristics and the recurrence pattern of the 26 December 2004 Indian Ocean earthquake. In *Proceedings of the Chapman Conference on Active Tectonics and Seismic Potential of Alaska.* Aleyeska.

2134 • 2006 Seismology Division, Department of Science and Technology. *A Consolidated Report on Global Positioning System Measurements of Coseismic Displacement due to 26 December 2004 Sumatra-Andaman Earthquake and their Analysis.* New Delhi: Government of India.

Impact: Flood

2135 ■ 1903 Inglis, J.W. *Narrative of the Principal Events Connected with Flood Embankments in Orissa and of the Origin and Development of the Orissa Canals, 1840–1990.* Kolkata: Bengal Secretariat Press.

2136 • 1931 Lahiri, R. Aftermath of North Bengal floods. *Modern Review*: 50 (5): 568–570.

2137 • 1933 Chaudhury, J. Orissa's grave situation. *Modern Review*: 54 (4): 436–438.

2138 • 1935 Commissariat, M.S. Famine, flood and pestilence in Gujarat during Aurangzeb's reign. In M.S. Commissariat (ed). *Studies in the History of Gujarat.*

2139 • 1937 Special Correspondent. Flood problem of Orissa. *Science and Culture*, 3 (10): 552–554.

2140 • 1953 Government of Bihar. *Preliminary Report on Floods in Bihar.* Patna.

2141 ▪ 1959 Bamzai, P.K. Flood problem in Kashmir. *Modern Review*. 106 (4): 307–309.

2142 • 1960 Economic Commission for Asia and Far East. *Final Report of the West Bengal Flood Enquiry Committee.* Kolkata.

2143 • 1962 Gulati, O.D. *Study of Rohtak Town with Special Refernce to Floods.* New Delhi: Department of Urban and Regional Planning, School of Planning and Architecture (Unpublished Postgraduate Dissertation).

2144 • 1964 Handa, C.L. The annual menace of floods. *Indian Journal of Power and River Valley Development*, 14 (11): 43–44.

2145 • 1967 Special Correspondent. Orissa: A heavy toll. *Economic and Political Weekly*, 2 (49): 1976.

2146 • 1968 Singhota, J.S. Flood control and assessment of agricultural damages. *Indian Journal of Power and River Valley Development*, 18 (1): 8.

2147 • 1970 Special Correspondent. Bihar: Cruel month. *Economic and Political Weekly*, 5 (17): 733–738.

2148 • 1971 Special Correspondent. Flood damage and protection. *Bhagirath*, (January 18): 1–3.

2149 • 1974 Special Correspondent. Maharashtra: Floods in Nanded. *Economic and Political Weekly*, 10 (43): 1654.

2150 • 1976 Special Correspondent. The floods. *India Today*, (June 5): 8.

2151 ▪ 1978 Borthapen, D.N. Problems of soil erosion and floods in the North-Eastern Region. In *Proceedings of the National Workshop on the Integrated Development of the Ganga Brahmaputra Barak Basin.* New Delhi:Gandhi Peace Foundation.

2152 • 1978 Chaterjee, P. West Bengal: In Birbhum after the floods. *Economic and Political Weekly*, 13 (40): 2001–2002.

2153 • 1979 Sain. K. *The Problem of Flood in India.* New Delhi: Birla Institute of Scientific Research.

2154 • 1980 Kayastha, S.L. and R.P. Yadav. Flood hazards in Lower Ghaghara plain (Uttar Pradesh): A case study in perception and impact on socio-economic development. *National Geographical Journal of India*, 26 (1–2): 19–28.

2155 ▪ 1982 Rao, M. Flood intensity and land evaluation of Cauvery delta (Tamil Nadu): Applying remote sensing techniques. *Photonirvachak, Journal of the Indian Society of Remote Sensing*, 10 (3): 23–30.

2156 • 1983 Kolarkar, A.S. *et al.* Impact of flash flood on agricultural land in Jaipur region of Rajasthan. *Journal of Indian Society of Soil Science*, 10: 580–586.

2157 • 1983 Menon, R. Flood of sorrow. *India Today*, (July 31): 46–48.

2158 ■ 1984 Amminaidu, E. *Environmental Impact of Flood: Regional Analysis of Vamsadhara River Basin.* Vishakhapatnam: Andhra Univeristy, Waltair (Unpublished Ph.D thesis).

2159 • 1984 Chitale, S.V. Kosi: The problem river of North Bihar. *Journal of Irrigation and Power*, 41: 197–202.

2160 • 1984 Rao, G.C. Damages from floods and cyclones in India. In B.R. Naidu (ed). *Disaster Management.* Tirupati: Sri Venkateswara University. 77–86.

2161 • 1984 Reddy, S.R. Flood damages of Vamsadhara river: An economic evaluation. In B.R. Naidu (ed). *Disaster Management.* Tirupati: Sri Venkateswara University. 55–62.

2162 • 1985 Ali, S.H. Tragedy of Kosi. *Economic and Political Weekly*, 20 (18): 794–795.

2163 • 1985 Kayastha, S.L. Flood induced population migration in India: A case study of Ghaghara Zone.In A.L. Kosinski and K.M. Etahi (eds). *Population Redistribution and Development in South Asia.* Jaipur: Rawat Publications. 79–88.

2164 ■ 1985 Sen, P.K. Flood problems: A challenge to environmental setting of a region. In H.H. Singh, *et al* (eds). *Geography and Environment: Issues and Challenges.* New Delhi: Concept Publishing Company. 77–90.

2165 • 1985 Special Correspondent. Nature's fury: A deadly deluge. *Frontline*, (Nov): 49–53.

2166 • 1985 Special Correspondent. State of siege. *India Today*, (March 8): 58–71.

2167 ■ 1986 Rath, B. and A. Tripathy. Flood damage estimates of the Rengali Multi-Purpose Project: Some methodological and conceptual issues. In *Proceedings of the Workshop on Flood Damage Assessment, in the Himalayan Region.*New Delhi: Central Board of Irrigation and Power. 71–79.

2168 • 1986 Venkateshwara, M. Caught in the swirl. *Frontline*, (September 16): 40–41.

2169 • 1987 Central Board of Irrigation and Power. *Flood Damage Assessment.* Delhi: Government of India.

2170 ■ 1987 Centre for Science and Environment. *The Wrath of Nature: Impact of Environmental Destruction on Floods and Drought.* New Delhi:

2171 • 1987 Menon, R. *et al.* Death and devastation. *India Today*, (September 30): 74–76.

2172 • 1987 Rao, D.K. Bearing the brunt. *Frontline*, 20 (September 6).

2173 • 1987 Venkataramani, G. A river of sorrow. *Frontline*, 20 (December 26): 42– 43.

2174 • 1988 Basu, S. and S. Santra. Flood problem in Howrah district. *Geographical Review of India*, 50 (4): 69–74.

2175 • 1988 Chawla, S.S. Punjab overwhelmed. *Frontline*, 21 (October 15): 23–34.

2176 • 1988 Mudgal, V. Watery woes. *India Today*, (October 31): 69–73.

2177 • 1988 Padmanathan, B.S. The fury of the floods. *Frontline*, 21 (October 15): 20–22.

2178 • 1988 Ramachandra, C.M. Bangalore: After the deluge. *Frontline*, 21 (Oct 1): 97–99.

2179 • 1988 Special Correspondent. Assam: Ordeal by water. *Frontline*, 21 (June 25): 105.

2180 • 1989 Goswami, D.C. Floods and their impact on agriculture of Assam. In P.C. Goswami (ed). *Agriculture in Assam*. Guwahati: Institute of Development Studies. 191–207.

2181 ■ 1989 Mookerjee, S. The flood hazard. *Science and Culture*, 55 (9): 345–353.

2182 • 1989 Prabhakara, M.S. Brahmaputra billows on. *Frontline*, 22 (Sept. 2): 41–51.

2183 • 1989 Singh, A.N. Assessing extent of damage caused by flooding and drought in dominantly rice cropland area using landsat data. In P. Singh (ed). *Problems of Wasteland and Forest Ecology in India*. New Delhi: Ashish Publishing House. 45–56.

2184 • 1989 Special Correspondent. Ravaged Andhra Pradesh. *Frontline*, 22 (August 19): 107–108.

2185 • 1990 Alam, M.J. Floods play havoc in Bihar. *New Age*, 38 (31): 11.

2186 ■ 1990 Pandit, A.K. and S.S. Quadri. Floods threatening Kashmir wetlands. *Journal of Environmental Management*, 31 (4): 299–311.

2187 • 1990 Satya, R. *Impact of Flood on Agriculture and Human Adjustments in Badlapur and Khurtalian Blocks (District Jaunpur)*. Varanasi: Banaras Hindu University (Unpublished Ph.D thesis).

2188 • 1990 Singh, S. *Effects of the 1988 Floods on the Rural Landscape of District Gurdaspur (Punjab State)*. Chandigarh: Punjab University (Unpublished Master of Philosophy Dissertation).

2189 • 1991 Center for Science and Environment. Nightmare in North Bihar. In *State of India's Environment, Third Citizen's Report: Floods, Flood Plains and Environmental Myths*. New Delhi. 95–25.

2190 • 1991 Center for Science and Environment. Annual mayhem in Assam. In *State of India's Environment, Third Citizen's Report: Floods, Flood Plains and Environmental Myths*. New Delhi. 71–94.

2191 ■ 1991 Centre for Science and Environment. The wrath of nature: The impact of environmental destruction on floods and drought. In *State of India's Environment, Third Citizen's Report: Floods, Flood Plains and Environmental Myths*. New Delhi.

2192 • 1991 Dey, S. and S. Mukherjee. Flood problem in North Bengal. *Indian Journal of Power and River Valley Development*, 41 (10): 193–195.

2193 • 1991 Pathak, M. *Flood Plains and Agricultural Occupance.* New Delhi: Deep and Deep Publications Pvt. Ltd.

2194 • 1991 Pereira, A. Devastating deluge. *India Today*, (October 15): 54.

2195 ■ 1991 Sen, P.K. Flood hazards and river bank erosion in the Lower Damodar basin. In S. Sharma (ed). *Indian Geomorphology.* New Delhi: Concept Publishing Company. 95–108.

2196 ■ 1991 Singh, A. *Environmental Degradation in Flood Prone Areas.* Jaipur: Pointer Publishers.

2197 • 1991 Yashwant, S. Mourning Mowad: Flood havoc in a Maharashtra village. *Frontline*, 24 (August 17): 37–39.

2198 • 1992 Krishna Kumar, R. Nature strikes back: The havoc in Kerala. *Frontline*, 25 (November 20–December 3): 86–88.

2199 • 1992 Mohan, M.M. Problem of plenty in Karnataka: A quirk of nature. *Frontline*, 25 (December 18–31): 128.

2200 • 1992 Sen, N. Impact of flood on crop damages: A case study of Lower Damodar valley region. *Indian Journal of Regional Science*, 24 (2): 65–72.

2201 • 1992 Thakural, E. *Rivers of Sorrow, Rivers of Change.* Delhi: Sage Publications.

2202 • 1993 Bhattacharjee, R. A journey of Majauli. *Frontline*, 26 (September 24): 66–74.

2203 ■ 1993 Chadha, S.K. Floods in India: An environmental concern. In S.K. Chadha (ed). *Understanding Environment.* Delhi: Amar Prakashan, 27–32.

2204 • 1993 Chadha, S.K. Fury of floods in Jammu and Kashmir state. In S.K. Chadha (ed). *Understanding Environment.* Delhi: Amar Prakashan. 181–184.

2205 • 1993 Mohan, M.M. The flood fury: A dream in Karnataka. *Frontline*, 26 (December 3): 129.

2206 • 1993 Patel, G. and S. Lokadhikar. Taken at the flood. *Down to Earth*, 2 (1): 13.

2207 • 1993 Singh, N. *et al.* Flash flood damage mapping in arid environment using satellite remote sensing: A case study of Pali region. *Photonirvachak, Journal of the Indian Society of Remote Sensing*, 21 (2): 75–86.

2208 • 1994 Mishra, P.R. Flooded out. *Down to Earth*, 3 (5): 11.

2209 ■ 1995 Gopalakrishnan, V. Impact of flood on environment ecology and population changes. *Journal of Human Ecology*, 6 (4): 255–260.

2210 ■ 1995 Sharma, K.K. Flood devastation: Politically volatile issue in Haryana. *New Age*, 43 (44): 4.

2211 ■ 1996 Ely, L.L. *et al.* Change in the magnitude and frequency of late Holocene monsoon floods on the Narmada river (Central India). *Bulletin of American Geological Soceity*, 108: 1134–1148.

2212 • 1996 Rao, M.V. Flood havoc: Death and devastation in Andhra Pradesh. *Frontline*, 29 (November 15): 32–33.

2213 • 1996 Sridhar, B. Flash floods: Unseasonal rain batters Chennai. *Frontline*, 29 (July 12): 117–118.

2214 • 1997 Bordoloi, N.K. *Impact of the Brahmaputra Flood and Erosion Hazard on Flood Plain Occupance in the Palasbari-Nagarpur Tract of Kamrup district.* Guwahati: Gauhati University (Unpublished Ph.D thesis).

2215 • 1997 Chakraborty, C. Floods, landslides and tidal waves affect West Bengal. *New Age*, 45 (38): 3.

2216 • 1997 Gogai, P. Monsoon mayhem. *Down to Earth*, 6 (6): 12.

2217 ■ 1997 Gupta, G.K. *et al.* An appraisal of flood and waterlogging hazards in Eastern Uttar Pradesh. In *Proceedings of the Symposium on the Earth Sciences in Environmental Assessment and Management.* Special Publication No. 48 (2). Kolkata: Geological Survey of India. 69–74.

2218 ■ 1997 Kumar, S. *et al.* Effect of floods on rural roads in Bihar. In T. Iwasaki *et al.* (eds). *Proceedings of the Group Seminar on Natural Disaster Reduction for Roads.*

2219 • 1997 Praveen, S. Crisis in the countryside. *Frontline*, 30 (October 17): 65–70.

2220 • 1997 Sharda, D. *et al.* Modelling flash floods hazard to a railway line: A geographic information systems approach. *Geocarto International*, 12 (3).

2221 • 1997 Swami, P. Northern misery in Rajasthan, Haryana and Jammu and Kashmir. *Frontline*, 30 (July 26–August 8).

2222 • 1997 Swami, P. Kashmir: Crisis in the countryside. *Frontline*, 30 (October 17): 65.

2223 • 1997 Venkatesan, V. Monsoon torrents. *Down to Earth*, 6 (8): 47–48.

2224 • 1998 Chaudhuri, K. *et al.* Swamped by floods. *Frontline*, 31 (October 9): 69–73.

2225 ■ 1998 Gupta, A. Geomorphological effects of floods in Indian rivers. In V.S. Kale (ed). *Flood Studies in India.* Bangalore: Memoirs of the Geological Society of India, No. 41, 143–154.

2226 • 1998 Jamwal, N. *Regional Patterns of Flood Caused Degradation and its Socio-economic Impact in Northwestern Himalaya from 1950–1992: A Geomorphological Study.* Jammu: Jammu University (Unpublished Ph.D thesis).

2227 • 1998 Singh, V.K. and R.K. Patel. A case study of socio-economic impact of floods in Haryana: A case study. In B.N. Verma (ed). *Flood Devastation and Agricultural Development in Eastern India.* Delhi: B.R. Publishing Corporation. 109–114.

2228 ■ 1998 Verma, B. N. *et al.* Eastern Region: The playground of floods. In B.N. Verma (ed). *Flood Devastation and Agricultural Development in Eastern India.* Delhi: B.R. Publishing Corporation. 25–40.

2229 • 1998 Verma, B. N.(eds). *Flood Devastation and Agricultural Development in Eastern India.* Delhi: B R Publishing Corporation.

2230 ■ 1999 Ahuja, B. S. Dimension of the problem-floods. *Shelter,* special issue, 5.

2231 • 1999 Biswas, R. *et al.* A community based study on health impact of flood in a vulnerable district of West Bengal. *Indian Journal of Public Health*, 43 (2): 89–90.

2232 • 1999 D' Souza, R. Floods: The problem. *Seminar*, 478: 12–17.

2233 • 1999 Gojoi, B. *Impact of Flood on Human Occupance in Sadiya Region, Assam.* Guwahati: Gauhati University (Unpublished Postgraduate Dissertation).

2234 • 1999 Krishnakumar, R. The crisis of Kosi. *Frontline*, 16 (20): 70.

2235 • 1999 Krishnakumar, R. The Kosi untamed. *Frontline*, 16 (20): 65–70.

2236 • 1999 Mitra, A. Kolkata Diary. *Economic and Political Weekly*, 34 (33): 3411.

2237 • 1999 Mitra, A. Kolkata Diary. *Economic and Political Weekly*, 34 (49): 3410–3411.

2238 • 1999 Tiwari, M. Breach of trust: The shifting sands, state of despair, no way out. *Down to Earth*, 8 (13): 30–45.

2239 • 2000 Agarwal, A. Now it is flood. *Down to Earth*, 9 (9): 4.

2240 ■ 2000 Chakraborti, A. K. Satellite remote sensing for near real time flood and drought impact assessment: Indian experience. In P.S. Roy, *et al* (eds). *Natural Disasters and their Mitigation: A Remote Sensing and Geographic Information Systems Perspective.* Dehradun: Indian Institute of Remote Sensing. 143–150.

2241 • 2000 Chakraborty, C. West Bengal flood devastation. *New Age*, 48 (43): 3.

2242 • 2000 Editorial. Bengal floods kill 893. *Down to Earth*, 9 (11): 11.

2243 • 2000 Singh, J. Flood hazard in the Middle Ganga plain: Impact, causes and prevention. In R.B. Singh (ed). *Disaster Management.* Jaipur: Rawat Publications, 84–93.

2244 • 2000 Sinha, A. Bihar floods: Captured on films. *Down to Earth*, 9 (8): 59.

2245 • 2000 Sur, D. *et al.* Severe cholera outbreak following floods in a northern district of West Bengal. *Indian Journal of Medical Research*, 112: 178–182.

2246 • 2001 Chattopadhyay, S. Flood fury in Orissa. *Frontline*, 18(16): 33–36.

2247 • 2001 Chattopadhyay, S. S. Living on the edge. *Frontline*, 18 (16): 35.

2248 • 2001 Das, B. Changing lives. *Down to Earth*, 10 (6): 20–24.

2249 • 2001 Fluitman, F. Orissa floods: 70 dead, over one crore affected: Government fails miserably in managing crisis. *New Age*, 49 (30): 13.

2250 • 2001 Mondal, N.C. *et al.* A risk factor of diarrhoea among flood victims: A controlled epidemiological study. *Indian Journal of Public Health*, 45 (4): 122–127.

2251 • 2001 Special Correspondent. Digha in troubled waters. *Down to Earth*, 10 (Dec.31): 15.

2252 • 2001 Special Correspondent. Flood fury. *Down to Earth*, 10 (10): 32–33.

2253 • 2002 Bavadam, L. Flood fury in Thane. *Frontline*, 19 (15): 40.

2254 ■ 2002 Bortamula, M. *Impact of Flood on Biodiversity in Kaziranga National Park*. Delhi: Department of Geography, Delhi School of Economics, University of Delhi (Unpublished Master of Philosophy Dissertation).

2255 • 2003 Editorial. Flood fury. *Down to Earth*, 12 (14): 17.

2256 ■ 2003 Jain, V. and R. Sinha. Geomorphological manifestations of flood hazards. *Geocarto International*, 18 (4): 51–60.

2257 ■ 2003 Kale, V.S. Geomorphic effects of monsoon floods on Indian rivers. *Natural Hazard Journal*, 28 (1): 65–84.

2258 • 2003 Singh, N.K. *Flood Impact and its Management in Bhagalpur District (Bihar).* Delhi: Department of Geography, Delhi School of Economics, University of Delhi (Unpublished Master of Philosophy Dissertation).

2259 • 2004 Chattopadhyay, S.S. Battling Bihar. *Frontline*, 21 (16): 46–47.

2260 • 2004 Chattopadhyay, S.S. Bengal's bane. *Frontline*, 21 (16): 47–48.

2261 • 2004 Talukdar, S. The tears of Assam. *Frontline*, 21 (16): 43–45.

2262 • 2005 Bobb, D. *et al.* Mumbai collapse. *India Today*, (August 8): 18–26.

2263 • 2005 Editorial. Mumbai: After the deluge. *Economic and Political Weekly*, 40 (32): 3555–3556.

2264 • 2005 Goyari, P. Flood damages and sustainability of agricuture in Assam. *Economic and Political Weekly*, 40 (26): 2723–2729.

2265 • 2005 Katakam, A, *et al.* High water and hell. *Frontline*, 22 (17): 4–11.

2266 • 2005 Menon, M. and A.K. Bhatt. The rural distress. *Frontline*, 22 (17): 20–24.

2267 • 2005 Sharma, R. and P. Menon. Bangalore's woes. *Frontline*, 22 (23): 24–26.

2268 • 2005 Viswanathan, S. Flooded out. *Frontline*, 22 (23): 20–23.

2269 • 2005 Viswanathan, S. Deluge and after. *Frontline*, 22 (26): 25–28.

2270 • 2006 Kshirsagar, N.A. *et al.* Floods in Mumbai: Impact of public health services by hospital staff and medical studies. *Journal of Postgraduate Medicine*, 52 (4):

2271 • 2006 Pegu, R. Drowned. *Down to Earth*, 14 (22): 50.

2272 • 2006 Thakurdesai, P. and B. Malini. Maroon drive. *India Today* (July): 26–27.

Impact: Heat and Cold Wave

2273 • 1999 Special Correspondent. Heat and death. *Down to Earth*, 8 (2): 13.

Impact: Heavy Rain

2274 • 1984 Naidu, K. M. Disaster management: Evaluation of damages due to heavy rains and floods in Andhra Pradesh. In B. R. Naidu (ed) *Disaster Management.* Tirupati Sri Venkateswara University. 63–70.

2275 • 1990 M. V. R. Rain village. *Frontline*, 23 (August 19): 109.

2276 • 1991 Venkateshwarlu, K. The direct hit. *Frontline*, 24 (May 26): 10–13.

2277 • 1997 Joshi, B. C. and J. Bhagwan. Cloudburst incidence in Bilaspur division of Himachal Pradesh. In T. Iwasaki, *et al.* (eds). *Proceedings of the Group Seminar on Natural Disaster Reduction for Roads*, 144–147.

2278 • 2000 Editorial. Rain claims 120 lives. *Down to Earth*, 9 (6): 13.

2279 • 2005 Bunsha, D. Rain havoc in Gujarat. *Frontline*, 22 (15): 34–36.

2280 • 2005 Ramachandran, R.Of clouds, winds and precipitation. *Frontline*, 22(17): 14–15.

2281 • 2005 Vaidyanathan, S. *et al.* Tragedy of agriculture. *Frontline,* 22 (26): 28–30.

2282 • 2006 Menon, P. Hopes and fears. *Frontline*, 22 (27): 4–5.

Impact: Landslide

2283 ■ 1949 Puri, G. S. The problem of land erosion and landslips in the Hoshiarpur Shiwaliks. *Indian Forester*, 75: 45–51.

2284 ■ 1952 Mukherjee, S. M. Landslides and sounds due to earthquakes in relation to upper atmosphere. *Indian Journal of Meteorology and Geophysics*, 3(4): 240–257.

2285 ■ 1967 Didwal, R. S. Nashri landslide. *Science and Culture*, 33 (4): 193–194.

2286 ■ 1975 Prasad, R. C. The landslides and erosion problem with special reference to the Kosi catchment. In *Proceedings of the Seminar on Landslides and Toe Erosion Problems with Special Reference to Himalayan Region.* Lucknow: Indian Society of Engineering Geology, 4.

2287 • 1978 Awasthi, D. Nainital sliding down. *India Today*, (December 15): 174.

2288 ■ 1979 Ravindra, R. and R. S. Negi. Observations on Alakananda tragedy of 1970. *Himalayan Geology*, 9 (2): 724–733.

2289 ■ 1979 Ray, S. *Geological Note on the Problem of Slope Stability at 30 kilometre along Shillong-Cherrapunji Road.* Kolkata: Geological Survey of India (Unpublished Report).

2290 ■ 1979 Saxena, P.B *et al.* Study of landsliding and land depletion in the Alaknanda valley of the Garhwal Himalaya. *Himalayan Geology*, 9 (2): 716–723.

2291 ■ 1980 Ghosh, D.K. Geological influence of landslides and other mass movements in the planning of Baira-Baledh Siul Link Project (Himachal Pradesh). In *Proceedings of the International Symposium on Landslides*. New Delhi: Sarita Pradhan. 19–27.

2292 ■ 1980 Prasad, C. and B.K. Verma. Slope-failure: A case study in the catchment areas of Alkananda and Bhagirathi Rivers. In *Proceedings of the International Symposium on Landslides*. New Delhi: Sarita Pradhan. 41–44.

2293 ■ 1986 Mukherjee, A.B. Landslides and man: Development and destruction in Western Himalaya. In M. Shafi and M. Raza (eds). *Spectrum of Modern Geography*. New Delhi: Concept Publishing Company. 397–428.

2294 ■ 1989 Loshali, D.C. *Certain Hydrologic Responses of Landslide Affected Forest Sites in Kumaon Himalayas*. Nainital: Kumaon University (Unpublished Ph.D thesis)

2295 ■ 1989 Sahni, K. Landsliding in the Western Himalaya and its environmental impact. In S.K. Chadha (ed). *Ecological Hazards in the Himalayas*. Jaipur: Pointer Publishers. 57–65.

2296 • 1992 Balu, S.V. The Meghamalai tragedy and the worry over landslides. *Frontline*, 25 (September 18): 126–127.

2297 ■ 1993 Jalote, P.M. Landslide blockade of Satluj river below km 288–289 on National Highway–22, Himachal Pradesh. *Journal of Engineering Geology*, 22 (1).

2298 ■ 1993 Pal, D. and M.P.S. Bisht. Landslides as an environmental hazard in the development of hilly areas. *Himalayan Geology*, 14: 157–170.

2299 ■ 1995 Bartarya, S.K. and M.P. Sah. Landslide induced river bed uplift in the Tal valley of Garhwal Himalaya. *Geomorphology*, 12 (2): 109–121.

2300 • 1995 Nagarajan, R. *Studies on Areas Affected by Landslides Using Remote Sensed Ancillary Data*. Mumbai: Indian Institute of Technology.

2301 • 1996 Bhagat, N.C. Problem of landslide hazard in Garhwal Himalayas. In K. Senneset (ed). *Proceedings of Seventh International Symposium on Landslides*. Rotterdam: Balkema. 1853–58.

2302 • 1996 Murlidharan, S. and S. Bukhari. Death on the Amarnath *yatra*. *Frontline*, 29 (September 20): 130–132.

2303 • 1996 Pradhan, S.R. Role of landslides in planning and execution of Lagyap Hydel Project, Sikkim: A case history. In *Proceedings of the International Symposium on Landslides*. New Delhi: Sarita Pradhan. 227–233.

2304 ■ 1996 Sah, M.P. The Molling slide of Kinnaur and its control on channel blocking and flash floods in the Lower Spiti valley (Himachal Pradesh). In A.P. Santhakumar, *et al.* (eds). *Proceedings of the International Conference on Disasters and Mitigation, (A-4).* Chennai: Anna University, 102–106.

2305 ■ 1996 Sah, M.P. *et al.* Slope failure and damming of river channels: Some examples from Satluj valley (Himachal Pradesh). *Himalayan Geology*, 17: 183–192.

2306 • 1996 Sair, J.S. Landslide problem on roads in Sikkim and North Bengal and measures adopted to control them. In *Proceedings of the International Symposium on Landslides.* New Delhi: Sarita Pradhan. 239–244.

2307 ■ 1997 Tolia, D.S. and K. Kumar *et al.* Disastrous landslides near Kullu town in Himachal Pradesh. In T. Iwasaki, *et al.* (eds). *Proceedings of the Group Seminar on Natural Disaster Reduction for Roads.* 95–100.

2308 ■ 1997 Unnikrishnan, P.V. Kerala landslides in a fragile ecosystem. *Economic and Political Weekly*, 32 (39):2445.

2309 • 1998 Ahmed, F. Hills of death. *India Today*, (April 12): 33–34.

2310 • 1998 Editorial. Landslide of terror. *Down to Earth*, 7 (8): 15–16.

2311 ■ 1998 Joshi, M.A. *Geographical Analysis of Slope Instability and its Environmental Impact in Saran Catchment of Almora District.* Nainital: Kumaon University (Unpublished Ph.D thesis).

2312 • 1998 Ramakrishnan, V. Survival battles: Uttar Pradesh landslides. *Frontline*, 31 (September 25): 43–45.

2313 • 1999 Bist, K.S. and M.P. Sah. The devastating landslides of August 1998 in Ukhimath area, Rudraprayag district . *Current Science*, 76 (4): 481–484.

2314 ■ 1999 Chattopadhyay, A. *A Study of Landslides and their Impact on Land use in Darjeeling Hills.* Burdwan: University of Burdwan (Unpublished Ph.D thesis).

2315 • 2000 Paul, S.K. *et al.* Catastrophic mass movement of 1998 monsoons at Malpa in Kali Valley, Kumaon Himalaya (India). *Geomorphology*, 35 (3–4):169–180.

2316 ■ 2001 Kandpal, G.C. *et al.* Problem of landslide/mass movement in parts of Sikkim Himalaya: A discussion. In *Proceedings of the Role of Earth Sciences in Integrated Development and Related Societal Issues.* Special Publication No. 65 (2). Kolkata: Geological Survey of India: 91–95.

2317 • 2001 Srivastava, D. *et al.* Avalanche impact around Kedarnath Township in Mandakini valley, Garhwal Himachal (Uttaranchal). In *Proceedings of the Role of Earth Sciences in Integrated Development and*

Related Societal Issues. Special Publication No. 65 (3). Kolkata: Geological Survey of India: 43–48.

2318 ■ 2001 Tolia, R.S. Landslide problems in the Uttarakhand region. In P. Sahni, *et al* (eds). *Disaster Mitigation: Experiences and Reflections.* New Delhi: Prentice-Hall of India Private Limited, 39–47.

2319 • 2003 Bhandari, R.K. Two great landslide tragedies of India. In H.K. Gupta (ed). *Disaster Management.* Hyderabad: University Press (India) Private Limited, 110–127.

2320 • 2003 Jayant, T.V. Debris flow. *Down to Earth*, 12 (11): 49–52.

2321 • 2003 Mishra, S. The angry hills. *India Today*, (October 27): 67–68.

2322 • 2003 Sah, M.P. *et al.* Cloudburst of 10 August 2002 and related landslides and debris flows around Budha Kedar (Thati Kathur) in Balganga valley (District Tehri). *Himalayan Geology*, 24 (2): 87–102.

2323 • 2004 Tripathi, P.S. Debris danger. *Frontline*, 21 (16): 49–52.

2324 • 2004 Uniyal, A. Landslides at Karnaprayag: Another Uttarkashi in the making. *Current Science*, 87 (8): 1031–1033.

2325 • 2005 Bukhari, S. Death under snow. *Frontline*, 22 (6): 43–46.

2326 • 2006 Pande, R.K. Landslide problems in Uttaranchal, India: Issues and challenges. *Disaster Prevention and Management*, 15 (2): 247–255.

Impact: Storm

2327 ■ 1976 Special Correspondent. North: Spin offs of hailstorm. *Economic and Political Weekly*, 11 (17): 622.

2328 ■ 1987 Dhar, O.N. and B.N. Mandal. Brief appraisal of severe rainstorms of Madhya Pradesh region for the optimum development of its water resources. *Indian Journal of Power and River Valley Development*, 37 (3 & 4): 88–93.

2329 • 1992 Narayanan, R.Y. Storm fury in southern district. *Frontline*, 25 (Dec. 18): 122.

Impact: Tsunami

2330 ■ 2005 Aglionby, J. and A. Meldrum. Deadly wall of water. *Frontline*, 22 (Jan): 12.

2331 • 2005 Balaraman, K. Malaria receptivity in the tsunami hit coastal villages of Southern India. *The Lancet*, 5 (9): 531–532.

2332 • 2005 Balaraman, K. *et al.* Risk of outbreak of vector-born diseases in the tsunami hit areas of Southern India. *The Lancet*, 5 (3): 128–129.

2333 • 2005 Bindra, S. *Tsunami: 7 Hours that Shook the World.* New Delhi: Harper Collins Publishers, India-Joint Venture with India Today Group.

2334 • 2005 Bunch, M. *et al.* Research in turbulent environments: Slums in Chennai, India, and the impact of the December 2004 tsunami on an eco-health project. *Eco-Health*, 2 (2): 150–154.

2335 ■ 2005 Chadha, R.K. The tsunami of the Great Sumatra earthquake of M 9.0 on 26 December 2004: Impact on the East Coast of India. *Current Science*, 88 (8): 1297–1300.

2336 • 2005 Chattopadhyay, S. Tsunami at Havelock: An eyewitness report. *Science and Culture*, 71 (1 & 2): 11–12.

2337 ■ 2005 Coyalkar, L.N. Perturbing. *Down to Earth*, 13 (20): 17.

2338 ■ 2005 Editorial. Waves of fury. *Economic and Political Weekly*, 40 (1): 3–4.

2339 • 2005 Editorial. Tsunami: The killer wave. *Journal of Indian Medical Association*, 75 (1).

2340 ■ 2005 Gokhale, V.A. Analytical study of living environment in the tsunami-affected areas of Tamil Nadu. *ISET Journal of Earthquake Technology*,42 (4): 219–226.

2341 • 2005 Greenhough, B. *et al.* Geographical encounters with the Indian Ocean tsunami. *Geographical Journal*, 171 (4): 369–371.

2342 • 2005 Gupta, R. Tsunami unravelled. *Down to Earth*, 14 (3): 38.

2343 • 2005 Jana, B. *Unprecedented Earthquakes and Dreadful Tsunami.* New Delhi: Mittal Publication.

2344 ■ 2005 Jayapal, P. *Geohydrographic Impact and Issues Post-Tsunami (India).* New Delhi: Advanced Professional Programme in Public Administration, Indian Institute of Public Administration (Unpublished Master of Philosophy Dissertation).

2345 • 2005 Jayaraman, K.S. India's nuclear debate hots after tsunami flood reactor. *Nature*, 433 (675).

2346 • 2005 Kar, S. The giant Asian tsunami. *Science and Culture*, 71 (1& 2): 7–10.

2347 • 2005 Krishnakumar, A. Orphans of waves. *Frontline*, 22 (2): 127–128.

2348 ■ 2005 Krishnakumar, R. An encounter at Kanyakumari. *Frontline*, 22 (2): 129.

2349 • 2005 Krishnakumar, R. A different battle at Colachal. *Frontline*, 22 (2): 13–14.

2350 • 2005 Krishnakumar, R. A view from sea. *Frontline*, 22 (2): 19.

2351 • 2005 Krishnakumar, R. Those eight minutes. *Frontline*, 22 (2): 20.

2352 • 2005 Krishnakumar, R. An ongoing tragedy. *Frontline*, 22 (3): 9–14.

2353 • 2005 Krishnakumar, R. Caught unaware in Kerala. *Frontline*, 22 (2): 17–22.

2354 • 2005 Kumar, A. *Oceanic Disaster.* Delhi: Daya Publishing.

2355 • 2005 Mehta, L. The tsunami and globalised uncertainty. *Economic and Political Weekly*, 40(3): 194–195.

2356 • 2005 Moszynski, P. Disease threatens millions in wake of tsunami. *British Medical Journal*, 326: 330–359.

2357 • 2005 Mukhopadhyay, S. *Tsunami: The Biggest Bane.* Delhi: Shipra Publication.

2358 • 2005 Murdoch, J. Heritage groups inventory: Tsunami's effect on cultural sites. *Architectural Record*, 193 (2): 26.

2359 • 2005 Nambath, S. A battered archipelago. *Frontline*, 22 (2): 23–24.

2360 • 2005 Nayak, K. and A. D. Sagar. South Asian tsunami. *The Lancet*, 365 (9463): 934–935.

2361 ■ 2005 Pal, A. Unstoppable tsunami. *Down to Earth*, 14 (March 15): 4.

2362 ■ 2005 Peterson, C. *et al.* Flood elevation inundation distance and flow competence of the 2004 Sumatra-Andaman tsunami, as recorded by tsunami deposits in thirteen shore-normal profiles from the Tamil Nadu coastline (India). *ISET Journal of Earthquake Technology*, 42 (4): 95–110.

2363 • 2005 Project Directorate. *Preliminary Assessment of the Impact of Tsunami from Selected Coastal Areas of India.* Chennai: Department of Ocean Development.

2364 • 2005 Radhakrishna, B. P. O, horrible-most horrible—devastating tsunami strikes coastline of India. *Journal of Geological Society of India*, 65 (2): 129–140.

2365 • 2005 Ramachandran, R. The tsunami phenomena. *Frontline*, 22 (1): 14–16.

2366 • 2005 Ramakrishnan, D. R. S. K. Trials of killer tsunami: A preliminary assessment using satellite remote sensing technique. *Current Science*, 88 (5): 709–711.

2367 ■ 2005 Ramanamurthy, M. V. *et al.* Inundation of sea water in Andaman and Nicobar Islands and parts of Tamil Nadu coast during 2004 Sumatra tsunami. *Current Science*, 88 (11): 1736–1740.

2368 • 2005 Ramesh, R. Will to disaster: Post-tsunami technical feasibility of Sethusamudram project. *Economic and Political Weekly*, 40 (26): 2648–2653.

2369 • 2005 Ramesh, R. Pre and post tsunami: Is the Sethusamudram shipping canal project technically feasible. *Economic and Political Weekly*, 40 (4): 271–275.

2370 ■ 2005 Reddy, H. R. V. *et al.* Coastal water quality off Dakshin Kannada after tsunami. *Current Science*, 88 (7): 1026–1027.

2371 • 2005 Sambanan, V. S. Now nature's turn *Frontline*, 22 (1): 11–12.

2372 • 2005 Shekhar, M. Tsunami waves. *Yojana*, 46: 20–24.

2373 • 2005 Subramanian, T. S. Chennai's black Sunday. *Frontline*, 22 (1): 9–10.

2374 • 2005 Subramanian, T. S. Killer waves. *Frontline*, 22 (1): 4–8.

2375 • 2005 Vishwanathan, S. Fisherman at sea. *Frontline*, 22 (3): 4–6.

2376 • 2005 Viswanathan, S. Breaking barriers. *Frontline*, 22 (2): 130–131.

2377 ■ 2006 Anilkumar N. *et al.* Post-tsunami oceanographic conditions in Southern Arabian Sea and Bay of Bengal. *Current Science*, 90 (3): 421–427.

2378 ● 2006 Arya, A.S. *et al.* Some aspects of tsunami impact and recovery in India. *Disaster Prevention and Management*, 15 (1): 51–66.

2379 ● 2006 Awasthi, K. Buried relics. *Down to Earth*, 15 (3): 48–49.

2380 ■ 2006 Chakraborti, P. *et al.* Science and technology input: Earthquake and tsunami effects, Andaman Islands, India. In S. M. Ramaswamy, *et al* (eds). *Geomatics in Tsunami.* New Delhi: New India Publishing Agency. 37–62.

2381 ■ 2006 Chandrasekar, N. *et al.* Geospatial assessment of tsunami 2004 damages along the rocky coast of Kanyakumari. In S. M. Ramaswamy, *et al* (eds). *Geomatics in Tsunami.* New Delhi: New India Publishing Agency. 135–146.

2382 ● 2006 Dhanapal, T. Tsunami: The citizen's perspective. In S. M. Ramaswamy, *et al* (eds). *Geomatics in Tsunami.* New Delhi: New India Publishing Agency. 203–207.

2383 ● 2006 Jayant, T. V. Killer factor. *Down to Earth*, 15 (8): 16.

2384 ● 2006 Jayant, T. V. Still life in waves. *Down to Earth*, 14 (16): 30–39.

2385 ■ 2006 Kannan, L *et al.* Reflection of boxing day tsunami on coral reef ecosystem of India. In S. M. Ramaswamy, *et al* (eds). *Geomatics in Tsunami.* New Delhi: New India Publishing Agency. 167–177.

2386 ■ 2006 Kumar, A. *et al.* 26 December 2004 Great tsunami: It's characteristics and coastal responses in Car Nicobar Islands. In S. M. Ramaswamy, *et al* (eds). *Geomatics in Tsunami.* New Delhi: New India Publishing Agency. 155–166.

2387 ● 2006 Manu Gupta, M. and A. Sharma. Compounded loss: The post tsunami recovery experience of Indian island communities. *Disaster Prevention and Management*, 15 (1): 67–78.

2388 ● 2006 Math, S. B. *et al.* Tsunami: Psychosocial aspect of Andaman and Nicobar Islands, assessments and intervention in the early phase. *International Review of Psychiatry*, 18 (3): 233–239.

2389 ■ 2006 Natesan, U. and S. Kalaivani. Tsunami induced water quality changes in Buckingham Canal, Chennai region (India). In S. M. Ramaswamy, *et al* (eds). *Geomatics in Tsunami.* New Delhi: New India Publishing Agency. 191–198.

2390 ■ 2006 Nirupama, N. *et al.* Persistent high-water levels around Andaman and Nicobar Islands following the 26 December 2004 tsunami. *Science of Tsunami Hazards*, 24 (3): 183–193.

2391 ● 2006 Raj, M. D. Photogrammetry applications for tsunami damage assessment. In S. M. Ramaswamy, *et al.* (eds). *Geomatics in Tsunami.* New Delhi: New India Publishing Agency. 17–30.

2392 • 2006 Rajamanickam,G. V. *et al.* Damage assessment in tsunami affected Pondicherry-Cuddalore coast (India). In S. M. Ramaswamy, *et al* (eds). *Geomatics in Tsunami.* New Delhi: New India Publishing Agency. 87–106.

2393 ■ 2006 Rajendran, A. and R. A. James. Physico-chemical perturbation on coastal system by tsunami events. In S. M. Ramaswamy, *et al* (eds). *Geomatics in Tsunami.* New Delhi: New India Publishing Agency. 147–153.

2394 ■ 2006 Rajendran, P. *et al.* Bacteriological analysis of water samples from tsunami hit coastal areas of Kanyakumari district. *Indian Journal of Medical Microbiology*, 24 (2).

2395 ■ 2006 Ramalingam, M., *et al.* Tsunami: 2004 damage assessment mapping (Chennai region). In S.M. Ramaswamy, *et al* (eds). *Geomatics in Tsunami.* New Delhi: New India Publishing Agency. 77–85.

2396 ■ 2006 Ramaswamy, S.M. *et al.* Geomatics in Tsunami damages, Nagapattinam coast (India). In S.M. Ramaswamy, *et al* (eds). *Geomatics in Tsunami.* New Delhi: New India Publishing Agency. 107–134.

2397 ■ 2006 Rasheed, A. *et al.* Tsunami impacts on morphology of beaches along South Kerala coast. *Science of Tsunami Hazard*, 24 (1): 24.

2398 • 2006 Reddy, K. M. *et al.* Damage assessment and mapping of tsunami affected areas in parts of Nellore coast (India). In S.M. Ramaswamy, *et al* (eds). *Geomatics in Tsunami.* New Delhi: New India Publishing Agency. 63–76.

2399 ■ 2006 Selvam, V. Reflections of recent tsunami on mangrove ecosystem, Cuddalore coast (India). In S. M. Ramaswamy, *et al* (eds). *Geomatics in Tsunami.* New Delhi: New India Publishing Agency. 179–190.

2400 ■ 2006 Shetye, S. R. The events associated with the Great tsunami of 26 December 2004: Sea level variation and impact on coastal region of India. In *Proceedings of the Brain Storming Session on the Great Tsunami Event of 26 December 2004.* New Delhi: Department of Science and Technology, Government of India. 57–64.

2401 • 2006 Tells, S. *et al.* Emotional impact following the tsunami in endogenous people and mainland settlers in Andaman. *Indian Journal of Medical Science*, 60 (2).

2402 • 2006 Venkatachalan, C. Syndrome of the tsunami (2004) ravaged society: A microscopic view from Nagapattinam. In S.M. Ramaswamy, *et al* (eds). *Geomatics in Tsunami.* New Delhi: New India Publishing Agency. 199–202.

Impact: Assorted

2403 • 1984 Naidu, B.R. The impact of natural disaster on agriculture in developing countries. In B.R. Naidu (ed). *Disaster Management.* Tirupati: Sri Venkateswara University, 27–34.

2404 • 1990 Ramdas, S. Gujarat. *Frontline*, 23.

2405 • 1992 Chandra Kantha. Caught in a trap. *Frontline*, 25: 65–70.

2406 • 1992 Mitra, A. Kolkata diary. *Economic and Political Weekly*, 27 (27): 1363–1364.

2407 • 1992 Ramanathan, R.Without a warning. *Frontline*, 25 (December 18): 123–125.

2408 • 1992 Special Correspondent. Poor pay for ecological neglect. *Down to Earth*, 1 (June 15): 5.

2409 • 1995 Bhatt, M. *Gender and Disaster: Perspectives on Women as Victims of Disasters.* Ahmedabad: Disaster Mitigation Institute.

2410 • 1995 Biswas, I.M. *A Study of Children Affected by Disaster 1995: Archival Material Post-traumatic Stress Disorder in Children.* New York: Department of Development Support and Management Services, Bank Street Collage of Education (Unpublished Master of Philosophy Dissertation).

2411 • 1996 Gardner, J.S. Tourism and risk from natural hazard: Manali (Himachal Pradesh). In R.B. Singh (ed). *Disasters, Environment and Development.* New Delhi: Oxford and IBH Publishing Co. Pvt. Ltd. 107–120.

2412 • 1996 Singh, R.B. and S. Ghai. Environmental disasters in marginal areas: A study of road construction induced impacts in a Himalayan Yamuna Catchment. In R.B. Singh and R. Majoral (eds). *Development Issues in Marginal Regions.* New Delhi: Oxford and IBH Publishing Co. Pvt. Ltd. 311–320.

2413 • 1997 Gardner, J.S. *et al.* Tourism and risk from natural hazards in the Manali area. In F. Berkes and J.S. Gardner (eds). *Sustainability of Mountain Environments in India and Canada.* Winnipeg: University of Manitoba.

2414 • 1997 Govindaswamy, M.V. (eds). *Proceedings of National Workshop on Psychological Consequences of Disasters.* Bangalore: National Institute of Mental Health and Neurosciences.

2415 • 1997 Gupta, A.K. Effect of natural disasters on roads in India. In T. Iwasaki, *et al* (eds). *Proceedings of the Group Seminar on Natural Disaster Reduction for Roads.* 1–16.

2416 • 1998 De, U.S. and K.S. Joshi. Natural Disasters and their Impact on Developing Countries. Geneva: *World Meteorological Organisation Bulletin*, 4 (47).

2417 • 2000 Banquer, A. Disability issues. In S. Parasuraman and P.V. Unnikrishnan (eds). *India Disasters Report: Towards a Policy Initiative.* New Delhi: Oxford University Press. 51–53.

2418 ■ 2000 Bhatt, C.P. Impact of natural disasters on the environment and development: Examples from Himalayas and Eastern and Western Ghats. *Science and Culture*, 66 (1–2): 20–25.

2419 • 2000 Jaiswal, S. Disaster and mental health. *Indian Journal of Social Work*, 61(4): 521–526.

2420 • 2000 Juvva, S. and P. Rajendran. Disasater and mental health: A current perspective. *Indian Journal of Social Work*, 61(4): 527–541.

2421 • 2000 Lohokare, M. *et al.* Women in disasters and mental health. *Indian Journal of Social Work*, 61(4): 565–580.

2422 • 2000 Murthy, R.S. Psycho-social consequences. In S. Parasuraman and P.V. Unnikrishnan (eds). *India Disasters Report: Towards a Policy Initiative.* New Delhi: Oxford University Press. 54–58.

2423 ■ 2000 Unnikrishnan, P.V. and M. Martin. Internal displacement due to disaster and conflicts. In S. Parasuraman and P.V. Unnikrishnan (eds). *India Disasters Report: Towards a Policy Initiative.* New Delhi: Oxford University Press. 282–286.

2424 • 2001 Agarwal, A. A tale of two disasters. *Down to Earth*, 9: 4.

2425 ■ 2001 Bhatt, C.P. Impact of natural disasters on environment and development: Examples from Himalaya and Eastern and Western Ghats. In V.K. Sharma (ed). *Disaster Management.* New Delhi: Indian Institute of Public Administration. 15–25.

2426 • 2001 Mahapatra, R. Net bill, gross disaster. *Down to Earth*, 9 (18): 22–23.

2427 • 2002 Aubrey, J. and S. Krishnadas. Psychological issues of disasters and the aftermath: A second victimisation. *Indian Journal of Social Work*, 63 (2): 206–222.

2428 • 2002 Chatterjee, M. *Hazards in Bhubaneshwar city: A Geographical Enquiry.* Delhi: Department of Geography, Delhi School of Economics, University of Delhi (Unpublished Master of Philosophy Dissertation).

2429 • 2002 Jaswal, S. Psychological aspects of disaster: An introduction. *Indian Journal of Social Work*, 63 (2): 139–151.

2430 • 2002 Mukhopadhyay, S. Natural disaster vis-à-vis agriculture. *Science and Culture*, 68 (7 and 8): 186–191.

2431 • 2002 Murthy, R.S. and M. Karki. Disaster and mental health: A psychological intervention. *Indian Journal of Social Work*, 63 (2): 173–182.

2432 • 2002 Special Correspondent. A heavy toll. *Economic and Political Weekly*, 37.

2433 • 2003 Special Correspondent. The shadow of a calamity. *Down to Earth*, 11 (24): 6.

2434 • 2004 Sarangi, P. and G.C. Penthoi. Economic implications of natural disasters in Orissa: A retrospective view. *Indian Journal of Regional Science*, 36 (1):

2435 • 2005 Aggarwal, M. Aftermath of disaster. In R.K. Sharma and G. Sharma (eds). *Natural Disaster: Institutional Building for Mitigating the Impact on Vulnerable Sections*. New Delhi: APH Publishing Corporation. 157–166.

2436 • 2005 Aryabandhu, M.M. and M. Wikramasinghe. *Gender Dimensions in Disaster Management: A Guide for South Asia*. New Delhi: Zubaan Publishers.

2437 • 2005 Kapur, A. *et al. Disasters in India: Studies of Grim Reality*. Delhi: Rawat Publication.

2438 ▲2005 Kapur, A. Delhi: A disaster prone national capital. In A. Kapur, *et al. Disasters in India: Studies of Grim Reality*. New Delhi: Rawat Publication. 33–76.

2439 ▲2005 Chopra, N. Greater Mumbai: Teaming with disasters. In A. Kapur, *et al. Disasters in India: Studies of Grim Reality*. Delhi: Rawat Publication. 77–124.

2440 ▲2005 Chakraborti, D. Kolkata: A city on the brink. In A. Kapur, *et al. Disasters in India: Studies of Grim Reality*. New Delhi: Rawat Publication. 125–160.

2441 • 2005 Mukhapadhyay, A.K. *Crisis and Disaster Management Turbulence and Aftermath*. New Delhi: New Age Publisher.

2442 • 2005 Srivastava, S. and D. Mukherji. Emerging cityscape and environmental issues. *Economic and Political Weekly*, 40 (36): 3905–3909.

2443 • 2006 Bhavani, S.T. and B. Kumar. Children's reaction to disaster. *School of Social Work*, 2 (12): 9–10.

2444 • 2006 Gandhi, S. and L. Krishnaswamy. Disaster and children. *School of Social Work*, 2 (12): 30–32.

2445 • 2006 Hultman, N.E. *et al.* The changing face of natural disaster: Risk resilience and natural security in a changing climate. *Journal of International Affairs*, 59 (2): 25–41.

2446 • 2006 Kumar, J. and K. Muralidharan. Disaster: Impact and preparedness. *School of Social Work*, 2 (12): 23–26.

2447 • 2006 Lala, M.K. and K.R. Lala. Survivors of natural disasters and post-traumatic stress disorder. *Journal of Indian Medical Association*, 76 (1).

2448 • 2006 Ranade, P.S. Economic cause and consequences of natural disasters. In P.S. Ranade (ed). *Natural Disasters: Management and Preparedness*. Hyderabad: Institute of Charted Finanacial Analysts of India University Press. 3–22.

2449 • 2006 Swain, M. and S. Mrutyunjay. *Gender and Disaster*. New Delhi: Serials Publications.

2450 • 2006 Visvesvaran, P.K. Impact of disasters on the school going child. *School of Social Work*, 2 (12): 3–4.

Response

The stimuli drawn by any action is referred to as a response. It is a decision or activity taken to correct an emerged situation. A response could vary from doing everything to doing nothing. References that research into the variety of reactions and actions disasters attract are gathered in the group 'response'. Be it the politicians, administration or the triforces a disaster draws a response from all these and others. Even local communities, *panchayats*, media and non-government organisations come to respond. Political bickering, peoples anger or providing camps for relief also are expressions of a response. Many an evaluative study of programmes and even reports of lessons learnt are a return on disaster Springing as an inherent need of disaster is the entire gamut of 'redoing'. Therefore references that carry the elements of rehabilitation, relocation, rebuilding, reconstruction and repair too are assembled within the group called 'response'. Among the 4004 references there are 538 on the theme of response. These form 13 per cent of the total (Tables 6.19, 6.20 and 6.21).

Table 6.19 References for each Category

Natural Disaster	References	Number
Cyclone	2451–2511	61
Drought	2512–2619	108
Earthquake	2620–2796	177
Floods	2797–2872	76
Heat and Cold Wave		0
Heavy Rain	2873–2874	2
Landslide	2875–2879	5
Storm		0
Tsunami	2880–2918	39
Assorted	2919–2992	74

Table 6.20 Research References on Response

'Natural' disaster	Book	Report	Thesis	Article[1]	Chapter[2]	Feature[3]	Total
Cyclone	6	11	3	18	7	16	61
Drought	5	6	4	38	17	38	108
Earthquake	3	23	11	47	9	84	177
Floods	1	4	9	28	11	23	76
Heat and Cold Wave	0	0	0	0	0	0	0
Heavy Rain	0	0	0	1	0	1	2
Landslide	0	0	0	1	2	2	5
Storm	0	0	0	0	0	0	0
Tsunami	0	1	1	10	4	23	39
General Hazard	5	8	2	24	19	16	74
Total	20	53	30	167	69	203	542

[1] in a journal; [2] in a book/report/proceedings; [3] in a magazine.

Table 6.21 Dominance of Human over Natural in Research References on Response

Phase	Natural	Human	Ecological	Total
1830–1947	0	3	0	3
1948–1989	0	78	5	83
1990–2006	3	449	3	456
Total	3	530	8	542

Response: Cyclone

2451 • 1977 Krishana, C.V. Andhra Pradesh: Politics of cyclone relief. *Economic and Political Weekly*, 12 (35): 2102.

2452 • 1977 Special Correspondent. Political use of cyclone. *Economic and Political Weekly*, 12 (28): 2118–2119.

2453 • 1978 Divi Seema Social Service Society. *Six Months after the Tidal Wave Tragedy of November 1977: Socio-Economic Survey Conducted at Divi Seema, May 1978*. Bhubaneswar: Government of Orissa.

2454 • 1978 Raghavalu, C.V. *The Andhra Cyclone: A Study of the Political and Administrative Process of Mass Death and Relief Effort*. New Delhi: Indian Council of Social Science (Project Report).

2455 • 1979 Caldwell, N. *et al.* An analysis of Indian press coverage of the Andhra Pradesh cyclone disaster of 19 November 1977. *Disasters: The Journal of Disaster Studies, Policy and Management*, 3 (2): 154.

2456 • 1979 Lewis, J. The Tamil Nadu cyclone November 1977: A comparison of newspaper reports. *Disasters: The Journal of Disaster Studies, Policy and Management*, 3 (2): 123.

2457 • 1979 Mishra, D.K. and H.R. Prakash. An evaluation of the Andhra Pradesh cyclone shelters programme: Guidelines for the Orissa programme. *Disasters: The Journal of Disaster Studies, Policy and Management*, 3 (4): 241–314.

2458 • 1979 Raghavulu, C.V. and P.S. Cohen. *The Andhra Cyclone of 1977: Individual and Institutional Responses to Mass Deaths*. Delhi: Vikas Publishing House.

2459 • 1979 Winchester, P. Disaster relief operations in Andhra Pradesh (Southern India) following the cyclone in November 1977. *Disasters: The Journal of Disaster Studies, Policy and Management*, 3 (2): 173–177.

2460 • 1979 Windass, M. Appropriate reconstruction training and information centre and the construction of houses in Andhra Pradesh following the cyclone and sea surge of November 1977. *Disasters: The Journal of Disaster Studies, Policy and Management*, 3 (2): 147.

2461 • 1979 Young, S.L. Oxfam's experience of shelter provision following the Andhra Pradesh cyclone: Some recent comments. *Disasters: The Journal of Disaster Studies, Policy and Management*, 3(2): 145.

2462 • 1980 Office of the District Collector. *Rehabilitation and Construction of Permanent Houses in Cyclone Affected Areas of the Krishna District.* Masulipatnam: Government of Andhra Pradesh.

2463 • 1981 Government of Andhra Pradesh. *The Andhra Pradesh Natural Calamities (Relief and Rehabilitation) Act 1981*. Hyderabad.

2464 • 1983 Rao, R. Cyclone and politics. *Economic and Political Weekly*, 18 (19): 1962.

2465 • 1990 Prasad, R.J.R. A political storm. *Frontline*, 23 (May 26): 116–118.

2466 • 1991 Special Correspondent. A lax response. *Frontline*, 24 (May 26): 14.

2467 • 1992 Sudan, R. Disaster management: Administrative response to cyclones. *Indian Journal of Public Administration (*special number*)*, 38 (1): 37–49.

2468 • 1992 Winchester, P. *Power, Choice and Vulnerability: A Case Study in Disaster Mismanagement in South India, during 1977–1988*. London: James and James Publishers.

2469 • 1995 Appa Rao, T.V.S.R. Braving the cyclones. *Down to Earth*, 4 (11): 11.

2470 ▲1995 Finance and Planning Department. *Report of the Cyclone Emergency Reconstruction Project.* Hyderabad Government of Andhra Pradesh.

2471 • 1996 TARU. *Cyclone Identification Mission following the Andhra Cyclone.* New Delhi.

2472 • 1998 Prasad, T.K. *Cyclone Disaster Management by the Prawn Growers in West Godavari District of Andhra Pradesh*. Hyderabad: Acharya N.G. Ranga Agricultural University (Unpublished Postgraduate Dissertation).

2473 • 1999 Banerjee, R. Nobody's child. *India Today*, (December 27): 34–38.

2474 • 1999 Banerjee, R. *et al.* The day hope died. *India Today*, (November 15): 18–28.

2475 • 1999 Boral, A. Save and reconstruct Orissa: Urges to save from disaster. *New Age*, 47 (49): 8–9.

2476 • 1999 Chakarvarthy, S. Bottle neck. *India Today*, (November 30): 32–35.

2477 • 1999 Chattopadhyay, S.S. Reaching relief. *Frontline*, 16 (24): 25–28.

2478 • 1999 Chattopadhyay, S.S. Succour for Orissa: Focus on the task before state government and non-government organisations towards rebuilding cyclone affected areas of Orissa. *Frontline*, 16(26): 45–46.

2479 • 1999 Chaudhuri, K. Political storm. *Frontline*, 16 (26): 44–45.

2480 • 1999 Government of Andhra Pradesh. *Cyclone Reconstruction Project.* Hyderabad.

2481 • 1999 Jayaraman, K.S. Scientists reel after Indian cyclone. *Nature*, 402 (November 18): 219.

2482 • 1999 Kumar, S. Cyclone exposes Indian government's lack of disaster preparation. *The Lancet*, 354 (9191):1709.

2483 • 1999 Mahalingam, S. A project in waiting. *Frontline*, 16 (24): 30.

2484 • 1999 Punetha, A.C. Preparedness for recurring cyclones in Andhra Pradesh and coping mechanisms for rural housing. *Shelter,* special issue, October: 65–67.

2485 • 1999 Roy, V. *et al.* Rehabilitation of Andhra Pradesh cyclone victims. *Shelter,* special issue, October: 95–96.

2486 • 1999 Sen, A. Rebuilding Orissa. *India Today*, (November 29): 18–23.

2487 • 2000 Banerjee, R. Castles in the air. *India Today*, (November 13): 66–68.

2488 • 2000 Bharat, S. and C.R. Chandrashekar. *Psycho-Social Care for Individuals: Information Manual I (Super Cyclone)*. Bangalore: Books for Change.

2489 • 2000 Bhattacharya, T. Experience in Gudari (Orissa). In S. Parasuraman and P.V. Unnikrishnan (eds). *India Disasters Report: Towards a Policy Initiative*. New Delhi: Oxford University Press. 158–159.

2490 • 2000 Das, S.K. and N. Manchanda. *Action Plan for Shelter and Settlement Reconstruction*. New Delhi: Orissa Reconstruction Campaign. Associated Architects (Mimeo).

2491 • 2000 Dhara, S. Lessons for Orissa cyclone. *Down to Earth*, 9 (8): 39.

2492 • 2000 India Disaster Report Team. Cyclones: Trapped in a spin cycle. In S. Parasuraman and P.V. Unnikrishnan (eds). *India Disasters Report: Towards a Policy Initiative*. New Delhi: Oxford University Press. 181–190.

2493 • 2000 Kumar, K.V.K. *et al. Psycho-Social Care for Community Level Helpers: Information Manual 2. (Super Cyclone).* Bangalore: Books for Change.

2494 ■ 2000 Kumar, M. *Post Disaster Reconstruction in Cyclone Affected Jamnagar District (Gujarat).* New Delhi: Deparment of Housing, School of Planning and Architecture (Unpublished Postgraduate Dissertation).

2495 • 2000 Mahapatra, R. A state in chaos: Orissa. *Down to Earth*, 9 (14): 30–33.

2496 • 2000 Merladet, J.S. The European Community Humanitarian Office in India. In S. Parasuraman and P.V. Unnikrishnan (eds). *India Disasters Report: Towards a Policy Initiative.* New Delhi: Oxford University Press. 139–140.

2497 • 2000 Panigrahi, N. *Super Cyclone in Orissa: Response of Welfare Agencies for Relief, Rehabilitation and Reconstruction.* Bhubaneswar: Nabakrishna Choudhury Centre for Development Studies Occasional Paper No. 51.

2498 • 2000 Prasad, P.V.V. Health service system response to cyclone disaster: Study of Guntur district. In R.B. Singh (ed). *Disaster Management.* Jaipur: Rawat Publications. 225–237.

2499 • 2000 Reddy, A.V.S. *et al. Cyclones in Andhra Pradesh: A Multidisciplinary Study to Profile Cyclone Response in Coastal Andhra Pradesh.* New Delhi: Oxfam.

2500 • 2000 Sarwal, T. and J. Gwynn. The Oxfam experience in India. In S. Parasuraman and P.V. Unnikrishnan (eds). *India Disasters Report: Towards a Policy Initiative.* New Delhi: Oxford University Press.

2501 • 2000 Srikala, B. *et al. Psychological Care for Individuals after the Orissa Super Cyclone.* Bangalore: Books for Change.

2502 • 2000 Winchester, P. *Cyclone Mitigation, Resource Allocation and Post-Disaster Reconstruction in South India: Lessons from Two Decades of Research.* London: Overseas Development Institution.

2503 • 2001 Kumar, K.V.K. *et al. Psyco-Social Care for Individuals: Information Manual 3 (Super Cyclone).* Bangalore: Books for Change.

2504 • 2001 Samal, K.C. Super Cyclone in Orissa: People's suggestions and reaction. In *National Symposium on Super Cyclone and Ecorestoration Efforts to Conserve Biodiversity of the State of Orissa.* Bhubaneswar: Postgraduate Department of Zoology, Utkal University.

2505 • 2001 Samal, K.C. *Coping Strategy of Weaker Sections (Dalits and Women) in Post-Super Cyclone Orissa: Case Study of Erasama Block.* New Delhi: Indian Council of Social Science, (Project Report).

2506 • 2001 Sharma, V.K. and C. Rout. Managing natural disasters: The lessons from Orissa Super Cyclone. In V.K. Sharma (ed). *Disaster Management.* New Delhi: Indian Institute of Public Administration. 208–221.

2507 • 2002 Adhikari, S.N. *Approach to Rehabilitation Process for Cyclone Affected Villages, Erasama Block (Orissa)*. New Delhi: School of Planning and Architecture (Unpublished Postgraduate Dissertation).

2508 • 2002 Das, K. Social mobalisation for rehabilitation and relief work in cyclone affected Orrisa. *Economic and Political Weekly*, 37 (10): 4859–4866.

2509 • 2002 Kanda, M. After the storm: A study of the response to the Orissa cyclone. In *Draft Working Paper prepared for Overseas Development Institute's Rural Livelihood Futures Workshop*, 17–19 October. New Delhi. 17–19.

2510 • 2002 Kumar, P. and Pareeda. Towards rebuilding a post disaster society: A case study of Super Cyclone affected coastal Orissa. *Indian Journal of Social Work*, 63 (2): 243–262.

2511 • 2005 Samal, K.C. The 1999 Super Cyclone in Orissa: Coping strategies of women in Ersama block. *Social Action*, 55 (1): 72–97.

Response: Drought

2512 • 1971 Limaye, B.V. and M.D. Rahalkar. Drought relief: A first hand account. *Economic and Political Weekly*, 6 (41): 2141–2143.

2513 • 1972 Surana, P. Maharashtra: Who cares about the drought? *Economic and Political Weekly*, 7 (37): 2269–2270.

2514 • 1973 Special Correspondent. Centre-state relations: Debt and drought relief. *Economic and Political Weekly*, 8: 1923–1932.

2515 • 1974 Mathur, K. and M. Bhattacharya. *Administrative Response to Emergency: A Study of Scarcity Administration in Maharashtra*. New Delhi: Indian Council of Social Science, (Project Report).

2516 • 1974 Mundle, S. *Planning and Resource Deployment in Drought Relief Operations: A Study of Aurangabad District (Maharashtra)*. New Delhi: Indian Institute of Public Administration.

2517 • 1974 Rangasami, A. The uses of drought. *Economic and Political Weekly*, 9 (50): 2069–2074.

2518 • 1978 Mullick, R.P. Rajasthan: Drought and election. *Economic and Political Weekly*, 14 (50): 2040–2041.

2519 • 1979 Ambhire, D.B. The drought prone areas of Maharashtra: A study of human response to the environment. In N.B.K Reddy (ed). *Proceedings of the All India Symposium on Drought Prone Areas of India*. Tirupati: Rayalseema Geographical Society. 167–174.

2520 ▲1979 Saibaba, G. Conceptual aspects of the strategy of development of drought prone areas. In N.B.K Reddy (ed). *Proceedings of the All India Symposium on Drought Prone Areas of India*. Tirupati: Rayalseema Geographical Society. 287–295.

2521 • 1979 Tewari, A.K. Perception of drought hazard among the people of drought prone areas in arid lands of India: A case study. In N.B.K Reddy (ed). *Proceedings of the All India Symposium on Drought Prone Areas of India.* Tirupati: Rayalseema Geographical Society.

2522 • 1981 Garcia, R.V. and J.C. Escudero. *Drought and Man, The 1972 Case History*. London: Pergamon Press.

2523 • 1981 Jodha, N.S. Role of credit in farmers adjustment against risk in arid and semi-arid tropical areas of India. *Economic and Political Weekly*, 16 (42 and 43): 1696–1709.

2524 • 1982 Special Correspondent. Drought: Policy failures. *Economic and Political Weekly*, 17 (43): 1710.

2525 • 1985 Dhawan, B.D. Irrigation performance during drought. *Economic and Political Weekly*, 20 (28): 1191–1196.

2526 • 1985 Quaraishi, S. Dry farming versus drought: A study in Western Rajasthan. In *Proceedings of the Sixth Indian Geographical Congress.* Department of Architecture and Regional Planning Kharakpur: Indian Institute of Technology.

2527 • 1986 Madan Mohan, M. Care for cattle. *Frontline*, (February 22).

2528 • 1986 Special Correspondent. Maharashtra fighting drought. *Economic and Political Weekly*, 21 (18): 769–770.

2529 • 1986 Special Correspondent. It's a way of life in Rajasthan. *Frontline*, (April 10): 36–39.

2530 • 1987 Bandopadhyay, J. *Ecology of Drought and Water Scarcity: Need for an Ecological Water Resource Policy*. Dehradun: Research Foundation for Science, Technology and Ecology.

2531 • 1987 Editorial. Disaster relief: Conveniently haphazard. *Economic and Political Weekly*, 22 (41): 1718.

2532 • 1987 Madan Mohan, M. A question of funding. *Frontline*, (May 2): 29.

2533 • 1987 Mitra, A. Drought relief: Belated, half hearted. *Economic and Political Weekly*, 22 (36 and 37):1529–1531.

2534 • 1987 Mitra, A. Government helping rich farmers: Profit from drought. *Economic and Political Weekly*, 22 (44): 1847–2112.

2535 • 1987 Patnaik, M. *Drought Proneness and Human Response: A Case Study of Orissa.* Kharagpur: Indian Institute of Technology, Kharagpur (Unpublished Ph.D thesis).

2536 • 1987 Special Correspondent. Money down the drain. *India Today*, (October 15): 150–152.

2537 • 1987 Rangasami, A. Financing scarcity relief planning and budgeting for relief expenditure: The problems of centre and states. In *Proceedings of the National Workshop on Management of Drought: Public Policies and Institutions.* New Delhi.

2538 ▲1988 Bhatia, V. Official drought relief measures: A case study of Gujarat. *Social Action*, 38 (2): 138–161.

2539 • 1988 Hubbard, M. Drought relief and drought proofing in the state of Gujarat. In D. Curtis (ed). *Preventing Famine: Policies and Prospects for Africa.* London: Routledge. 120–130.

2540 • 1988 Prasad, P. *Famines and Droughts, Survival Strategies.* Jaipur: Rawat Publication.

2541 • 1988 Rangasami, A. The limits of drought. *Seminar* (special issue), 346 (1): 30–35.

2542 • 1988 Sen, S. R. Coping with drought. *Economic and Political Weekly*, 23 (30): 1523–1524.

2543 • 1989 Jain, L. C. Drought public distribution system and cooperatives. *Economic and Political Weekly*, 24 (14): 502–503.

2544 • 1989 Special Correspondent. Ordeal in Orissa: Fighting the drought. *Frontline*, 22 (April 29): 90–92.

2545 • 1989 Special Correspondent. A paralysed administration. *India Today*, (October 16): 63–69.

2546 • 1990 Clairmonte, F. F. Harvest of the holocaust. *Economic and Political Weekly*, 25 (42 and 43): 2365–2368.

2547 • 1990 Kulkarni, S. N. *Famines, Droughts and Scarcities in India: Relief Measures and Policies.* Allahabad: Chugh Publishers.

2548 ■ 1990 Sharma, R. K. *Role of Farm Extension in Drought Management.* Delhi: Agricultural Economics Research Centre.

2549 • 1991 Chen, M. A. *Coping with Seasonality and Drought.* New Delhi: Sage Publications.

2550 • 1991 Karanth, G. K. *Farmers Survival Strategies in a Drought Prone District: A Case Study of a Village in Chitradurga District.* Bangalore: Institute for Social and Economic Change.

2551 • 1991 Nadkarni, M. V. Coping with droughts: Then and now. *Mainstream*, 29 (January 17): 21–25.

2552 • 1992 Acharya, S. S. Rajasthan drought and response of rural families. *Economic and Political Weekly*, 27 (36): 1894–1896.

2553 • 1992 Agarwal, B. Gender relations and food security: Coping with seasonality, drought and famine in South Asia. In L. Beneria and S. Feldman (eds). *Unequal Burden: Economic Crisis, Persistent Poverty and Women's Work.* Boulder: Westview Press.

2554 • 1992 Dhanagare, D. N. 1992 drought in Maharashtra: Misplaced priorities, mismanagement in water resources. *Economic and Political Weekly*, 27 (27): 1421–1425.

2555 • 1992 Kale, M. S. Drought relief in Maharashtra. *Economic and Political Weekly*, 27 (46): 2446.

2556 • 1992 Mitra, A. Policy watch: Waiting for the rains. *Economic and Political Weekly*, 27 (20–21): 1043–1046.

2557 • 1992 Mitra, A. and G. Pangare. Chittaur kept head above water in drought. *Down to Earth*, 1 (8): 44–45.

2558 • 1992 Mohanty, M. Kalahandi awaits. *Mainstream*, 30 (April 18): 7–9.

2559 • 1992 Phadke, A. R. S. Left response to drought in Mahrashtra. *Economic and Political Weekly*, 27 (6): 253–254.

2560 ▲ 1992 Special Correspondent. Drought action. *Down to Earth*, 1 (2): 49.

2561 • 1993 Balagopal, K. Andhra Pradesh: A year of drought. *Economic and Political Weekly*, 28 (38): 1970–1972.

2562 • 1993 Mahajan, A. Human situation in Kalahandi: An eco-developmental analysis. In P. K. Nayak and A. Mahajan (eds). *Kalahandi-Human Encounter with Drought.* New Delhi: Reliance Publishing House. 11–51.

2563 • 1993 Mishra, D. Hunger in Kalahandi: Blinkered understanding. *Economic and Political Weekly*, 28 (24 and 25): 1261–72.

2564 • 1994 Phadke, A. Anti-drought movement in Sangli district. *Economic and Political Weekly*, 29 (48): 3024–3025.

2565 • 1994 Suresh, B. R. *Water Balance Drought Climatology, Cropping Pattern and Land Use and Water Development.* Anantapur: Sri Krishna Devaraya University (Unpublished Ph.D Thesis).

2566 • 1995 Prasad, N. P. *Socio-cultural Adaptations in a Drought Prone Village of Rayalseema.* Hyderabad: Osmania University (Unpublished Ph.D thesis).

2567 • 1995 Reddy, K. P. A study of adaptation of Drought Prone Area Programme in Rangareddy district (Andhra Pradesh). *Journal of Rural Development*, 14 (4): 367–383.

2568 • 1996 Prasad, C. and N. P. Purendra. Response to drought: An overview. *Man in India*, 76 (1): 35–49.

2569 • 1997 Bhatt, M. Maintaining families in drought India: The fodder security system of the Banaskantha women. In P. Fernando and V. Fernando (eds). *South Asian Women: Facing Disasters, Securing Life.* Columbo, Sri Lanka: Intermediate Technology Development Group and Duryog Nivaran. 35–44.

2570 • 1997 Editorial. Helping hand. *Down to Earth*, 6 (3): 12.

2571 • 1997 Fisher, R. J. *If Rain Doesn't Come: An Anthropological Study of Drought and Human Ecology in Western Rajasthan.* New Delhi: Manohar Publications.

2572 • 1997 Prasad, C. and N. P. Purendra. State response to famines and drought: A case of Anantpur district, 1800–1947. *Man in India*, 77 (2 and 3): 219–231.

2573 • 1997 Prasad, N. Adaptations of peasants in stress environment. *Economic and Political Weekly*, 32 (37): 2315–2317.

2574 • 1997 Ray, T. When the rain failed: Mass migration from Orissa. *Frontline*, 30 (January 10): 91–92.

2575 • 1998 Acharya, M. Dark area. *Down to Earth*, 6 (20): 16–17.

2576 • 2000 Avijit. A. Parched hopes. *India Today*, (May 15): 50–55.

2577 • 2000 Brijnath, R. Depths of despair. *India Today*, (May 8): 36–39.

2578 • 2000 Dash, J. and N.R. Bahura. Crisis management lessons from the past: A case study of drought situation. *Man in India*, 80 (1): 75–88.

2579 • 2000 George, T.J.S. Drought and *babudom*. *Mainstream*, 38 (June 3): 4.

2580 • 2000 Madhurkar, U. The good neighbours. *India Today*, (March 13): 54–55.

2581 • 2000 Madhurkar, U. and N.K. Singh. Path breakers. *India Today*, (May 8): 57–58.

2582 • 2000 Menon, A.K. Dying hopes. *India Today*, (May 8): 51.

2583 • 2000 Ramachandran, R. Dealing with drought. *Frontline*, 17 (12): 108–112.

2584 • 2000 Sainath, P. Everybody loves a good drought. In S. Parasuramanand P.V. Unnikrishnan (eds). *India Disasters Report: Towards a Policy Initiative*. New Delhi: Oxford University Press. 173–176.

2585 • 2000 Sengupta, S. Drought: Reaping scarcity. In S. Parasuraman and P.V. Unnikrishnan (eds). *India Disasters Report: Towards a Policy Initiative*. New Delhi: Oxford University Press. 165–172.

2586 • 2000 Sharma, P.P. Man in desert: Drought, desertification and indigenous knowledge for sustainable development. *Man in India*, 80 (3 and 4): 225–226.

2587 • 2000 Special Correspondent. Managing a crisis. *Down to Earth*, 9 (13): 22–23.

2588 • 2000 Special Correspondent. Government criticised. *Down to Earth*, 9 (2): 12.

2589 • 2000 Tiwari, M. Standing the test of drought. *Down to Earth*, 8 (16): 30–40.

2590 • 2000 Yoga, R. Gujarat drought: A lesson in water mismanagement. *Down to Earth*, 9 (8): 14.

2591 • 2001 Agarwal, A. Drought of 1999–2000: Jolted into action. In A. Agarwal, *et al* (eds). *Making Water Everybody's Business: Practice and Policy of Water Harvesting*. New Delhi: Centre for Science and Environment. 361–362.

2592 • 2001 Agarwal, A. Fixing the drought. *Down to Earth*, 9 (17): 5.

2593 • 2001 Bhatia, B. In pursuit of relief. *Frontline*, 18 (16): 42–44.

2594 • 2001 Bhatia, B. When hope is lost. *Frontline*, 18 (16): 43.

2595 • 2001 Das, B. Drought of 1999–2000: Miles to go. In A. Agarwal, *et al* (eds). *Making Water Everybody's Business: Practice and Policy of Water Harvesting.* New Delhi: Centre for Science and Environment. 382–396.

2596 • 2001 Disaster Management Institute. *Agenda for Drought Relief: Community Based Action Review of Drought Relief 2000 in Gujarat some Lessons for Asia.* Ahmedabad:

2597 • 2001 Hirway, T. Vicious circle of drought and scarcity works: Why not break it? *Indian Journal of Agricultural Economics*, 56 (4): 708–721.

2598 • 2001 Kumkum, D. The Madhya Pradesh government and the people work together to fight drought. *Down to Earth*, 9: 22–23.

2599 • 2001 Mahapatra, R. Drought of relief. *Down to Earth*, 10 (2): 30–34.

2600 • 2001 Mahapatra, R. *et al.* Drought of 1999–2000: Second generation mission. In A. Agarwal, *et al* (eds). *Making Water Everybody's Business: Practice and Policy of Water Harvesting.* New Delhi: Centre for Science and Environment. 397–414.

2601 • 2001 Maheshwari, U. L. *et al.* Pattern of drought and survival strategies in farm households in a drought prone district in Tamil Nadu. *Indian Journal of Agricultural Economics*, 56 (4): 683–695.

2602 • 2001 Saha, A. Out of focus: An analysis. *Down to Earth*, 10 (2): 35–39.

2603 • 2001 Special Correspondent. Community initiative. *Down to Earth*, 9 (19): 15.

2604 • 2002 Aiyar, S. Thirst aid. *India Today*, (June 9): 28–45.

2605 • 2002 Chari, K. B. and S. A. Abasi.Kaliveli wetland: A typical bulwark against drought. In R. Nagarajan (ed). *National Workshop Proceedings on Recent Trends in Drought Assessment Monitoring and Management*, (June 28–30). Mumbai: Centre of Studies in Resources Engineering Indian Institute of Technology. 115–123.

2606 • 2002 Editorial. Despair on drought. *Economic and Political Weekly*, 36 (31): 3215–3216.

2607 • 2002 Hassan, I. U. Left high and dry. *Down to Earth*, 10 (17): 9.

2608 • 2002 Mahapatra, R. Time to get cracking. *Down to Earth*, 10 (17): 22–24.

2609 • 2002 Muralidharan, S. Too little, too late. *Frontline*, 19 (19): 25–26.

2610 • 2002 Muralidharan, S. Drought challenges. *Frontline*, 19(17): 4–8.

2611 • 2002 Special Correspondent. A mid-summer dream. *Down to Earth*, 11 (3): 39–43.

2612 • 2002 Singh, R. V. Drought 2002: What do people do? *Down to Earth*, 11 (13).

2613 • 2003 Pandey R. K. *Empowerment of Dalit Women through Government Programme in Bihar (A Case Study of Drought Prone Area of Nawada District).* New Delhi: Indian Council of Social Science, (Project Report).

2614 • 2003 Purie, A. Painting the village green. *India Today*, (October 20): 72.

2615 • 2003 Special Correspondent. Too much soothsaying: How much rain? *Down to Earth*, 12 (2): 60.

2616 • 2004 Appasaheb, T. A. *Response to Drought in the Rain Shadow Zone of Maharashtra*. Delhi: Department of Geography, Delhi School of Economics, University of Delhi (Unpublished Master of Philosophy Dissertation).

2617 • 2004 Bavadam, L. Delayed, flawed strategy. *Frontline*, 21 (6): 45–46.

2618 • 2004 Chandrakanth, M. *et al.* Combating negative externalities of drought. *Economic and Political Weekly*, 39 (11): 1164–1170.

2619 • 2004 Sivakumar, S. and F. Kerbart. Drought sustenance and livelihood: *Akal* survey in Rajasthan. *Economic and Political Weekly*, 39 (3): 285–294.

Response: Earthquake

2620 • 1935 Andrews, C. F. *The Indian Earthquake*. London: George Allen and Unwin Publishers.

2621 • 1935 Brett, W. B. *Report on the Bihar Earthquake and on the Measures Taken in Consequence their of upto 31 December 1934*. Patna: Supretendent Government Printing.

2622 • 1951 Naik, D. Assam earthquake relief. *Modern Review*: 89 (2): 148–150.

2623 • 1968 Special Correspondent. Maharashtra: Politics of earthquake relief. *Economic and Political Weekly*, 3 (36): 1885–1888.

2624 • 1969 Maishi, P. G. *Replanning and Rehabilitation in Earthquake Devastated Areas: A Case Study of Koyna*. New Delhi: School of Planning and Architecture (Unpublished Postgraduate Dissertation).

2625 • 1970 Deuskar, V. R. and S. M. Bhalerao. Koyna earthquake and after. *Bhagirath*, 17 (3): 124–128.

2626 • 1978 Mukherjee, T. *Economic Analysis of Natural Hazards: A Preliminary Adjustments to Earthquakes and their Costs*. Roorkee: Natural Hazards Research Working Group,17 (Working Papers).

2627 • 1988 Building Material Technology Promotion Council. *Assessing Maharashtra Earthquake Rehabilitation Programme*. New Delhi: Ministry of Urban Development.

2628 • 1988 Chengappa, R. *et al.* Waiting for a breakthrough. *India Today*, (September 15): 44–45.

2629 • 1991 Ahmed, F. Looting misery.*India Today*, (September 15): 139.

2630 • 1991 Kumar, A. Garhwal cries for relief, Bhartiya Janta Party fiddles. *New Age*, 39 (44): 1–15.

2631 • 1991 Mishra, D. K. and A. Ravindra. Coping with the Garhwal earthquake. *Economic and Political Weekly*, 26 (47): 2667–2669.

2632 • 1991 Pathak, R. The relief report. *India Today*, (November 30): 165–170.

2633 • 1991 Pathak, R. Uttarkashi: The relief rip-off. *India Today*, (February 15): 165–170.

2634 • 1991 Srinivas, K.After the quake: Apathy adds to Garhwal's misery. *Frontline*, 24 (November 9): 4–5.

2635 • 1992 Building Material Technology Promotion Council. *Reconstruction of Earthquake Resistant Houses in Garhwal Region*. New Delhi: Ministry of Urban Development.

2636 • 1992 Reddy, T. No rehabilitation yet to Garhwal quake victims. *New Age*, 40 (22): 11.

2637 • 1993 Bhaskar, B. It took us by surprise. *Frontline*, 26 (October 22): 131.

2638 • 1993 Editorial. Earthquake: Politics of relief. *Economic and Political Weekly*, 28 (45): 2424.

2639 • 1993 Editorial. Communist Party of India (M) team visits quake hit villages. *New Age*, 41 (41): 1.

2640 • 1993 Kumar, S. N. Hooked to doles? *Frontline*, 26 (November 5).

2641 • 1993 Lakshmipathi, T. Post-quake riddles. *Frontline*, 26 (December 3): 117–118.

2642 • 1993 Menon, P. Brick by brick: Challenges on the housing front. *Frontline*, 26 (November 5): 14.

2643 • 1993 Padmanabhan, R. Rebuilding debates. *Frontline*, 26 (December 3): 118–119.

2644 • 1993 Padmanabhan, R. The rebuilding effort. *Frontline*, 26 (November 5): 13.

2645 • 1993 Padmanabhan, R. With great effort: Testing time for state administration. *Frontline*, 26 (October 22): 122–125.

2646 • 1993 Pathak, R. Time is the enemy. *India Today*, (November 15): 83–88.

2647 • 1993 Rao, U. S. and D. P. Rao. Latur earthquake: Selection of sites for rehabilitation of the affected villages in Latur and Osmanabad districts of Maharashtra, India (A remote sensing based study). *Current Science*, 65 (10): 772–774.

2648 • 1993 Revenue and Forests Department. *Proposal for Maharashtra Earthquake Rehabilitation Programme*. Mumbai: Government of Maharashtra.

2649 • 1993 Shankar, U. Shattered: Chaotic response to calamity. *Down to Earth*, 2 (October 13): 5–14.

2650 • 1993 Sharma, R. The living fields: Few fatalities in Karnataka but fear pervades. *Frontline*, 26 (October 22): 13.

2651 • 1993 Sharma, R. The shattered and the unmoved. *Frontline*, 26 (October 22): 1.

2652 • 1993 Special Correspondent. Wising up after the quake. *Down to Earth*, 2 (November 15): 14.

2653 • 1993 Swami, T. Lessons of 91: Uttarkashi revisited. *Frontline*, 26 (November 5): 118–119.

2654 • 1994 Bilham, R. The 1737 Kolkata earthquake and cyclone evaluated. *Bulletin of the Seismological Society of America*, 84 (5): 1650–1657.

2655 • 1994 Bokil, M.S. Maharashtra earthquake: Institutional tremors. *Economic and Political Weekly*, 29 (1 and 2): 17–20.

2656 • 1994 Desarda, H.M. Rehabilitation of earthquake victims: Problems of housing. *Economic and Political Weekly*, 29 (6): 291–292.

2657 • 1994 Koppikar, S. At ground zero still. *India Today*, (October 31): 205–207.

2658 • 1994 Menon, P. Living it down: The task of resettlement. *Frontline*, 27 (Dec 3): 1.

2659 • 1994 Muralidharan, S. Central succour. *Frontline*, 27 (October 22): 5.

2660 • 1994 Parikh, R. *et al.* Mechanism of coping and psychopathology following the Latur earthquake: The profile study. *Archives of Indian Psychiatry*, 1 (2): 97–101.

2661 • 1994 Patel, S.B. Marathwada earthquake: Soft credit, soft appraisal. *Economic and Political Weekly*, 29 (8): 401–405.

2662 • 1994 Project Management Unit. *Guidelines for Repair, Strengthening and Reconstruction of Houses Damaged in the 30 September Earthquake in Maharashtra.* Mumbai: Government of Maharashtra.

2663 • 1994 Ranjan, A. *Rehabilitation of an Earthquake Affected Village: Malkundji (Latur).* New Delhi: School of Planning and Architecture (Unpublished Postgraduate Dissertation).

2664 • 1994 Ravi, S. A tour in Latur for the Prime Minister. *Frontline*, 27 (Nov 5): 18–20.

2665 • 1994 Revenue and Forest Department. *Rehabilitation Action Plan for Earthquake affected People.* Mumbai: Government of Maharashtra.

2666 • 1994 Revenue and Forest Department. *Earthquake Rehabilitation Review Committee of the Government of Maharashtra.* Mumbai: Government of Maharashtra.

2667 • 1994 Special Correspondent. Latur: Diminishing interest. *Economic and Political Weekly*, 29 (7): 653–656.

2668 • 1994 Srivastava, R. The earthquake capital rises again. *Down to Earth*, 2 (24): 6.

2669 • 1995 Ahmedabad Study Action Group. *Repairs and Retrofitting: A Manual for People of Marathwada, Latur.* Ahmedabad.

2670 • 1995 Khan, A.A. *Post-disaster Reconstruction in Earthquake Affected Area: Case Study of Latur, Maharashtra.* New Delhi: School of Planning and Architecture (Unpublished Postgraduate Dissertation).

2671 • 1995 Parasuraman, S. Relief assistance and rehabilitation following the Maharashtra earthquake. *Disasters: The Journal of Disaster Studies, Policy and Management*, 19(2): 178.

2672 • 1995 Sen Gupta, S.N. Lessons on Latur earthquake. *Indian Journal of Geology*, 67 (1): 26–33.

2673 • 1995 TARU. *Marathwada Earthquake: Rehabilitation and Reconstruction Work Two Years After*. New Delhi.

2674 • 1996 Government of Maharashtra. *Maharashtra Emergency Earthquake Rehabilitation Program*. Mumbai.

2675 • 1996 Kaur, G.P. *Evaluation of Rehabilitation Programme for Housing in Earthquake Affected Areas: Latur and Osmanabad*. New Delhi: School of Planning and Architecture (Unpublished Postgraduate Dissertation).

2676 • 1996 Oxford Committee for Famine Relief. *Strategy for Housing Rehabilitation in Earthquake Affected Villages of Marathwada*. Mumbai.

2677 • 1996 Parasuraman, S. Relief assistance and rehabilitation lessons in disaster management. *Economic and Political Weekly*, 31 (27): 1654.

2678 • 1996 Tubbesing, S.K. *Post-Earthquake Investigation Field Guide: Learning from Earthquake*. Oakland: Earthquake Engineering Research Institute.

2679 • 1997 Earthquake Rehabilitation Cell. *Mantralaya Plan for Economic Rehabilitation of People Affected due to Land Acquisition for Resettlement of Earthquake Victims*. Mumbai: Government of Maharashtra.

2680 • 1997 Mangnale, M. Bhukampayan: *The Story of Earthquake and Rehabilitation*. Latur: Latur Press.

2681 • 1997 Rajalakshmi, T.K. Coping with calamity. *Frontline*, 30 (June 13): 31–32.

2682 • 1997 Special Correspondent. Quake queries. *Down to Earth*, 6 (4): 12.

2683 • 1997 Thapliyal, G.B. *A Case Study of Earthquake Disaster in Uttarkashi: Response to Crisis*. New Delhi: Advanced Professional Programme in Public Administration, Indian Institute of Public Administration (Unpublished Master of Philosophy Dissertation).

2684 • 1998 Svetlana, N. and S.S. Momin. Post earthquake housing rehabilitation project in Maharashtra. *Indian Concrete Journal*, 72 (11).

2685 • 1999 Arya, A.S. Post-earthquake relocation, reconstruction, repair and retrofitting of buildings. *Shelter*, special issue, *October*: 97–99.

2686 • 1999 Earthquake Engineering Research Institute. *Innovative Earthquake Rehabilitation in India: Lessons Learnt over Time*. Oakland.

2687 • 2001 Jain, V. *et al.* Pediatric surgical emergencies in the setting of a natural disaster: Experiences from the 2001 earthquake in Gujarat. *Journal of Pediatric Surgery*, 38 (5): 663–667.

2688 • 1999 Joglekar, M.N. Community involvement in planning and design of a rehabilitation programme: A case of Latur Rehabilitation Programme. *Shelter,* special issue, October: 43–48.

2689 • 1999 Roy, V. *et al.* Rehabilitation of Jabalpur earthquake victims. *Shelter,* special issue, October: 93–95.

2690 • 1999 Suresh, V. Rehabilitation of the housing stock in the earthquake affected areas of Chamoli and Rudraprayag. *Shelter,* special issue, October: 89–92.

2691 • 1999 Verma, A. *Evaluation of Rehabilitation Project in Earthquake Affected Areas: A Case Study of Jabalpur Tehsil.* New Delhi: Deparment of Housing, School of Planning and Architecture (Unpublished Postgraduate Dissertation).

2692 • 2000 Bapat, A. Earthquake rehabilitation efforts. *Journal of Human Ecology*, 11 (6): 463–469.

2693 • 2000 Bhatt, C.P. and N. Juyal. Living with earthquake: Lesson from Chamoli and Rudraprayag (Central Himalaya). *Science and Culture*, 66 (1 and 2): 16–19.

2694 • 2000 Gopalan, P. Community participation in Latur. In S. Parasuraman and P.V. Unnikrishnan (eds). *India Disasters Report: Towards a Policy Initiative.* New Delhi: Oxford University Press. 210–213.

2695 • 2000 Krishnadas, S. Socio-legal intervention in Latur. In S. Parasuraman and P.V. Unnikrishnan (eds). *India Disasters Report: Towards a Policy Initiative.* New Delhi: Oxford University Press. 217–218.

2696 • 2000 Pandey, N.R. *et al.* Mental health care in Marathwada earthquake disaster 1: Organisation of services. *Indian Journal of Social Work*, 61(4): 631–639.

2697 • 2000 Pandey, N.R. *et al.* Mental health care in Marathwada earthquake disaster : Short term out-reach counselling. *Indian Journal of Social Work*, 61(4): 640–651.

2698 • 2001 Aiyar, V.S. *et al.* Surviving the government: Earthquake rescue and relief. *India Today*, (February 12): 40–46.

2699 • 2001 Arya, A.S. *Reconstruction and New Construction of Houses in Kachchh Earthquake Areas of Gujarat.* Gandhinagar: Gujarat State Disaster Management Authority.

2700 • 2001 Bavadam, L. Rebuilding Kutch. *Frontline*, 18 (5): 118–119.

2701 • 2001 Bidwai, P. Gujarat's unlearnt lessons. *Frontline*, 18 (5): 120.

2702 • 2001 Chakravarti, S. Hope in hell. *India Today*, (February 26): 22–28.

2703 • 2001 Corieggia. M. Helping hands for quake victims. *Down to Earth*, 9 (20): 3.

2704 • 2001 Dasgupta, S. Is Keshu *Bhai* upto it? *India Today*, (February 12): 47–49.

2705 • 2001 Earthquake Engineering Research Institute. *Earthquake Rebuilding in Gujarat India: An Earthquake Engineering Research Institute Recovery Reconnaissance Report.* Oakland.

2706 • 2001 Editorial. Comic relief. *Down to Earth*, 9 (20): 5.

2707 • 2001 Editorial. No sign of science. *Down to Earth*, 9 (19): 43.

2708 • 2001 Editorial. Gujarat killer earthquake: Relief bungling needs thorough probe. *New Age*, 49 (7): 1–2.

2709 • 2001 Editorial. Cashing Gujarat misery. *New Age*, 49 (6): 2.

2710 • 2001 Editorial. Feverish chaos. *Down to Earth*, 10 (12): 5.

2711 • 2001 Faizee, S. Biased administration compounds miseries. *New Age*, 49 (6):1–2.

2712 • 2001 Faizee, S. Gujarat quake: A story of nature's fury, man's greed and government's inefficiency. *New Age*, 49 (5): 1–2.

2713 • 2001 *Frontline* Team. Picking up the pieces. *Frontline*, 18 (4): 131–134.

2714 • 2001 *Frontline* Team.Vignettes of relief. *Frontline*, 18 (4): 128–130.

2715 • 2001 Ghosh, A. and G.S. Mudur. Politicians, bureaucrats and builders responsible for the disaster but not eligible for aid. *New Age*, 49 (6): 8–9.

2716 • 2001 Goel, A. Rehabilitation of Gujarat areas affected by earthquake. In S.L. Goel and R. Kumar (eds). *Disaster Management.* Delhi: Deep and Deep Publications Pvt. Ltd. 238–264.

2717 • 2001 Humanitarian Initiatives and Disaster Mitigation Institute. *Independent Evaluation of Expenditure of Disaster Emergency Committee India Earthquake Appeal Funds, January to October 2001,* Volume 2 (Evaluation Report). United Kingdom.

2718 • 2001 International Federation of Red Cross and Red Crescent Society. Local response and coordination key to Gujarat recovery. In *World Disasters Report 2001*. Geneva. 21.

2719 • 2001 Iyer, L. and F. Ahmed. Earthquake: Politics of pearl grain reapers. *India Today*, (February 12): 51–52.

2720 • 2001 Joshi, V.H. Socio-economic aspects of earthquake disaster relief and rehabilitation. In H.R. Wason, *et al* (eds). *Proceedings of the Workshop on Recent Earthquakes of Chamoli and Bhuj.* Roorkee. 467–478.

2721 • 2001 Katiyar, S.K. and R. Khandelwal. Relocation in Kutch: Bridging old fault lines. *Economic and Political Weekly*, 36 (44): 2319–2321.

2722 • 2001 Kumar, S. India tries to rebuild 600000 lives after earthquake. *The Lancet*, 357 (9254): 447.

2723 • 2001 Louis, P. Gujarat: Earthquake and after, *Economic and Political Weekly*, 36 (11): 908–910.

2724 • 2001 Madhurkar, U. Kangra safe corners: An analysis. *Down to Earth*, 9 (24): 31.

2725 • 2001 Madhurkar, U. And greed hits home. *India Today*, (February 19): 20–24.

2726 • 2001 Madhurkar, U. Leader in paralysis. *India Today*, (March 5): 25–28.

2727 • 2001 Mahadevia, D. Privatising earthquake rehabilitation. *Economic and Political Weekly*, 36 (39): 3670–3673.

2728 • 2001 Mahapatra, R. Now where Gujarat. *Down to Earth*, 10 (1): 20–21.

2729 • 2001 Mathew, G. Gram panchayats: Where are they in Gujarat? *Kurukshetra*, 50 (7): 18–21.

2730 • 2001 Mehta, L. Reflections on the Kutch earthquake. *Economic and Political Weekly*, 36 (31): 2931–2936.

2731 • 2001 Menon, P. Rehabilitation as a dream. *Frontline*, 18 (March 17): 112–113.

2732 • 2001 Menon, P. Relief, delayed or denied. *Frontline*, 18 (4): 9–15.

2733 • 2001 Mitra, S. Earthquake: Uninsured property, leave it to God. *India Today*, (February 12): 67–68.

2734 • 2001 Murlidharan, S.Unbounded confusion. *Frontline*, 18 (4): 20–22.

2735 • 2001 Renuka, M. Deathquake: Unknown heroes. *India Today*, (February 12): 76.

2736 • 2001 Roonwal, G.S. The Gujarat earthquake catastrophe: Lessons to be learnt for disaster preparedness. *Current Science*, 80 (8): 911–912.

2737 • 2001 Rutten, M. Involvement and indifference: Gujarat earthquake in wider perspective. *Economic and Political Weekly*, 36 (35): 3358–3361.

2738 • 2001 Sharma, R. India plans massive hospital rebuilding after earthquake. *British Medical Journal*, 322 (7284): 451.

2739 • 2001 Sharma, V.K. and C. Rout. Gujarat earthquake: Emerging issues and the lessons for future. In V.K. Sharma (ed). *Disaster Management*. New Delhi: Indian Institute of Public Administration. 177–188.

2740 • 2001 Special Correspondent. Shaken not stirred. *Down to Earth*, 9 (20): 20.

2741 • 2001 Special Correspondent. Beyond financial assistance. *Economic and Political Weekly*, 36 (11): 880.

2742 • 2001 Special Correspondent. Deathquake: Good similarities, state of the heart. *India Today*, (February 12): 74–75.

2743 • 2001 Special Correspondent. Foundation for tomorrow. *India Today*, (Feb 12): 61.

2744 • 2001 Special Correspondent. Competitive chaos. *India Today*, (February 12): 4.

2745 • 2001 Special Correspondent. Post-paralysis. *India Today*, (February 12): 16.

2746 • 2001 Special Correspondent. New foundation. *India Today*, (February 19): 43.

2747 • 2001 Special Correspondent. Shame on you. *India Today*, (February 19): 17.

2748 • 2001 Special Correspondent. Strategy partners. *India Today*, (March 5): 65.

2749 • 2001 Special Correspondent. Relief buildings need thorough probe. *New Age*, 49 (7): 1–2.

2750 • 2001 Spodek, H. Crisis and response: Ahmedabad 2000. *Economic and Political Weekly*, 36 (19): 1627–1638.

2751 • 2001 Sud, N. Earthquake response: Beyond bricks and mortar. *Economic and Political Weekly*, 36 (35): 3353–3354.

2752 • 2001 Suresh, V. Natural disaster rehabilitation: Housing and structural mitigation, case study for earthquake affected areas. In V.K. Sharma (ed). *Disaster Management*. New Delhi: Indian Institute of Public Administration. 126–156.

2753 • 2001 Swami, P. Far from the focus. *Frontline*, 18 (4): 16–20.

2754 • 2001 Thekaekara, M.M. To rebuild Kutch. *Frontline*, 18 (April 14): 75–78.

2755 • 2001 Unnithan, S. Earthquake: Looking back, Latur still shaken. *India Today*, (February 12): 72–73.

2756 • 2001 Vatsa, K.S. *The Bhuj Earthquake of District Kutch, State of Gujarat, (January 26) 2001: Identification of Priority Issues* (A Reconnaissance Report). Alexandria: World Institute for Disaster Risk Management.

2757 • 2001 Venkatesan, V. and T. K. Rajalakshmi. Losses and lessons. *Frontline*, 18 (4): 22.

2758 • 2002 Bavadam, L. Kutch, a year after. *Frontline*, 19 (2): 45–48.

2759 • 2002 Chandhari, S. *Rehabilitation Plan for Earthquake Affected Rural Settlements in Gujarat*. New Delhi: School of Planning and Architecture (Unpublished Postgraduate Dissertation).

2760 • 2002 Chatterjee, P. One year after Gujarat earthquake. *The Lancet*, 359 (9303): 327.

2761 • 2002 Faizee, S. Gujarat one year after the quake. *New Age*, 50 (5): 6–15.

2762 ■ 2002 Gupta, A. *Post-Disaster Reconstruction Strategy for Earthquake Affected Areas of Anjar (Kutch)*. New Delhi: School of Planning and Architecture (Unpublished Postgraduate Dissertation).

2763 • 2002 Krinitzsiky, E.L. and M.E. Hynes. The Bhuj (India) earthquake: Lessons learnt for earthquake safety of dams on alluvium. *Engineering Geology*, 66 (3–4): 163–196.

2764 • 2002 Kropec, M. *Bhuj Reconstruction Project.* Ahmedabad: Disaster Mitigation Institute.

2765 • 2002 Madhurkar, U. On firmer ground. *India Today*, (February 4): 50–52.

2766 • 2002 Navalli, S. and R. Mahapatra. Stumbling blocks. *Down to Earth*, 10 (20): 22–24.

2767 • 2002 Puri, B. and M.M. Harjai. Gujarat earthquake pediatric surgery perspective. *Journal of Indian Association of Pediatric Surgeons*, 7 (1).

2768 • 2002 Shaw, R. *et al.* International cooperation in a post disaster scenario: A case study from Gujarat (India). *Journal of Natural Disaster Science*, 24 (2).

2769 • 2002 Vora, B. Gujarat squandering in the name of quake relief. *New Age*, 50 (7): 16.

2770 • 2002 Yadav, V. *Disaster and Response: A Geographical Interpretation of Killari Earthquake in Latur District (Maharashtra).* Delhi: Department of Geography, Delhi School of Economics, University of Delhi (Unpublished Master of Philosophy Dissertation).

2771 • 2003 Abhiyaan and United Nations Development Programme. *Coming Together: A Brief Documentation of the Post-Earthquake Rehabilitation Efforts by Various Non-Government Organisations.* New Delhi: Government of India.

2772 • 2003 Bose, C.P.B. Non-government organisations in the context of natural disaster in India: Some issues. *Loyola Journal of Social Sciences*, 17(2): 171–191.

2773 • 2003 Shaw, R. *et al.* Community recovery and its sustainability: Lessons from Gujarat earthquake of India. *Australian Journal of Emergency Management*, 18 (2): 28–34.

2774 • 2004 International Federation of Red Cross and Red Crescent Society. Gujarat's poor invest to recover. In *World Disasters Report 2004.* Geneva: 21.

2775 • 2004 Khare, R. *Kutch: Triumph of the Spirit.* New Delhi: Rupa and Company.

2776 • 2004 Madhurkar, U. The high road to recovery. *India Today*, (Sept. 13): 50–51.

2777 • 2004 Mishra, P.K. *The Kutch Earthquake 2001: Recollections, Lessons and Insights.* New Delhi: National Institute of Disaster Management.

2778 • 2004 Priya, K.R. Post-quake recovery in urban Kachchh. *Economic and Political Weekly*, 39 (37) 4229–4231.

2779 • 2005 Ahmad, M. Post-disaster emergency health management. *Indian Journal for Practising Doctors*, 2 (4).

2780 • 2005 Bidwai, P. Shaken not stirred. *Frontline*, 22 (22): 103–104.

2781 • 2005 Chokshi, H. *Livelihood Rehabilitation for Affected Areas of Kutchh (Gujarat).* New Delhi: School of Planning and Architecture (Unpublished Postgraduate Dissertation).

2782 • 2005 Editorial. Another disaster-another reminder. *Indian Journal for Practising Doctors*, 2 (4).

2783 • 2005 Husain, A. Open ticket to goodwill. *India Today*, (November 21): 44–46.

2784 • 2005 Husain, A. India's disaster: Deadly delay. *India Today*, (October 24): 72–74.

2785 • 2005 Husain, A. Left in the cold. *India Today*, (October 31): 50–52.

2786 • 2005 Mishra, P. K. Natural calamities: Reconstruction and rehabilitation. *Yojana*, 46: 93–96.

2787 • 2005 Navlakha, G. Crisis as an oppurtunity. *Economic and Political Weekly*, 40 (47): 4874–4876.

2788 • 2005 Ozerdem, A. and T. Jacoby. The Gujarat earthquake. In A.Ozerdem and T. Jacoby (eds). *Disaster Management in Civil Society: Earthquake Relief in Japan, Turkey and India.* York: International Library of Postwar Reconstruction and Development. 71–92.

2789 • 2005 Special Correspondent. The great shake up. *India Today*, (October 24): 60–61.

2790 • 2005 Suvhajyoti, S. *Participatory Approach for Post-Earthquake Reconstruction in the Village of Kachchh, Gujarat.* New Delhi: School of Planning and Architecture (Unpublished Postgraduate Dissertation).

2791 • 2005 Venkatesh, S. A. C. *et al.* Post-disaster public health issue. *Indian Journal for Practising Doctors*, 2 (4).

2792 • 2005 Zahir-ud-din. Jammu and Kashmir earthquake: Civilian and government responses. *Economic and Political Weekly*, 40 (44 and 45): 4666–4667.

2793 • 2005 Zaidi, H. Where is the help? *India Today*, (October 24): 68 –70.

2794 • 2005 Zaidi, H. Kashmir quake: Utterly helpless. *India Today*, (October 31): 45–48.

2795 • 2006 Department of Science and Technology. *National Programme on Seismicity: Highlights and Achievements.* New Delhi: Government of India.

2796 • 2006 Swami, P. Spring of hope in Tangdhar. *Frontline*, 22 (4): 64.

Response: Flood

2797 • 1944 Special Correspondent. Damodar Flood Enquiry Committee. *Science and Culture*, 9 (8): 338.

2798 • 1958 Chakravarty, A.K. *et al.* A plea for floods. *Indian Journal of Meteorology and Geophysics*, 8 (1): 16–18.

2799 ▲1962 Borwankar, P.G. *Poona: Redevelopment of Flood Affected Areas.* New Delhi: Department of Urban and Regional Planning, School of Planning and Architecture (Unpublished Postgraduate Dissertation).

2800 • 1968 Special Correspondent. Unmoved by tragedy. *Economic and Political Weekly*, 3 (43): 1665.

2801 • 1974 Thakur, C. *Human Perception and Adjustment to Flood Hazard in the Ganga Flood Plains*. Delhi: Unpublished Department of Geography, Delhi School of Economics, University of Delhi (Unpublished Ph.D thesis).

2802 • 1975 Ramachrandran, R. and S.C. Thakur. Human perception and adjustment to flood hazard in the Ganga *khadar*. In S.P. Dasgupta (ed). *The Upper Gangetic Flood-Plain: A Regional Survey*. Kolkata: National Atlas and Thematic Mapping Organisation. 86–95.

2803 • 1976 Special Correspondent. Flood relief: Power relations in action. *Economic and Political Weekly*, 11 (38): 1518–1519.

2804 • 1977 Kayastha, S.L. and R.P. Yadav. Human perception and adjustment to envi onmental hazards: A case study of three villages of Ganga flood plain. In C. Robert (ed). *Man, Culture and Settlement*. New Delhi: Kalyani Publishers.

2805 • 1978 Bhattacharya, M. *Bureaucratic Response to Emergency: The West Bengal Floods of 1978*. Kolkata: Department of Political Science, University of Calcutta (Unpublished Ph.D thesis).

2806 • 1978 Special Correspondent. West Bengal flood relief: The gasping rich and the helpless poor. *Economic and Political Weekly*, 13 (45 and 46): 1894.

2807 • 1980 Yadav, R.P. *Human Adjustment to Flood and Strategy for Development in Eastern Uttar Pradesh and Related Area Development Strategy*. Varanasi: Banaras Hindu University (Unpublished Ph.D thesis).

2808 • 1981 Bhattacharya, M. *Bureaucratic Response to Emergency: A Study of West Bengal Flood Relief 1978*. New Delhi: Indian Council of Social Science. (Project Report).

2809 • 1981 Gandhi, K.R.K. Uttar Pradesh: Politics of flood relief vs flood control. *Economic and Political Weekly*, 16 (47): 1889–1891.

2810 • 1982 Editorial. Orissa: Army assistance during floods. *Economic and Political Weekly*, 17 (18): 1722–1723.

2811 • 1986 Mathew, P.J. A flood of fury. *Frontline*, (September 6): 84–89.

2812 • 1986 Pathak, M. *Types of Agricultural Occupance of Flood Plains along the Ganga River between Varanasi and Ballia*. Burdwan: University of Burdwan (Unpublished Ph.D thesis).

2813 • 1986 Special Correspondent. Tragedy: Tempo of relief work on Godavari floods. *Frontline*, (September 6): 41–44.

2814 ▲1987 Government of Bihar. *Flood in Bihar, July and August 1987: A Memorandum Submitted to Central Team, Government of India*. Patna.

2815 • 1988 Dobal, M.K. Drama in Delhi. *Frontline*, 21 (October 15): 25–27.

2816 • 1988 Roy, P.K. Flooded by neglect. *Frontline*, 21 (October 5): 98–102.

2817 • 1989 Nayyar, K.P. and R. Madhavan. Kerala: Flood rituals. *Frontline*, 22 (August 19): 111–113.

2818 • 1990 Agarwal, B. Social security and the family: Coping with seasonality and calamity in rural India. *Journal of Peasant Studies*, 17 (3): 341–412.

2819 • 1990 Kattelmann, R. Conflicts and cooperation over floods in the Himalayan Ganges region. *Water International*, 15 (4): 189–194.

2820 • 1990 Maithani, B.P. Combating the flood menace. In K.M. Gupta (ed). *Himalayas: Man and Nature*. New Delhi: Lancers Books. 171–75.

2821 • 1990 Singh, J.P. *Human Response to Flood Hazard in the Lower Rapti-Saraju Doab.* Varanasi: Banaras Hindu University (Unpublished Ph.D thesis).

2822 • 1991 Panda, G.K. Human perception and community response to the flood hazards of Orissa coastal zone: An Indian experience of human adaptation to the natural hazards. In V.S. Mahajan (ed). *Environmental Planning: Machinery and Management.* New Delhi: Deep and Deep Publications Pvt. Ltd. 331–348.

2823 • 1991 Sen, N. *Flood, Man and Settlement: Case Study of Lower Damodar Valley Region*. New Delhi: School of Planning and Architecture (Unpublished Ph.D thesis).

2824 • 1992 Dutta, M.K. *Human Adjustment to Flood, Study of People's Perception: A Case Study of Kokilamukh Village in Jorhat District (Assam).* Delhi: Department of Geography, Delhi School of Economics, University of Delhi (Unpublished Master of Philosophy Dissertation).

2825 • 1992 Warier, K.U. Centre disappoints Kerala in relief. *New Age*, 40 (51): 6.

2826 • 1993 Dhawan, B.D. Coping with floods in Himalayan rivers. *Economic and Political Weekly*, 28 (18): 849–852.

2827 • 1993 Jain, R. and A. Goswami. Where have we gone wrong? *Down to Earth*, 2 (7): 10–13.

2828 • 1993 Satapathy, S. *Floods and Flood Control Policies: An Analysis with Reference to Mahanadi Delta in Orissa.* Trivandrum: Centre for Development Studies.

2829 • 1995 Rao, M.S. Online relief plan simulation of regular floods in Orissa. In *Proceedings of the World Congress on Natural Hazards Reduction.* 295–306.

2830 • 1996 Satya Ram. Rural cropping pattern in flood prone area of Jaunpur district, Uttar Pradesh. *National Geographical Journal of India*, 42 (1 and 2): 82–93.

2831 • 1997 Editorial. Scheme watch. *Down to Earth*, 6 (4): 10.

2832 • 1997 Editorial. End discrimination in relief work: Communist Party of India leaders visit flood affected areas of Punjab. *New Age*, 45 (38): 4.

2833 • 1997 Sinha, R. and D. Choudhary. Drained dry. *Down to Earth*, 7 (March 15): 22.

2834 • 1998 Khan, N. A. The boat people. *Down to Earth*, 7 (10): 41–42.

2835 • 1998 Yadav, C. P. Impact of flood on agricultural innovations and adjustments under flood prone agriculture. In B. N. Verma (ed). *Flood Devastation and Agricultural Development in Eastern India.* Delhi: B. R. Publishing Corporation. 141–148.

2836 • 1999 Editorial. Flood fury. *Down to Earth*, 8 (7): 13.

2837 • 1999 Krishnakumar, R. A transplanted village. *Frontline*, 16 (20): 66–67.

2838 • 1999 Mishra, D. K. Flood protection that never was: Case of Mahananda basin of North Bihar. *Economic and Political Weekly*, 34 (24): 2013.

2839 • 1999 Yadav, R. P. and H. L. Yadav (eds). *Floods in Eastern Uttar Pradesh: Human Adjustment and Related Area Development Strategy.* New Delhi: Radha Publications.

2840 • 2000 Basu, K. The business of faith. In S. Parasuraman and P. V. Unnikrishnan (eds). *India Disasters Report: Towards a Policy Initiative.* New Delhi: Oxford University Press. 354–355.

2841 • 2000 Dasgupta, K. Rising the phoenix. *Down to Earth*, 9 (9): 46.

2842 • 2000 Editorial. Mumbai: Watery tales. *Economic and Political Weekly*, 35 (28): 2513–2515.

2843 • 2000 Editorial. Floods: Water wheels. *Economic and Political Weekly*, 35 (40): 3540.

2844 • 2000 Lochan, K. and S. Awasthi. Lessons from Eastern Uttar Pradesh. In S. Parasuraman and P. V. Unnikrishnan (eds). *India Disasters Report: Towards a Policy Initiative.* New Delhi: Oxford University Press. 161–163.

2845 • 2000 Mishra, D. K. Inundated by excuse. *Down to Earth*, 9 (8): 55.

2846 • 2000 Misra, D. K. Breaches of promises. In S. Parasuraman and P. V. Unnikrishnan (eds). *India Disasters Report: Towards a Policy Initiative.* New Delhi: Oxford University Press. 155–157.

2847 • 2000 Ray, C. N. Mismanaging floods. *Economic and Political Weekly*, 35 (46): 3998–3999.

2848 • 2000 Roy, D. Flood: A small matter of history. In S. Parasuraman and P. V. Unnikrishnan (eds). *India Disasters Report: Towards a Policy Initiative.* New Delhi: Oxford University Press. 147–150.

2849 • 2000 Shrivastava, A. K. and R. B. Singh. Government response to flood disaster preparedness, relief and mitigation in Haryana. In R. B. Singh (ed). *Disaster Management.* Jaipur: Rawat Publications. 308–318.

2850 • 2001 Mishra, D.K. Living with floods: Peoples perspective. *Economic and Political Weekly*, 36 (29): 2756–2761.

2851 • 2001 News Report. Flood devastation in Orissa: Communist Party of India with other parties stage *dharna* before assembly. *New Age*, 49 (32): 7.

2852 • 2001 Special Correspondent. Orissa ill-prepared again. *Economic and Political Weekly*, 36 (29): 2724.

2853 • 2003 Jairath, J. Drought of sanity and flood of the absurd: Politics of water discourse. *Economic and Political Weekly*, 38 (7): 665–668.

2854 • 2003 Rafique, A. Floods, poverty and seasonal migration. *Economic and Political Weekly*, 38.

2855 • 2004 Editorial. Floods: All talk, no action. *Economic and Political Weekly*, 39 (33): 3660–3662.

2856 • 2004 Moench, M. and A. Dixit. *Adaptive Capacity and Livelihood Resilience for Responding to Floods and Droughts in South Asia.* Colorado: International Institute for Social and Environmental Transition.

2857 • 2005 Chatterjee, P. Mopping up in Mumbai. *The Lancet*, 366 (9488): 795.

2858 • 2005 Date, V. Mumbai floods: The blame game begins. *Economic and Political Weekly*, 40 (34): 3714–3716.

2859 • 2005 Katakam, A. A service outside the church. *Frontline*, 22 (17): 8.

2860 • 2005 Katakam, A. A question of reliance. *Frontline*, 22 (17): 10–11.

2861 • 2005 Pai, A. Troubled course. *India Today*, (December 19): 56–58.

2862 ■ 2005 Prasad, R. and Thakur, B. Floods, embankment and development in the lower Kosi plain. In B. Thakur *et al* (eds). *Urban and Regional Development in India: Essays in Honour of Professor L. N. Ram.* New Delhi. Concept Publishing Company. 632–657.

2863 • 2005 Ramachandran, R. Lessons not learnt. *Frontline*, 22 (17): 12–16.

2864 • 2005 Ray, C.N. Mismanagement of flood in Gujarat. *Economic and Political Weekly*, 40 (28): 3004–3006.

2865 • 2005 Revi, A. Lessons from the deluge: Priorities for multi-hazard risk mitigation. *Economic and Political Weekly*, 40 (36): 3911–3916.

2866 • 2005 Srinivasan, S. After the floods: The health services responsibilities. *Indian Journal of Medical Ethics*, 13 (4).

2867 • 2005 Viswanathan, S. Politics in troubled waters. *Frontline*, 22 (26): 31–32.

2868 • 2006 Despande, A.A. When disaster strikes: Our experience at King Edward Memorial Hospital, tertiary referral center in Mumbai. *Journal of Postgraduate Medicine*, 52 (4).

2869 • 2006 Editorials. Mumbai floods: A year later. *Economic and Political Weekly*, 41(30): 3264–3265.

2870 ■ 2006 Mahanta, C. and A. Mahanta. Bridge over Brahmaputra. *Economic and Political Weekly*, 41(7): 579–581.

2871 • 2006 Silove, D. *et al.* Understanding community psychosocial need after disasters: Implications for mental health services. *Journal of Postgraduate Medicine*, 52 (2).

2872 • 2006 Srinivasan, R.K. Wrong target. *Down to Earth*, 14 (17): 28–29.

Response: Heavy Rain

2873 • 2004 Mazari, R.K. and M.P. Sah. Pulia Nala cloudburst of (July 16) 2003, District Kullu (Himachal Pradesh): Lessons for policy implementation. *Himalayan Geology*, 25 (2): 153–161.

2874 • 2004 Jayant, T.V. Water woes wet Kerala. *Down to Earth*, 13 (1): 26–33.

Response: Landslide

2875 • 1993 Phekaekara, M.M. Mountain tragedy: Lessons from the Nilgiri landslip. *Frontline*, 26 (December 31): 130–132.

2876 • 1998 Ramakrishnan, V. Lessons from landslides. *Frontline*, 31 (October 9): 74–76.

2877 • 1998 Sarkar, S. *Landslide Hazard Zonation and Slope Stability Assessment Techniques: Applications to Srinagar-Rudraprayag (Garhwal Himalayas).* Roorkee: University of Roorkee (Unpublished Ph.D thesis).

2878 • 2004 Thigale, S.S. and A.S. Khandge. Coping with landslide disasters in the Western Ghats. In K.S. Valdiya (ed). *Coping with Natural Hazards: Indian Context.* Hyderabad: Orient Longman. 153–164.

2879 • 2005 Rautela, P. and R.K. Pande. Implication of ignoring the old disaster management plans: Lessons learnt from the Amparav tragedy of 23 September 2004 in the Nainital district of Uttaranchal. *Disaster Prevention and Management*, 14 (3): 388–394.

Response: Tsunami

2880 • 2005 Banerjee, S. Reflections on the tsunami. *Economic and Political Weekly*, 40 (2): 97–98.

2881 • 2005 Bidwai, P. Coping with the tsunami crises. *Frontline*, 22 (2): 100–102.

2882 • 2005 Chatterjee, P. Mental health care for India's tsunami survivors. *The Lancet*, 365 (9462): 833–834.

2883 • 2005 Cherian, J. A crucial contribution. *Frontline*, 22 (2): 133.

2884 • 2005 Chinnaraj, S. *Reconstruction and Resettlement Strategy for Tsunami Affected Areas of Nagappatanam*. New Delhi: School of Planning and Architecture (Unpublished Postgraduate Dissertation).

2885 • 2005 Editorial. Coping with the future. *Economic and Political Weekly*, 40 (3): 163–164.

2886 • 2005 Gupta, R. Waves of unawarness. *Journal of Geological Society of India*, 65 (4): 512.

2887 • 2005 Human Rights Watch. *After the Deluge: India's Reconstruction following the 2004 Tsunami*. New York: Human Rights Watch.

2888 • 2005 Krishnakumar, R. Relief and camps. *Frontline*, 22 (2): 18.

2889 • 2005 Kurien, J. A good occasion for change. *Frontline*, 22 (3): 18–20.

2890 • 2005 Linnerooth, B. J. *et al.* Refocusing disaster aid. *Science*, 309 (5737): 1044–1046.

2891 • 2005 Menon, P. Rebuilding lives. *Frontline*, 22 (2): 4–8.

2892 • 2005 Mohan, N. *et al.* Relief and rehabilitation: Ensuring inclusion. *Economic and Political Weekly*, 40 (15): 1493–1495.

2893 • 2005 Ramachandran, R. For a reliable system. *Frontline*, 22 (3): 21–23.

2894 • 2005 Sana, S. *et al.* We did our bit. *Down to Earth*, 13 (19): 55.

2895 • 2005 Silove, D. and A. Zwi. Translating compassion into psycho-social aid after the tsunami. *The Lancet*, 366: 269–271.

2896 • 2005 Srivivasam, S. Tsunami rehabilitation. *Yojana*, 46: 65–67.

2897 • 2005 Subramanian, T. S. A majestic survivor. *Frontline*, 22 (2): 132.

2898 • 2005 Suryanarayan, P. S. Relief and unease. *Frontline*, 22: 43–45.

2899 • 2005 Swaminathan, M. S. Towards sustainable rehabilitation. *Frontline*, 22 (4): 40.

2900 • 2005 Varde-Naqvi, S. Car Nicobar resilience. *Frontline*, 22 (13): 95–97.

2901 • 2005 Viswanathan, S. Signs of hope. *Frontline*, 22 (2): 9–12.

2902 • 2006 Ahluwalia, A. D. Geoethical audit of tsunami, December 26 2004: Challenge before leaders, media and scientists. *Current Science*, 90 (6): 762–764.

2903 • 2006 Balaji, C. Tsunami fallout. *Down to Earth*, 14 (17): 14.

2904 • 2006 Fertig, B. *et al.* Tsunamis and the international response: Economic, social and environmental dimensions. In P. S. Ranade (ed). *Natural Disasters: Management and Preparedness*. Hyderabad: Institute of Charted Finanacial Analysts of India University Press. 90–110.

2905 • 2006 George, A. Partners in action: Non-government organisations in tsunami relief and rehabilitation. In S. M. Ramaswamy, *et al* (eds). *Geomatics in Tsunami*. New Delhi: New India Publishing Agency. 209–213.

2906 • 2006 Giles, J. India's ban on foreign boats hinders tsunami research. *Nature*, 439 (January 26): 380–380.

2907 • 2006 Jayant, T. V. Studied delay. *Down to Earth*, 14 (19): 14.

2908 • 2006 Krishnakumar, R. A question of accountability. *Frontline*, 22 (27): 13–16.

2909 • 2006 Menon, P. and H. Appanaswamy. Rising to the challenge. *Frontline*, 22 (27): 8–10.

2910 • 2006 Radhakrishna, M. G. *et al.* Tsunami, one year after the tragedy. *India Today*, (January 2): 26.

2911 • 2006 Santhanam, R. The tsunami disaster and Tamil Nadu's response. In S. M. Ramaswamy, *et al.* (eds). *Geomatics in Tsunami.* New Delhi: New India Publishing Agency. 215–220.

2912 • 2006 Shaw, R. Indian Ocean tsunami and aftermath: Need for environment-disaster synergy in the reconstruction process. *Disaster Prevention and Management*, 15 (1): 5–20.

2913 • 2006 Siva Kumar, B. R. and Singh, B. Indian tsunami (2004): Initiatives by Natural Resource Data Management Systems, Department of Science and Technology, India. In S. M. Ramaswamy, *et al* (eds). *Geomatics in Tsunami.* New Delhi: New India Publishing Agency. 31–36.

2914 • 2006 Sridhar, V. Partners in rebuilding. *Frontline*, 22 (27): 11–12.

2915 • 2006 Stipe, J. Rebuilding lives: The aftermath of the South Asia earthquake. *Journal of International Affairs*, 59(2):298–312.

2916 • 2006 Subramanian, T. S. A long wait for relief. *Frontline*, 22 (27): 17–20.

2917 • 2006 Ugra, S. Islands of renewal: Counting the cost. *India Today*, (January 2): 42

2918 • 2006 Vasanthi. Riders on the storm. *India Today*, (January 2): 38.

Response: Assorted

2919 • 1967 Special Correspondent. How a calamity was averted. *Economic and Political Weekly*, 2 (16): 1316–1317.

2920 • 1967 Special Correspondent. Government caught unprepared. *Economic and Political Weekly*, 2 (20): 916–917.

2921 • 1978 Jodha, N. S. Effectiveness of farmers adjustment to risk. *Economic and Political Weekly*, 13 (25): A38–A48.

2922 • 1982 Geipal, R. *Disaster and Reconstruction.* London: Allen and Unwin

2923 • 1984 Bose, C. P. B. *Disaster, Politics and Administration: A Study of Three Disasters.* Vishakhapatnam: Andhra University, Waltair (Unpublished Ph.D thesis).

2924 • 1987 Bhoi, P. K. *Perception and Human Response to Natural Hazards in Coastal Orissa (The Lower Mahanadi-Devi interfluve): A Case Study.* Bhubaneswar: Department of Geography, Utkal University (Unpublished Master of Philosophy Dissertation).

2925 • 1990 Ministry of Agriculture. *Contingency Action Plan for Natural Calamities.* New Delhi: Government of India.

2926 • 1990 Special Correspondent. Height of ruthlesssness. *India Today*, (May 8): 65–6

2927 • 1990 Winchester, P. Economic power and response to risk: A case study of disaster. In J. Handmer and A. N. Rowsell (eds). *Hazards and the Communication of Risk*. London: Gower.

2928 • 1991 Special Correspondent. What can be done right away? *Frontline*, 24: 93.

2929 • 1991 Special Correspondent. Fatal indifference. *India Today*, (May 8): 62.

2930 • 1991 Valery, A. Facing disaster on your own terms. *Social Welfare*, 38 (1): 10–13.

2931 • 1992 Bose,P. K. Mass media and natural disaster. *Science and Culture*, 58 (5): 98–99.

2932 • 1992 Dasgupta, S. Natural disaster reduction: Media role. *Science and Culture*, 58 (5–8): 100–104.

2933 • 1992 Inter-Ministerial Task Force. *Coping with Natural Disasters: An Integrated Approach*. New Delhi: Department of Science and Technology, Government of India.

2934 • 1992 Mohommaed, I. and M. I. El–Sabh. Coping with natural disasters is everyone's concern. In D.K. Sinha (ed). *Natural Disaster Reduction for Nineties: Perspectives, Aspects and Strategies*. Kolkata: International Journal Services. 112–124.

2935 • 1992 Mukhopadhyay, S. P. Media and reduction of natural disaster. *Science and Culture*, 58 (5–8): 93–94.

2936 • 1992 Padmanabhan, B.S. Disaster reduction and sustainable development. *Kurukshetra*, 41(3): 35–36.

2937 • 1992 Rattien, S. The role of media in hazard mitigation and disaster management. *Science and Culture*, 58 (5–8): 85–88.

2938 • 1992 Sensarma, A. K. Role of media in natural disaster reduction: Post-disaster phase. *Science and Culture*, 58 (5–8): 89–92.

2939 • 1992 Special Correspondent. Fighting natural disasters. *Kurukshetra*, 41(3): 7–44.

2940 • 1993 Mohan, M. M. Politics of relief: Game of one-upmanship in Karnataka. *Frontline*, 26. 5 November: 17–18.

2941 • 1993 Special Correspondent. The East Coast road to ecological disaster. *Down to Earth*, 2. 13 January: 8–9.

2942 • 1994 Prakash, I. *Disaster Management: Societal Vulnerability to Natural Calamities and Man-Made Disasters, Preparedness and Response*. Ghaziabad: Rashtra Prahari Prakashan.

2943 • 1995 Pant, J.C. Disaster relief in India. In G.K. Mishra and G.C. Mathur (eds). *Natural Disaster Reduction*. New Delhi: Reliance Publishing House. 79–96.

2944 • 1995 Pathak,Y.P. and M.A. Sinare. Military bridging equipment and boats for disaster management. In G.K. Mishra and G.C. Mathur (eds). *Natural Disaster Reduction*. New Delhi: Reliance Publishing House. 28–66.

2945 ● 1996 Singh, R.B (ed). *Disaster Environment and Development.* Proceedings of the International Geographical Union.Delhi: Oxford and IBH Publishing Ltd.

2946 ● 1997 Biswas, R. *et al.* Community perception about disaster preparedness and response programme in four states of India. *Indian Journal of Public Health*, 41(1): 11–15.

2947 ▲ 1997 Datta, P. *et al. Documentation on Disaster Preparedness Work in North Bengal and North Bihar.* Kolkata: Oxford Committee for Famine Relief-India Trust.

2948 ● 1998 Jeyanth, K. *et al.* Crisis intervention by community participation. *International Journal of Mass Emergencies and Disasters*, 16 (3): 363–370.

2949 ● 1998 Rahim, K. Abatement of man-made hazards vis-a-vis natural hazards: Role of collective action. *Indian Journal of Adult Education*, 53 (9): 18–26.

2950 ● 1999 Sinha, A. Relief administration and capacity building for coping mechanism towards disaster reduction. *Shelter,* special issue, October: 9–12.

2951 ● 2000 Baker, L. Shelter during disasters. In S. Parasuraman and P.V. Unnikrishnan (eds). *India Disasters Report: Towards a Policy Initiative.* New Delhi: Oxford University Press. 191–194.

2952 ● 2000 Das, S.N. Financing disaster response. In S. Parasuraman and P.V. Unnikrishnan (eds). *India Disasters Report: Towards a Policy Initiative.* New Delhi: Oxford University Press. 115–117.

2953 ● 2000 Editorial. Disaster management: Mostly non-existent. *Economic and Political Weekly*, 35 (34): 3353–3354.

2954 ● 2000 George, T. Responding with faith. In S. Parasuraman and P.V. Unnikrishnan (eds). *India Disasters Report: Towards a Policy Initiative.* New Delhi: Oxford University Press. 127–129.

2955 ● 2000 India Disasters Report Team. How the Parliament debates disaster issues. In S. Parasuraman and P.V. Unnikrishnan (eds). *India Disasters Report: Towards a Policy Initiative.* New Delhi: Oxford University Press. 118–123.

2956 ● 2000 Lynch, S. and P.V. Unikrishanan. A profile of agencies involved in disaster responses. In S. Parasuraman and P.V. Unnikrishnan (eds). *India Disasters Report: Towards a Policy Initiative.* New Delhi: Oxford University Press. 130–137.

2957 ● 2000 Murthy, R.S. Disaster and mental health: Responses of mental health professionals. *Indian Journal of Social Work*, 61(4): 675–692.

2958 ● 2000 Pant, N.K. Armed forces and natural disasters. *Down to Earth*, 8 (9): 3.

2959 • 2000 Parasuraman, S. and P.V. Unnikrishnan. Disaster response in India: An overview. In S. Parasuraman and P.V. Unnikrishnan (eds). *India Disasters Report: Towards a Policy Initiative.* New Delhi: Oxford University Press. 3–20.

2960 • 2000 Prasannan, R. Role of the Indian Armed Forces in disaster relief. In S. Parasuraman and P.V. Unnikrishnan (eds). *India Disasters Report: Towards a Policy Initiative.* New Delhi: Oxford University Press. 102–103.

2961 • 2000 Press Council of India. *Preparing People to Cope with Disasters: A Report on Role of Media.* New Delhi.

2962 • 2000 Singh, P.M. *et al. Disaster Response in India.* New Delhi: A Joint Study Initiative by the Center of Excellence in Disaster Management and Humanitatian Assistance and the Foreign Military Studies Office.

2963 • 2000 Tully, M. Media reportage, public perceptions and policies. In S. Parasuraman and P.V. Unnikrishnan (eds). *India Disasters Report: Towards a Policy Initiative.* New Delhi: Oxford University Press. 42–45.

2964 • 2000 Unnikrishnan, P.V. and Sarwal. The Sphere project's humanitarian charter. In S. Parasuraman and P.V. Unnikrishnan (eds). *India Disasters Report: Towards a Policy Initiative.* New Delhi: Oxford University Press. 142–144.

2965 • 2000 Valdiya, K.S. Free access to topographical maps: Coping with natural hazards, assessing development potential and monitoring environmental changes. *Current Science*, 79 (4): 479–480.

2966 • 2001 Dasgupta, K. Availing in delays. *Down to Earth*, 9: 22.

2967 • 2001 Editorial. Solid foundation. *Down to Earth*, 9 (19): 47.

2968 • 2001 Jayaraman, K.S. Indian science benefits from sizeable funding increase. *Nature*, 410 (March 8): 134.

2969 • 2001 Lok Sabha Secretariat. *Natural Calamities in the Country Lok Sabha Debates Under Rule 193.* New Delhi: Government of India.

2970 • 2001 Medury, U. Coping with disasters: A community based approach. In P. Sahni, *et al* (eds). *Disaster Mitigation: Experiences and Reflections.* New Delhi: Prentice-Hall of India Private Limited. 155–170.

2971 • 2001 Nanda, P. (ed). *Civil Society and Natural Disasters.* Bhubaneswar: Orissa Disaster Mitigation Mission.

2972 • 2001 Satendra. Dealing with disasters. *Down to Earth*, 9 (23): 53.

2973 • 2002 D' Souza, D. The attitude of relief. *Seminar*, 509: 90–94.

2974 • 2002 Enarson, E. *'We Want Work': Rural Women in the Gujarat Drought and Earthquake.* Quick Response Research Report No. 135. Boulder: Natural Hazards Research and Applications Information Center, University of Colorado.

2975 • 2002 Parasuraman, S. and P.V. Unnikrishnan. Disaster response in India: An overview. *Indian Journal of Social Work*, 63(2): 151–172.

2976 • 2002 Swami, S.K. *Trigger Mechanism: A Concept for Emergency Response Plans*. New Delhi: Department of Agriculture and Cooperation, Natural Disaster Management Division, Government of India.

2977 • 2004 Diaz, J.O.P. and R. Lakshminarayana. Psychological support in event of mass destruction: Challenges and lessons. *Economic and Political Weekly*, 39 (21): 2121–2127.

2978 • 2004 Sharma, M. and H. Abdi. Biotechnology interventions in rehabilitation of victims of natural disasters: Some instances. In K.S. Valdiya (ed). *Coping with Natural Hazards: Indian Context.* Hyderabad: Orient Longman. 263–266.

2979 • 2004 Sur, D. and S.K. Bhattacharya. Sanitation medicare in the hazard-affected areas. In K.S. Valdiya (ed). *Coping with Natural Hazards: Indian Context.* Hyderabad: Orient Longman. 257–262.

2980 • 2004 Valdiya, K.S. (eds). *Coping with Natural Hazards: Indian Context.* Hyderabad: Orient Longman.

2981 • 2005 Bhatt, M.R. Keep the poor in focus. *Down to Earth*, 14 (7): 50.

2982 • 2005 Iyer,S.D. Post-disaster reflections. *Science and Culture*, 88 (10): 1532.

2983 • 2005 Jha, T. and Achal, A. Coping with multiple natural hazards in north Bihar. In B. Thakur *et al* (eds). *Urban and Regional Development in India: Essays in Honour of Professor L.N. Ram.* New Delhi. Concept Publishing Company. 621–631.

2984 • 2005 Mahapatra, R. *et al.* What? Why? Our disaster preparedness. *Down to Earth*, 14: 27–36.

2985 • 2005 Saikia, A. Disasater unpreparedness in Assam. *Economic and Political Weekly*, 40 (24): 2389–2391.

2986 • 2006 Aravind, R.E and S. Jeyaram. Needs of special children in disaster. *School of Social Work*, 2 (12): 5–8.

2987 • 2006 Kotoch, A. The responders' cauldron: The uniqueness of international disaster response. *Journal of International Affairs*, 59 (2): 153–172.

2988 • 2006 Latta, K.S. Children's response to disasters: Role of parents. *School of Social Work*, 2 (12): 27–29.

2989 • 2006 Prabhu,T and E. Sinu. Psycho-social care for children in disaster. *School of Social Work*, 2 (12): 11–18.

2990 • 2006 Sharma, K.K, *et al* (eds). *Environmental Geohazards: Science and Society*. New Delhi: Research India Press.

2991 • 2006 Singh, R.B. A perspective post-mortem of natural disaster. In R.B. Singh (ed). *Natural Hazards and Disaster Management Vulnerability and Mitigation*. Jaipur Rawat Publications, 290–305.

2992 • 2006 Visvesvaran, P.K. Crisis intervention when calamity strikes. *School of Social Work*, 2 (12): 30–32.

Forecasting

The catchment of research on forecast is defined by those who raise the inquiry 'what is likely to occur or happen'. Keen to project the future, such research is about probability and prediction. All research that focuses its inquiry on that which is forthcoming, be it the use of sophisticated models, simulations and even traditional methods of investigation, falls under the rubric of 'forecasting'. The reference space of future could be a long-term projection like the analysis of swarms as precursors of earthquakes or of a short term trajectory of a cyclone veering towards a coast, both fall within the group of 'forecasting'. Inbuilt within this category are also references that are concerned with communication about the future. Warnings that announce the possible arrival of a disaster are also capped within those on forecasting. The number of research references on the theme of 'forecasting' were 246. This is 6 per cent of the total (Tables 6.22, 6.23 and 6.24).

Table 6.22 References for each Category

Natural Disaster	References	Number
Cyclone	2993–3050	58
Drought	3051–3074	24
Earthquake	3075–3159	85
Floods	3160–3191	32
Heat and Cold Wave	–	0
Heavy Rains	3192	1
Landslide	3193–3202	10
Storm	3203–3207	5
Tsunami	3208–3226	19
Assorted	3227–3236	10

Table 6.23 Research References on Forecasting

'Natural' disaster	Book	Report	Thesis	Article[1]	Chapter[2]	Feature[3]	Total
Cyclone	1	4	3	40	7	3	58
Drought	0	3	1	3	14	3	24
Earthquake	1	3	2	51	11	17	85
Floods	0	5	1	15	9	2	32
Heat and Cold Wave	0	0	0	0	0	0	0
Heavy Rain	0	0	0	1	0	0	1
Landslide	1	0	0	1	8	0	10
Storm	0	0	0	5	0	0	5
Tsunami	0	2	0	5	2	10	19
General Hazard	1	1	0	4	3	1	10
Total	**4**	**18**	**7**	**125**	**54**	**36**	**244**

[1] in a journal; [2] in a book/report/proceedings; [3] in a magazine.

Table 6.24 Dominance of Natural over Human in Research References on Forecasting

Phase	Natural	Human	Ecological	Total
1830–1947	2	0	0	2
1948–1989	56	2	0	58
1990–2006	157	27	0	184
Total	**215**	**29**	**0**	**244**

Forecasting: Cyclone

2993 ■ 1972 Das, P.K. Prediction model for storm surges in the Bay of Bengal. *Nature*, 239 (September 22): 211–213.

2994 ■ 1975 Ghosh, S.K. Cyclone tracking and storm warning. *Geographical Review of India*, 37 (3): 207–209.

2995 ■ 1975 Sikka, D.R. Forecasting the movement of tropical cyclones in the Indian seas by non-divergent barotropic model. *Indian Journal of Meteorology, Hydrology, and Geophysics*, 26 (4)323–325.

2996 ■ 1977 Ghosh, S.K. Prediction of storm surges on the East Coast of India. *Indian Journal of Meteorology and Geophysics*, 28 (1): 157–168.

2997 ■ 1979 Kivganov, A.F. and U.C. Mohanty. Hydrodynamical method of forecasting the trajectories of tropical cyclones in the Indian seas. *Indian Journal of Meteorology, Hydrology, and Geophysics*.15: 9–15.

2998 ■ 1981 Johns, B. *et al.* Numerical simulation of the surge generated by the 1997 Andhra cyclone. *Quarterly Journal of the Royal Meteorological Society*, 107: 919–934.

2999 ■ 1983 Johns, B. *et al.* Simulation of storm surges using a three dimensional numerical model: An application to the 1977 Andhra cyclone. *Quarterly Journal of the Royal Meteorological Society*, 109: 211–224.

3000 ■ 1984 Mandal, G.S. Cyclone warning and its limitations. *Vayumandal*, 14: 49–55.

3001 ● 1984 Sinha, A.K. and S.U. Avrani. The disaster warning process: A study of the 1981 Gujarat cyclone. *Disasters: The Journal of Disaster Studies, Policy and Management,* 8 (1): 67–73.

3002 ■ 1985 Ghosh, S.K. *Probable Maximum Storm Surge on the Coasts of India and Bangladesh.* New Delhi: South Asian Publications.

3003 ■ 1989 Pal, P.K. *et al.* Cyclone track prediction using Indian National Satellite data. *Proceedings of Indian Academy of Sciences, Earth and Planetary Sciences*, 98 (4): 353–364.

3004 ■ 1991 Mandal, G.S. *Tropical Cyclones and their Forecasting and Warning Systems in the North Indian Ocean.* Geneva: World Meteorological Organisation.

3005 ■ 1991 Raj, Y.E.A. and P.V. Pillai. Prediction of movement of post monsoon depressions and cyclonic storms over Arabian Sea. *Vayumandal*, 21 (3 and 4): 120–130.

3006 ● 1992 Sensarma, A.K. Andhra cyclones of 1977 and 1990: A study of development in a disaster warning-response system in India. In *Proceedings of the World Congress on Natural Hazards Reduction.* 165–183.

3007 ■ 1994 Bansal, R.K. *et al.* Impact of European Remote Sensing Satellite-1 scatterometer data on the analysis and prediction of a severe cyclonic storm in the Bay of Bengal. *Vayumandal*, 24 (3 and 4): 59–66.

3008 ■ 1994 Das, H.P. *The Incidence, Prediction, Monitoring and Mitigation Measures of Tropical Cyclones and Storm Surges.* Pune: Agricultural Meteorological Division of India.

3009 ■ 1994 Das, P.K. On the prediction of storm surges. *Sadhana: Academy Proceedings in Engineering Sciences*, 19 (4): 583–595.

3010 ■ 1994 Joseph, P.V. Tropical cyclone hazards and warning and disaster mitigation system in India. *Sadhana: Academy Proceedings in Engineering Sciences*, 19 (4): 551–566.

3011 ■ 1994 Mohanty, U.C. Tropical cyclones in the Bay of Bengal and deterministic methods for prediction of their trajectories. *Sadhana: Academy Proceedings in Engineering Sciences*, 19 (4): 567–582.

3012 ■ 1995 Mandal, G.S. Tropical cyclones and their warning systems. In G.K. Mishra and G.C. Mathur (eds). *Natural Disaster Reduction.* New Delhi: Reliance Publishing House. 128–155.

3013 ■ 1995 Mandal, G.S. *Tropical Cyclones and their Damage Potential: Status of Wind Engineering in India.* Nagpur: Indian Society For Wind Engineering.

3014 ■ 1997 Dube, S.K. *et al.* Storm surge in the Bay of Bengal and Arabian Sea: The problem and its prediction. *Mausam*, 48 (2): 283–304.

3015 ■ 1997 Krishna Rao, A.V.R.K. Tropical cyclone synoptic method of forecasting. *Mausam*, 48 (2): 239–256.

3016 ■ 1997 Mohanty, U.C and A. Gupta. Deterministic methods for prediction of tropical cyclone tracks. *Mausam*, 48 (2): 257–252.

3017 ■ 1997 Rao, Y.V.R. *et al.* Storm surge prediction and frequency analysis for the Andhra coast of India *Mausam* 48 (4): 555–556.

3018 ■ 1998 Balasubramanian, K.V. Cyclone detection radar Chennai: 25 years of meritorious service. *Vayumandal*, 28 (3 and 4): 89–97.

3019 • 1998 Banerjee, R. Wasted warning. *India Today*, (April 27): 58–59.

3020 ■ 1998 Krishnamurti, T.N. and B. Jha Cyclone track prediction. *Sadhana-Academy Proceedings in Engineering Sciences*, 23 (5–6): 653.

3021 ■ 1998 Kumar, A. *Some Numerical Simulation Studies of Tropical Cyclone Circulation over Bay of Bengal.* Vishakhapatnam: Andhra University. Waltair (Unpublished Ph.D thesis).

3022 • 1999 Editorial. Cyclone warning system. *Down to Earth*, 8 (15): 11.

3023 ■ 1999 Mandal, G.S. Forecasting and warning systems for cyclones in India. *Shelter,* special issue, *October*: 24–26.

3024 ■ 2001 Ghose, S.K. Prediction of storm surges on the East Coast of India. *Mausam*, 52 (1): 137–148.

3025 ■ 2001 Mandal, G.S. Tropical cyclones and their forecasting and warning system in India. In P. Sahni, *et al* (eds). *Disaster Mitigation: Experiences and Reflections.* New Delhi: Prentice-Hall of India Private Limited. 171–185.

3026 • 2002 Dash, B. Lessons from Orissa Super Cyclone: Need for integrated warning system. *Economic and Political Weekly*, 37 (45): 4270–4271.

3027 • 2002 Dash, B. *Warning and Response: A Case Study of Orissa Super Cyclone 1999.* New Delhi: Jawarahal Nehru University (Unpublished Ph.D Thesis).

3028 ■ 2002 Indian Meteorological Department. *Damage Potential of Tropical Cyclones.* Pune: Office of the Additional Director General of Meteorological Research

3029 ■ 2002 Kalsi, S.R. Early warning system for tropical storms in India. In J. Zschau and A.N. Kuppers (eds). *Early Warning Systems for Natural Disaster Reduction.* Netherlands: Springer. 191–198.

3030 ■ 2002 Kaur, U. Warning and forecasting of tropical cyclones in India. In S.K. Tandon and B.Thakur (eds). *Recent Advances in Geomorphology, Quaternary Geology and Environmental Geosciences: Indian Case Studies.* New Delhi: Manisha Publications. 409–422.

3031 ■ 2002 Trivedi, D.K. *et al.* Numerical simulation of a super cyclonic storm Orissa.: Impact of intial conditions. *Meteorological Applications*, 9 (3): 367–376.

3032 ■ 2003 Nagaratna, K. *Numerical Simulations of Tropical Cyclones over Bay of Bengal using Betis Miller Cumulus Parameterisation Scheme.* Vishakhapatnam: Andhra University, Waltair (Unpublished Ph.D thesis).

3033 ■ 2004 Choudhury, S. and S. Chattopadhyay. Identification of coasts vulnerable for severe tropical cyclone: A statistical evaluation. *Mausam*, 55 (3): 502–507.

3034 ■ 2004 Dube, S.K. A strategy for cyclone disaster reduction: Real-time storm surge prediction. In K.S. Valdiya (ed). *Coping with Natural Hazards: Indian Context.* Hyderabad: Orient Longman. 197–203.

3035 ■ 2004 Mandal, M. *et al.* Study on the impact of parameterization of physical processes on prediction of tropical cyclones over the Bay of Bengal with National Centre for Atmospheric Research/ Pennsylvania State University mesoscale model. *Natural Hazard Journal*, 31 (2): 391–414.

3036 ■ 2004 Mohanty, U.C. and M. Mandal. Coping with tropical cyclone disaster: The forecasting aspect. In K.S. Valdiya (ed). *Coping with Natural Hazards: Indian Context.* Hyderabad: Orient Longman. 180–196.

3037 ■ 2004 Mohanty, U.C. *et al.* Simulation of Orissa Super Cyclone 1999 using Pennsylvania State University/ National Centre for Atmospheric Research meso-scale model. *Natural Hazard Journal*, 31 (2): 373–390.

3038 ■ 2006 Agnihotri, N.P. *et al.* A bay river coupled model for storm surge prediction along the Andhra Coast of India. *Natural Hazard Journal*, 39 (1): 83–101.

3039 ■ 2006 Ashrit, R.G. *et al.* Multilattice microcannonical simulation of the 1999 Orissa Super Cyclone: Impact of bogus vortex on track and intensity prediction. *Mausam*, 57 (1): 129–134.

3040 ■ 2006 Goswami, P. *et al.* Advance forecasting of cyclone track over North Indian Ocean using a global circulation model. *Mausam*, 57 (1): 111–118.

3041 ■ 2006 Gupta, A. Current status of tropical cyclone track prediction: Techniques and forecast errors. *Mausam*, 57 (1): 151–158.

3042 ■ 2006 Mandal, M. and U.C. Mohanty. Numerical experiments for improvement in mesoscale simulation of Orissa Super Cyclone. *Mausam*, 57 (1): 79–96.

3043 ■ 2006 Rambabu, C.G. Quick scat scatterometer wind data: Impact on tropical cyclone forecasting by a meso-scale model. *Mausam*, 57 (1): 141–150.

3044 ■ 2006 Rao, D.V.B. and D.H. Prasad. Numerical prediction of the Orissa Super Cyclone (1999): Sensitivity to the convection boundary layer and explicit moisture processes. *Mausam*, 57 (1): 61–78.

3045 ■ 2006 Rao, Y.V.R. *et al.* Tropical cyclone prediction by numerical models in Indian Meteorological Department. *Mausam*, 57 (1): 47–60.

3046 ■ 2006 Sandeep, S. *et al.* The impact of assimilation of advanced microwave sounding unit data for the prediction of a tropical cyclone over India using a mesoscale model. *International Journal of Remote Sensing*, 27 (20): 4621–4653.

3047 ■ 2006 Shivare, R.P. and V.S. Srinivas. Tropical cyclonic storms: Modelling studies in Indian Air Force. *Mausam*, 57 (1): 135–140.

3048 ■ 2006 Sikka, D.R. Major advances in understanding and prediction of tropical cyclones over North Indian Ocean: A perspective. *Mausam*, 57 (1): 165–196.

3049 ■ 2006 Singha, U.N. *et al.* Preliminary results on the simulation of the 1999 Orissa Super Cyclone using a general circulation model with a new boundary layer code. *Mausam*, 57 (1): 119–128.

3050 ■ 2006 Trivedi, D.K. *et al.* Impact of physical parameterisation schemes on the numerical simulation of Orissa Super Cyclone 1999. *Mausam*, 57 (1): 97–110.

Forecasting: Drought

3051 ■ 1970 Pandey, S.N. Ground water investigation by resistivity method in the drought affected area near Markundi, (District Mirzapur). *Annals of Arid Zone*. 977–84.

3052 ■ 1979 Sarkar R.P. Droughts in India and their predictability. In *Proceedings of the International Symposium on Hydrological Aspects of Droughts*. New Delhi: Indian Institute of Technology: 33–39.

3053 ■ 1979 Seshagiri, N. Stochastic vector flow graph: Analysis of meteorological cause-effect relationships for drought prediction. In *Proceedings of the International Symposium on Hydrological Aspects of Drought*. New Delhi: Indian Institute of Technology.

3054 ■ 1984 Ramakrishna, Y.S. *et al.* On prediction of droughts in the Indian arid region. *Mausam*, 35 (3): 349–354.

3055 • 1984 Reddy, D.N. Towards a drought watch and management system in Andhra Pradesh: A proposal. In B.R. Naidu (ed). *Disaster Management*. Tirupati: Sri Venkateswara University. 49–54.

3056 ■ 1985 Mooley, D.A. and B. Parthasarathy. A probability model for areal extent of drought and flood over India during summer monsoon season. *Annals of National Association of Geographers, India*, 5 (1): 34–44.

3057 ■ 1988 Thiruvengadachari, S. *National Agricultural Drought Assessment and Monitoring System in India: The Satellite Contribution.* (Special Publication on Space and Drought Management). Bangalore: Indian Space Research Organisation. 44–89.

3058 ■ 2000 Das, H.P. Monitoring the incidence of large scale drought in India. In D.A.Wilhite (ed). *Drought: A Global Assessment (Volume 1).* London: Routledge, 181–195.

3059 ● 2000 Joshi, Y.G. Prediction and perception of drought: The indigenous vision of a tribal society in an overexploited ecosystem. In R.B. Singh (ed). *Disaster Management* Jaipur: Rawat Publications. 132–138.

3060 ■ 2000 Suri, S. Amateur radios: Right frequencies. In S. Parasuraman and P.V. Unnikrishnan (eds). *India Disasters Report: Towards a Policy Initiative.* New Delhi: Oxford University Press. 91–92.

3061 ■ 2001 Roy Sinha, K.C. and M.P. Shewal. Probability of occurrence of drought in various subdivisions of India. *Mausam*, 52 (3): 541–546.

3062 ■ 2001 Special Correspondent. Drought warning for Northwestern region. *New Age*, 49(7): 11.

3063 ■ 2002 Center of Studies in Resource Engineering. *Recent Trends in Drought Assessment, Monitoring and Management.* Mumbai: Indian Institute of Technology.

3064 ■ 2002 Jeyaseelam, A.D. Recent trends in the remote sensing applications to drought assessment and monitoring. In R. Nagarajan (ed). *National Workshop Proceedings on Recent Trends in Drought Assessment Monitoring and Management.* Mumbai: Centre of Studies in Resources Engineering Indian Institute of Technology. 38–46.

3065 ■ 2002 Muralidharan, S. Troubled record, poor outlook. *Frontline*, 19(21): 50.

3066 ● 2002 Nagarajan, R. Integrated drought vulnerability assessment and monitoring for micro-level planning. In R. Nagarajan (ed). *National Workshop Proceedings on Recent Trends in Drought Assessment Monitoring and Management.* Mumbai: Centre of Studies in Resources, Engineering Indian Institute of Technology. 47–57.

3067 ■ 2003 Editotial. Cloud watch. *Down to Earth.* 11 (18): 20.

3068 ■ 2003 Jat, M.L. *Prediction of Drought under Different Agro-Climatic Zones of Rajasthan.* Udaipur: Udaipur University (Unpublished Ph.D Thesis).

3069 ■ 2003 Shewale, M.P. and D.M. Pase. On the long term drought prediction over Saurashtra and Kutch. *Mausam*, 54 (3): 754–757.

3070 ■ 2003 Sikka, D.R. Evalution of monitoring and forecasting of summer monsoon over India: A review of monsoon drought of 2002. *Proceedings of Indian National Science Academy*, 69 (A5): 479–504.

3071 ■ 2003 Singh, R.K.S. Vegetation and temperature condition indices from National Oceanic and Atmospheric Administration (US)-Advanced Very High Resolution Radiometer data from monitoring drought over India. *International Journal of Remote Sensing*, 24 (22): 4393–4402.

3072 ■ 2003 Sridharan, V.K. *et al.* Geographic information systems approach to drought forecasting using parametric probability distribution functions. *In Proceedings of the Second Conference on Disaster Management Case Histories*. Pilani: Birla Institute of Technology. 172–178.

3073 ■ 2004 Samra, J.S. *Review and Analysis of Drought Monitoring, Declaration and Impact Management in India*. Colombo: International Water Management Institute.

3074 ■ 2005 Prakash, V.S. Karnataka state takes lead in the country in establishing drought monitoring system. In R.K. Sharma and G. Sharma (eds). *Natural Disaster: Institutional Building for Mitigating the Impact on Vulnerable Sections*. New Delhi: APH Publishing Corporation. 167–176.

Forecasting: Earthquake

3075 ■ 1936 Davison, C. Future earthquake in Assam. *Science and Culture*, 1(11): 708–709

3076 ■ 1956 Tandon, A.N. Zone in India liable to earthquake damage. *Indian Journal of Meteorology and Geophysics*, 7 (1): 93.

3077 ■ 1969 Pakisep, L.C. *et al.* Earthquake prediction and control. *Science,* 166 (3912): 1467–1474.

3078 ■ 1974 Gupta, H.K. and B.K. Rastogi. Will another damaging earthquake occur in Koyna? *Nature*, 248 (March 15): 215–216.

3079 ■ 1979 Rao, S.P. and R.B. Rao. Estimated earthquake probabilities in Northeast India, Andaman and Nicobar Islands. *Mausam*, 30:167–273.

3080 ■ 1980 Guha, S.K. *et al. Probabilistic Estimate of Seismic Risk in India and Surrounding Vicinity*. Chichester: John Wiley and Sons.

3081 ■ 1980 Ramalingeswara, R.B. *Investigations of Probabilities and Risk of Earthquakes on the Indian Subcontinent*. Vishakhapatnam: Andhra University, Waltair (Unpublished Ph.D thesis).

3082 ■ 1981 Seeber, L. and J.G. Ambruster. Great detachment earthquakes along the Himalayan arc and long term forecasting. In D.E. Simpson and

P.G. Richards (eds). *Earthquake Prediction: An International Review, Maurice Ewing, Series 4*. The American Geophysical Union. 259–277.

3083 ■ 1982 Goswami, D.C. and S.K. Sharma. Probabilistic earthquake expectancy in the Northeast Indian region. *Bulletin of the Seismological Society of America*, 72: 999–1009.

3084 ■ 1984 Chakrabarty, S.K. *et al.* The usefulness of geomagnetic observations for predicting the earthquakes. *Mausam*, 35 (2): 213–218.

3085 ■ 1985 Gupta, H.K. Cachar earthquake of 31 December 1984: Is it a signal for the beginning of seismic activity? *Journal of Geological Society of India*, 26 (2): 145–140.

3086 ■ 1986 Gupta, H.K *et al.* Seismicity of the Northeast India region: Earthquake swarms precursory to moderate magnitude to great earthquakes. *Journal of Geological Society of India*, 28 (5): 367–406.

3087 ■ 1987 Gupta, A.K. Seismicity of Northeast India region: An earthquake forecast. *Indian Journal of Geology*, 59 (3): 165–178.

3088 ■ 1987 Khattri, K.N. Great earthquakes, seismicity gap and potential for the earthquake disaster along the Himalaya plate boundary. *Tectonophysics*, 138 (1): 79–92.

3089 ■ 1987 Saha, S.N. Micro earthquakes and earthquake prediction. *Geographical Review of India*, 49 (4): 37–46.

3090 ■ 1989 Bhatia, S.C. *et al.* On intermediate-term prediction of strong earthquakes in the Himalayan arc region using pattern recognition algorithm M 8. *Proceedings of Indian Academy of Sciences, Earth and Planetary Sciences*, 98 (1): 111–124.

3091 ■ 1989 Gupta, H.K. and H.N. Singh. Earthquake swarms precursory to moderate magnitude to great earthquakes in the Northeast India region. *Tectonophysics*, 167 (2–4): 285–298.

3092 ■ 1989 Padale, J.G. *et al.* Feasibility of earthquake prediction in Koyna region (India) *Mausam*, 40 (2): 281

3093 ■ 1989 Saha, S.N. Earthquakes: Prediction and mitigation of hazards. *Science and Culture*, 55 (9): 332–338.

3094 ■ 1989 Virk, H.S. Forecasting of earthquakes. *Science and Culture*, 55 (9): 338–343.

3095 • 1990 Arya, A.S. Damage scenario of a hypothetical 8.0 magnitude earthquake in the Kangra region of Himachal Pradesh. *Bulletin of Indian Society of Earthquake Technology*, 27 (3): 121–132.

3096 ■ 1992 Gupta, H.K. Earthquake prediction related studies in India. *Current Science*, 62 (1 and 2): 34–36.

3097 ■ 1992 Nanda, A.R. and K.S. Natarajan. Population projection for earthquake prone zones of India 1981–2001. In *Proceedings of the World Congress on Natural Hazards Reduction*. 35–39.

3098 • 1993 Iyengar, R.N. How safe is the proposed Tehri dam to earthquake? *Current Science*, 65 (5): 384–392.

3099 ■ 1993 Khattri, K.N. Seismic gaps and likelihood of occurrence of larger earthquake in Northeast India. *Current Science*, 65 (11 and 12): 885–888.

3100 • 1993 Padmanabhan, R. Warnings unheeded. *Frontline*, 26 (October 22): 11.

3101 ■ 1994 Guha, S.K. Earthquake prediction: Present status and future prospects. *Indian Journal of Geology*, 66 (1): 1–14.

3102 ■ 1994 Khattri, K.N *et al.* Seismic hazard estimation using modelling of earthquake strong ground motions: A brief analysis of 1991 Uttarkashi earthquake (Himalaya) and prognostication for a great earthquake in the region. *Current Science*, 67: 343–353.

3103 ■ 1994 Subramanyam, V. Potential quakes. *Down to Earth*, 3 (15): 12.

3104 ■ 1994 Tagare, G.V. Predictability of earthquakes. *Deccan Geographer*, 32 (1): 13–18.

3105 ■ 1995 Chander, V.K. and R.K. Gahalaut. A simulation of upper crustal stresses for great and moderate thrust earthquakes of the Himalaya. *Proceedings of Indian Academy of Sciences, Earth and Planetary Sciences*, 104 (1): 115–129.

3106 ■ 1996 Chander, R. and V.K. Gahalaut. On some likely hazards of the next great earthquake of the Garhwal, Kumaon and Western Nepal Himalaya. *Himalayan Geology*, 17: 11–16.

3107 ■ 1996 Guptasharma, D. Is the seismic risk similar everywhere along the Himalayan collision belt? *Himalayan Geology*, 17: 1–10.

3108 ■ 1996 Khattri, K.N. Numerical simulation of earthquake strong ground motion. *Himalayan Geology*, 17: 45–52.

3109 ■ 1996 Maru, V.M. *et al.* Earthquakes: How much are they amenable to prediction? *Himalayan Geology*, 17: 147–156.

3110 ■ 1996 Nagarajan, R. Rapid assessment procedure to demarcate areas susceptible to earthquake induced ground failure for environment management-A case study from part of Northeast India. *Bulletin of Engineering Geology and the Environment*, 61 (2): 99–119.

3111 ■ 1996 Rajarathnam, S. Identification of seismic-prone faults using remote sensing data: A case study of Tamil Nadu . In R.B. Singh (ed). *Disasters, Environment and Development.* Delhi: Oxford and IBH Publishing Co. Pvt. Ltd. 219–227.

3112 ■ 1996 Virk, H.S. Radom studies for earthquake prediction *Himalayan Geology*, 17: 91–104.

3113 ■ 1997 Joshi, A. Modelling of active lineament for predicting earthquakes around Dehradun and Garhwal Himalayas. *Tectonophysics*, 283 (1–4): 289–310.

3114 ■ 1997 Mukherjee, S. *Re-evaluation of Seismic Potentialilty of Delhi Rohtak Area using Remote Sensing and Seismological Data.* New Delhi: School of Environmental Sciences, Jawaharlal Nehru University. (Unpublished Project Report).

3115 ■ 1997 Parvez, I. A. and A. Ram. Probabilistic assessment of earthquake hazards in the Northeast Indian Peninsula and Hindukush regions. *Pure and Applied Geophysics*, 149 (4): 731–746.

3116 ■ 1998 Ahmed, K. Warning quakes. *Down to Earth*, 7 (15): 22–23.

3117 ■ 1998 Bilham, R. Geodetic constraints on the translation and deformation of India: Implication for the future Great Himalayan earthquakes. *Current Science*, 74 (3): 213–229.

3118 ■ 1999 Burgmann, R. Model inversion of global positioning systems and leveling measurements across the Himalaya: Implications for earthquake hazards and future geodetic networks. *Himalayan Geology*, 20 (1): 57–72.

3119 ■ 1999 Khattri, K. N. Probabilities of occurrence of great earthquakes in Himalaya. *Current Science*, 77 (7): 967–972.

3120 ■ 1999 Khattri, K. N. Probabilities of occurrence of great earthquakes in the Himalayas *Proceedings of Indian Academy of Sciences, Earth and Planetary Sciences*, 108 (2): 87–92.

3121 ■ 1999 Kumar, D. *et al.* Modelling of accelerograms of two Himalayan earthquakes using a novel semi-empirical method and estimation of accelerograms for a hypothetical great earthquake in Himalaya. *Current Science*, 76 (6): 819–830.

3122 ■ 1999 Saikia, C. K. *et al.* Ground motion time histories from large Himalayan earthquakes for earthquake loss reduction in India. *Himalayan Geology*, 20 (1): 131–148.

3123 ■ 1999 Sharma, S. K. The probability of occurrence of a high magnitude earthquake in Northeast India. *Journal of Geophysics*, 20 (3): 129–135.

3124 ■ 1999 Sinha, R. and N. Adarsh. A postulated earthquake damage scenario for Mumbai. *ISET Journal of Earthquake Technology*, 36 (2–4): 169.

3125 ■ 2000 Bhattacharya, S. N. Recent advances in seismic instrumentation and data interpretation in India. *Current Science*, 79 (9): 1347–1358.

3126 ■ 2001 Bilham, R. Future shocks. *Down to Earth*, 10 (9): 9.

3127 ■ 2001 Gupta, H. K. Medium term forecast of the 1988 Northeast India earthquake. *Tectonphysics*, 338 (3–4): 281–286.

3128 ■ 2001 Kerr, R. A. Tectonic design combine for Indian disaster, more coming. *Science*, 291: 802–803.

3129 ■ 2001 Kumaraswamy, K. Earthquakes of South India and animal based forecasting. In Ganesh *et al.* (eds). *Proceedings of the Workshop on*

Popular Lectures On Facts and Myths of Earthquake, Indian Context. Tiruchirapalli: Bharathidasan University. 51–58.

3130 ■ 2001 Rajamanickam, G.V. Earthquake potential in Tamil Nadu. In Ganesh, *et. al* (eds). *Proceedings of Workshop on Popular Lectures on Facts and Myths of Earthquake, Indian Context.* Tiruchirapalli: Bharathidasan University. 40–41.

3131 ■ 2001 Ramanamurthy, M.V. Forcasting earthquakes and volcanic activity. *Journal of Geological Society of India*, 58 (3)272.

3132 ■ 2001 Romanowicz, B. and D. Giardini. The future of permanent seismic networks. *Science*, 293 (5537): 2000–2001.

3133 ■ 2001 Sharma, M.L. and D. Shanker. Estimation of seismic hazard parameters in the Himalayas and its vicinity from mixed data file. *ISET Journal of Earthquake Technology*, 38 (2–4): 93–102.

3134 ■ 2001 Singh, R.B. *et al.* Geo-informatics for land resource monitoring and magnitude of earthquakes in India. In *Proceedings of the 88 Annual Session of the Indian Science Congress Association.* Indian Science Congress. 82–83.

3135 ■ 2001 Suresh, V. Deathquake: Where next? *India Today*, (February 12): 63–65.

3136 • 2001 Vaux, T. *The Earthquake in Gujarat,: Report of a Monitoring Visit.* London.

3137 ■ 2002 Bhattacharya, S.N. Monitoring of earthquakes in India: Recent advances. In V. Subramaniam (ed). *Environmental Hazards in South Asia.* New Delhi: Concept Publishing Company. 11–16.

3138 ■ 2002 Majumdar, T.J. and D. Massonnet. Differential synthetic aperture radar interferometry applications for monitoring geological hazard, with special reference to Latur earthquake 1993. *Current Science*, 83 (4): 502–508.

3139 ■ 2002 Singh, S.K. *et al.* Ground motion in Delhi from future great earthquake in the central seismic gap of the Himalayan arc. *Bulletin of the Seismological Society of America*, 92 (2): 555–569.

3140 ■ 2003 Agarwal, S.K. and A. Chourasia. Application of geographic information systems in assessment of building damage vulnerability due to earthquake in Jabalpur area. In M. Mohanty and S. Aggarwal (eds). *Applications of Geographic Information Systems in Earthquake and Landslide Hazard Assessment.* New Delhi: Department of Science and Technology, Government of India. 12–14.

3141 • 2003 Bapat, A. Ready to quake? Delhi is completely earthquake prone. *Down to Earth*, 11 (20): 54.

3142 ■ 2003 Behera, L. *Seismic Imaging of Mahanadi Delta and its Tectonic Significance.* Hyderabad: Osmania University (Unpublished Master of Philosophy Dissertation).

3143 ■ 2003 Joshi, A. Predicting strong motion parameters for the Chamoli earthquake of 28 March 1999, Garhwal Himalaya (India) from simplified finite fault model. *Journal of Seismology,* 7 (1) :1–17.

3144 ■ 2003 Singh, S.K. *et al.* Estimation of ground motion for Bhuj M 7.6 (26 January 2001) and for future earthquakes in India. *Bulletin of the Seismological Society of America*, 93: 353–370.

3145 ■ 2003 Sinha, R. Geographic information systems application in earthquake risk assessment of urban areas. In M. Mohanty and S. Aggarwal (eds). *Applications of Geographic Information Systems in Earthquake and Landslide Hazard Assessment.* New Delhi: Department of Science and Technology, Government of India. 15–16.

3146 ■ 2004 De, R. and J.R. Kayal. Seismotectonics model of the Sikkim Himalaya: Constraint from micro earthquake surveys. *Bulletin of the Seismological Society of America*, 94 (3): 1395–1400.

3147 ■ 2004 Petersen, M.D. *et al.* Sensitivity analysis of seismic hazard for the Northwestern portion of Gujarat . *Tectonophysics*, 390 (1–4): 105–115.

3148 ■ 2004 Shanker, D. *et al.* Regional time-predictable modeling in the Hindu Kush Pamir-Himalayas region. *Tectonophysics*, 390 (1–4): 129–140.

3149 ■ 2004 Tiwari, R.K. *et al.* Characterization of earthquake dynamics in Northeastern India region: A modern non-linear forecasting approach. *Pure and Applied Geophysics*, 161(4): 865–880.

3150 ■ 2005 Kerr, R.A. and P. Bagla. A seismic murmur of what's ahead for India. *Science*, 310 (5746): 208.

3151 ■ 2005 Menon, A.K. Seisomology: The next big one. *India Today*, (Oct 24): 80.

3152 ■ 2005 Radhakrishna, B.P. Further threats of earthquakes and tsunami. *Journal of Geological Society of India*, 65 (4): 512.

3153 ■ 2005 Seismology Division, Department of Science and Technology. *National Programme on Global Positioning System for Earthquake Hazard Assessment.* New Delhi: Governtment of India.

3154 • 2005 Sridhar, V. Living on the edge. *Frontline*, 22 (3): 15–18.

3155 ■ 2005 Unnithan, S. India: Aftermath what lies ahead. *India Today*, (October 24): 76–79.

3156 ■ 2006 Gaur, V.K. Earthquake and tsunami threat to South Asia. In *Proceedings of the Brain Storming Session on the Great Tsunami Event of 26 December 2004*. New Delhi: Department of Science and Technology, Government of India. 9–24.

3157 ■ 2006 Jaiswal, K. and R. Sinha. Probabilistic modelling of earthquake hazard in stable continental shield of the Indian Peninsula. *ISET Journal of Earthquake Technology*, 43 (3).

3158 ■ 2006 Pandey, Y. and P.K.S. Chauhan. Seismic vulnerability of Delhi. In R.B. Singh (ed). *Natural Hazards and Disaster Management Vulnerability and Mitigation*. Jaipur: Rawat Publications, 124–140.

3159 ■ 2006 Rajendran, K. *et al.* Do the historic ruptures along the Andaman-Sumatra and Makran subduction zones signal future great earthquakes. In *Proceedings of the Chapman Conference on Active Tectonics and Seismic Potential of Alaska Aleyeska Resort*, Girdwood.

Forecasting: Flood

3160 ■ 1938 Bose, N.K. Floods and prediction of flood levels by river models. *Science and Culture*, 4 (3): 195–196.

3161 ■ 1955 Davar, K.S. Flood forecasting. *Indian Journal of Power and River Valley Development*, 5 (3): 19–22.

3162 ■ 1955 Sircar, P.K. Probability of high floods in the Rupnarayan, Kansai and Subarnrekha. *Geographical Review of India*, 17 (1): 3–5.

3163 ■ 1965 Bose, N.K. and A.K. Sinha. Prediction of flood stages in the Yamuna at Delhi. *Indian Journal of Power and River Valley Development*, 15 (1): 1–8.

3164 ■ 1965 Ministry of Irrigation and Power. *Report of the Technical Committee on Scientific Flood Forecasting and Warning*. New Delhi: Government of India.

3165 ■ 1965 Shenoy, R.C. Flood forecasting and warning. *Bhagirath*, (July 12): 79–81.

3166 ■ 1976 Jena, A.B. and S.P. Rout. Flood forecasting in river Mahanadi below Hirakud by unit hydrograph method. *Indian Journal of Power and River Valley Development*, 26(1): 108–114.

3167 ■ 1976 Narasimha Rao, D.V.L. *On the Development of Scientific Flood Forecasting System, with Special Reference to India Catchments*. Vishakhapatnam: Andhra University, Waltair (Unpublished Ph.D thesis).

3168 ■ 1977 Sinha, B. Flood forecasting in India. *Bhagirath*, (October 24): 151–158.

3169 ■ 1978 Guha, M.K. *et al.* A linear model for flood forecasting. *Mausam*, 29 (1): 136–142.

3170 ■ 1979 Sinha, B. Flood forecasting in India. *Bhagirath*, 26 (October): 144–150.

3171 ■ 1980 Chender, S. *et al.* Flood stage forecasting in rivers. In M.J. Lighthill (ed*). Monsoon Dynamics*. London: Cambridge University Press.

3172 ■ 1980 Kathuria, S.N. A numerical model for flood routing. *Mausam*, 31 (4): 523–528.

3173 ■ 1981 Abdi, S.D.S. Flash floods and their warning. In *Proceedings of the International Conference on Flood Disaster*. New Delhi: Indian National Science Academy. 238–50.

3174 ■ 1981 Ramaswamy, C. Meteorological aspects of severe and catastrophic floods in India during the period 1923–1979 and their importance in flood forecasting on the issue of early warning. In *Proceedings of the International Conference on Flood Disaster*. New Delhi: Indian National Science Academy.

3175 ■ 1984 Chander, S. and P. N. Kapoor. *Newer Techniques in High Flow Range Forecasting*. New Delhi: Report of the Indian National Committee for the International Hydrological Programme, Civil Engineering Department, Indian Institute of Technology.

3176 ■ 1984 Irrigation and Power Department. *Flood Forecasting for Mahanadi Basin for Operation of Hirakud Reservoir*. Bhubaneswar: Government of Orissa.

3177 ■ 1986 Abbi, S. D. S. Flood estimation in Himalayan catchment. In *Proceedings of Workshop on Flood Estimation in Himalayan Region*. Central Board of Irrigation and Power, New Delhi.

3178 ■ 1986 Special Correspondent. Flood forecasting and warning set: Uttar Pradesh. *Bhagirath*, 33: 191.

3179 ■ 1989 Central Water Commission. *Manual on Flood Forecasting*. New Delhi: Government of India.

3180 ■ 1989 Verma, R. *et al.* Flood forecasting in Jhelum basin. *Bhagirath*, 36 (October–December): 216–218.

3181 ■ 1990 Garde, R. J., and U. C. Kothyari. Flood estimation in India catchments. *Journal of Hydrogeology*, 113: 135–146.

3182 ■ 1991 Central Water Works. *Flood Forecasting and Warning Network: Performance Appraisal*. New Delhi: Government of India.

3183 ■ 1992 Singh. R. The flood in North India in 1988: Forecasting aspects. *Mausam*, 43 (2): 131–136.

3184 • 1997 Editorial. Delhi under duress. *Down to Earth*, 6 (5): 10.

3185 ■ 1998 Kukreja, P. B. Flood forecasting and warning system in India. In B. N. Verma (ed). *Flood Devastation and Agricultural Development in Eastern India*. Delhi: B. R. Publishing Corporation. 377–388.

3186 ■ 2002 Goswami, D. C. Flood forecasting in the Brahmaputra river, India: A Case Study. In S. R. Chalise and M. Shrestha (eds). *Report of the Regional Cooperation for Flood Disaster Mitigation in the Hindukush-Himalayas*. Kathmandu: The International Centre for Integrated Mountain Development, (Internal Report).

3187 ■ 2002 Sharma, A. Early warning, community interpretations and perceptions: A case of recurrent floods and their warnings in Delhi. In J. Zschau and A. N. Kuppers (eds). *Early Warning Systems for Natural Disaster Reduction*. Netherlands: Springer. 77–80.

3188 ■ 2003 Kalsi, S.R. Monitoring and forecasting of tropical cyclones and their associated effects. In H.K. Gupta (ed). *Disaster Management.* Hyderabad: University Press (India) Private Limited. 48–84.

3189 ■ 2004 Kshirsagar, M.M. and P. Vijaygopal. Flood warning system for a thermal power project in flood prone area. *Indian Journal of Power and River Valley Development*, 54 (5 and 6): 5–6.

3190 ■ 2004 Singh, R.B. and D. Kumar. Monitoring, mapping and mitigation of flood disaster in India using remote sensing and geographic information systems: A case Study. In K. Takara, *et al* (eds). *Monitoring, Prediction and Mitigation of Disasters by Satellite Remote Sensing.* Kyoto: Disaster Prevention Research Institute.

3191 ■ 2006 Singh, R.B. and D. Kumar. Monitoring, mapping and mitigation of flood disaster in India using remote sensing and geographic information systems: A case study. In R.B. Singh (ed). *Natural Hazards and Disaster Management Vulnerability and Mitigation.* Jaipur: Rawat Publications. 206–217.

Forecasting: Heavy Rain

3192 ■ 2004 Dimri, A.P. *et al.* Simulation of heavy precipitation with an intense western disturbance over Western Himalayas. *Natural Hazard Journal*, 31 (2): 499–519.

Forecasting: Landslide

3193 ■ 1992 Gerhard, J. Landsliding and Sediment Prediction in the Himalaya: Contrasting Conceptual Frameworks. In R.B. Singh (ed). *Dynamics of Mountain Geosystems.* New Delhi: Rawat Publications. 40.

3194 ■ 1992 Haigh, M.J. Estimating landslide outfalls in mountain environments. In R.B. Singh (ed). *Dynamics of Mountain Geosystems.* New Delhi: Rawat Publications.

3195 ■ 1996 Sambandan, S.T., and V. Natarajan. Evaluation of landslide susceptibility in central parts of Kerala. In A.P. Santhakumar, *et al* (eds). *Proceedings of the International Conference on Disasters and Mitigation, (A–4).* Chennai: Anna University. 107–111.

3196 ■ 2004 Nagarajan, R. What to do in landslide probable event situation? In R. Nagarajan (ed). *Lanslide Disaster: Assessment and Monitoring.* New Delhi: Anmol Publications. 314–318.

3197 ■ 2004 Nagarajan, R. Assessment and monitoring of Konkan landside. In R. Nagarajan (ed). *Lanslide Disaster: Assessment and Monitoring.* New Delhi: Anmol Publications. 202–221.

3198 ■ 2004 Nagarajan, R. Landslide classification, monitoring and risk assessment: An overview. In R. Nagarajan (ed). *Lanslide Disaster: Assessment and Monitoring.* New Delhi: Anmol Publications. 1–21.

3199 ■ 2004 Nagarajan, R. (eds). *Lanslide Disaster: Assessment and Monitoring.* New Delhi: Anmol Publications.

3200 ■ 2004 Neaupane, K.M., and S.H. Achet. Use of back propagation neural network for landslide monitoring: A case study in the Higher Himalaya. *Engineering Geology*, 74 (3–4): 213–226.

3201 ■ 2004 Rao, K.S. Synthetic aperture radar inferometry and its application to landslides. In R. Nagarajan (ed). *Lanslide Disaster: Assessment and Monitoring.* New Delhi: Anmol Publications. 113–122.

3202 ■ 2006 Shukla, A.K. *et al.* Avalanche risk estimation for approach road from Manali to south portal of proposed Rohtang tunnel. In K.K. Sharma, *et al* (eds). *Environmental Geohazards: Science and Society.* New Delhi: Research India Press. 128–137.

Forecasting: Storm

3203 ■ 1980 Misra, P.K. *et al.* Forecasting hailstorms over India. *Mausam*, 31 (3): 385–396.

3204 ■ 1999 Ghosh, S. *et al.* Identification of significant parameters of the prediction of pre-monsoon thunderstorms at Kolkata. *International Journal of Climatology*, 19 (6): 673–681.

3205 ■ 2001 Chaudhuri, S. and S. Chattopadhyay. Measure of line: A relevant parameter for forecasting pre-monsoon thunderstorm over Greater West Bengal. *Mausam*, 52 (4): 679–684.

3206 ■ 2001 Madan, O.P. and N. Ravi. Forecasting peak surface dust wind in association with thunderstorm activity during pre-monsoon season in Delhi. *Mausam*, 52 (1): 385–396.

3207 ■ 2004 Mukhopadhyay, P. Idealized simulation of a thunderstorm over Kolkata using regional atmospheric modelling system. *Journal of Indian Geophysical Union*, 8 (4): 253–266.

Forecasting: Tsunami

3208 • 2005 Acharyya, S.K. Scientists failed to warn vulnerable coastal people in time. *Journal of Geological Society of India*, 65 (4): 506.

3209 ■ 2005 Bagla, P. A dead spot for the tsunami network. *Science*, 310: 1604.

3210 • 2005 Cyranoski, D. Solo efforts hamper tsunami warning efforts. *Nature*, 433 (January 27) 343.

3211 • 2005 Dash, B. Tsunami warning system: Beyond technological fixes. *Economic and Political Weekly*, 40 (4): 269–271.

3212 ■ 2005 Hamsapriya, T. and S. Sumathi. Fault tolerant tsunami warning system using mirrored servers. *Indian Journal of Power and River Valley Development*, 55 (10 and 11): 238–247.

3213 ■ 2005 Jayaraman, K.S. India pledges to fund alert system in wake of disaster. *Nature*, 433 7021, (January 6).

3214 ■ 2005 Jayaraman, K. S. India makes waves over tsunami warning system. *Nature*, 438 (December 22): 1060–1061.

3215 ● 2005 Kerr, R. A. Failure to gauge the quake crippled the warning effort. *Science*, 307 (5707): 201–201.

3216 ■ 2005 Kintisch, E. Global tsunami warning system takes shape. *Science*, 307: 331.

3217 ● 2005 Marris, E. Inadequate warning system left Asia at the mercy of tsunami. *Nature*, 433 (January 6): 3–5.

3218 ■ 2005 Nanang, T. P. and I. Gunawan. Tsunami sources in the Sumatra region and simulation of the 26 December 2004 Aceh tsunami. *ISET Journal of Earthquake Technology*, 42 (4): 111.

3219 ■ 2005 Ramachandran, R. The Pacific system. *Frontline*, 22 (3): 24–25.

3220 ● 2005 Subramanian, T. S. Their own warning system. *Frontline*, 22 (2): 16.

3221 ■ 2006 Gupta, H.K. Need of the hour. In *Proceedings of the Brain Storming Session on the Great Tsunami Event of 26 December 2004*. New Delhi: Department of Science and Technology, Government of India. 53–56.

3222 ■ 2006 Murty, T. S. *et al.* Numerical modelling concepts for tsunami warning systems. *Current Science*, 90 (8): 1073–1081.

3223 ■ 2006 Murty, T. S. *et al.* Tsunami warning system for the Indian Ocean. In *Proceedings of the Brain Storming Session on the Great Tsunami Event of 26 December 2004*. New Delhi: Department of Science and Technology, Government of India. 25–32.

3224 ● 2006 Ramachandran, R. Working towards early warning. *Frontline*, 22 (27): 21–24.

3225 ● 2006 Singh, R. B. Tsunami disaster in India: Warning, responses, impact and mitigation. In R.B. Singh (ed). *Natural Hazards and Disaster Management Vulnerability and Mitigation*. Jaipur: Rawat Publications Jaipur. 62–79.

3226 ● 2006 Symonds, P. The Asian tsunami: Why there were no warnings. In P.S. Ranade (ed). *Natural Disasters: Management and Preparedness*. Hyderabad: Institute of Charted Finanacial Analysts of India University Press. 8–155.

Forecasting: Assorted

3227 ■ 1991 Seshagiri, N. The distributed control system: Theoretic model of natural hazard propagation and reduction. In *Proceedings of the World Congress on Natural Hazards Reduction*. 40–52.

3228 ● 1992 Reddy, A. V. S. Integrated early warning systems for disaster management and long term development. *Journal of Rural Development*, 11 (3): 265–289.

3229 • 1993 James, M. Panchayati Raj and early warning of disasters. *Economic and Political Weekly*, 28 (21): 1019–1020.

3230 ▪ 1994 Singh, R. B (ed). *Space Technology for Disaster Monitoring and Mitigation in India.* Tokyo: International Center for Disaster-Mitigation Engineering, Report 6.

3231 ▪ 1996 Singh, R.B. Disaster monitoring and mitigation. In R.B. Singh (ed). *Disasters, Environment and Development.* New Delhi: Oxford and IBH Publishing Co. Pvt. Ltd. 31–62.

3232 ▪ 1996 Singh, R. B (ed). *Research in Geography of Disaster and Environment Monitoring and Forecasting.* Delhi: A P H Publishing Corporation.

3233 • 1999 Gadhok, T.K. Urban risk in India. *Shelter,* special issue, *October.* 7–8.

3234 ▪ 2000 Singh, A. K. Disaster mapping and forecasting. In S. Parasuraman and P.V. Unnikrishnan (eds). *India Disasters Report: Towards a Policy Initiative.* New Delhi: Oxford University Press. 88–90.

3235 ▪ 2005 Schiermeier, Q. Natural disaster: The choas to come. *Nature,* 438 (December 15): 903–906.

3236 • 2006 Sarkar, S. Disaster Management Act 2005: A disaster in waiting. *Economic and Political Weekly*, 41 (35): 3760–3763.

Management

It is to the Latin word *maneggiare,* which means to handle, as derived from the hand, that the etymology of management can be traced. The goals of research on management are to lay down the principles of handling, directing and controlling an event. To achieve this end the research could identify, critique and suggest ground rules for efficient management. A number of stake holders are engaged in the management of disasters. All references that assess, evaluate or give importance to the role of communities, armed forces, government, non-government, administration and even modern information technology belong to the arena of this group of research. Research that comments on the level and degree of preparedness, and lists the dos and don'ts, and provides the code of standards as in the case of construction of buildings in seismic regions or land use plans in a flood prone area are also part and parcel of the references on 'management'. Amidst the 4004 there are 766 references on the theme of 'management', this is 19 per cent of the total (Tables 6.25, 6.26 and 6.27).

Table 6.25 References for each Category

Natural Disaster	References	Number
Cyclone	3237–3284	48
Drought	3285–3411	127
Earthquake	3412–3542	131
Floods	3543–3698	156
Heat and Cold Wave	-	0
Heavy Rain	-	0
Landslide	3699–3727	29
Storm	3728	1
Tsunami	3729–3746	18
Assorted	3747–4004	258

Table 6.26 Natural and Human in Research References on Management

Phase	Natural	Human	Ecological	Total
1830–1947	2	0	0	2
1948–1989	68	50	45	163
1990–2006	138	308	157	603
Total	**208**	**358**	**202**	**768**

Table 6.27 Research References on Management

'Natural' disaster	Book	Report	Thesis	Article[1]	Chapter[2]	Feature[3]	Total
Cyclone	1	18	4	10	15	0	48
Drought	13	24	6	30	44	10	127
Earthquake	1	16	15	69	22	8	131
Floods	3	28	17	66	35	7	156
Heat and Cold Wave	0	0	0	0	0	0	0
Heavy Rain	0	0	0	0	0	0	0
Landslide	0	7	0	6	15	1	29
Storm	1	0	0	0	0	0	1
Tsunami	0	1	0	8	6	3	18
General Hazard	34	54	7	68	86	9	258
Total	**53**	**148**	**49**	**257**	**223**	**38**	**768**

[1] *in a journal;* [2] *in a book/report/proceedings;* [3] *in a magazine.*

Management: Cyclone

3237 ▲1971 Office of the Chief Engineer. *Final Report of the Cyclone Disaster Mitigation Committee.* Hyderabad: Government of Andhra Pradesh.

3238 ▲1972 Special Correspondent. Cyclone distress mitigation committee. *Indian Journal of Meteorology and Geophysics*, 23 (1): 115–118.

3239 ● 1979 Andhra Pradesh State Housing Corporation. *Cyclone Prone Area Management Housing Programme: A Feasibility Report.* Hyderabad Andhra Pradesh State Housing Corporation.

3240 ■ 1979 Ressler, E.M. Observations on the development of educational materials following the Andhra Pradesh cyclone 1977. *Disasters: The Journal of Disaster Studies*, Policy and Management, 3 (3): 283–287.

3241 ● 1980 Revenue Department. *Community Cyclone Construction Programme for Cyclone Prone Areas of Andhra Pradesh.* Hyderabad: Government of Andhra Pradesh.

3242 ● 1981 Revenue Department. *Cyclone Contingency Plan of Action.* Hyderabad: Government of Andhra Pradesh.

3243 ● 1984 Narayan, N. An approach to cyclone disaster preparedness. In B. R. Naidu (ed). *Disaster Management.* Tirupati: Sri Venkateswara University. 87–90.

3244 ■ 1984 Ramesh, K. S., and G. V. Sunder Sheshu. *Cyclone Disaster Management in Vellore District.* Hyderabad: Administrative Staff College of India.

3245 ● 1985 Raghavulu, N.B. Disaster preparedness in agriculture, with particular reference to cyclones in Andhra Pradesh. *Vayumandal,* 15 (1and 2).

3246 ▲1985 Ramesh, K. S. *Management of Cyclone Disasters: A Systems Approach.* Hyderabad: Administrative Staff College of India, Mime.

3247 ■ 1986 Saha, A.K. Recommendations of Cyclone Review Committee. New Delhi: Department of Science and Technology, Government of India.

3248 ● 1986 Winchester, P. *Cyclonic Vulnerability and Housing Policy in the Krishna Delta, South India* (1977 to 1983). Norwich: School of Development Studies, University of East Angilia (Unpublished Ph.D Thesis).

3249 ● 1988 Building Material and Technology Promotion Council. *Cyclone Resistant Housing Guidelines for Do's and Dont's while Constructing a House in the Cyclone Prone Area.* New Delhi: Housing and Urban Development Corporation.

3250 ▲1990 Sinha, D.K. *Some Remarks on Management of Natural Disasters Pertaining to Coastal Areas in Coastal Management of West Bengal.* Kolkata: Sea Explorers' Institute.

3251 • 1991 Rao, S.M.V. *From Neoscience to Science and Beyond: A Critical Investigation of Buildings in Cyclone-Prone Areas.* Master of Building Science, University of Southern California (Unpublished Postgraduate Dissertation).

3252 ▲ 1994 Suresh, K.M. *A Study of Cyclone Disaster Management of Coastal Tracts of Guntur District (Andhra Pradesh).* Hyderabad: Acharya N.G Ranga Agricultural University (Unpublished Postgraduate Dissertation).

3253 • 1995 Shanmugasundaram, J. *et al.* Engineering of structures for cyclone disaster mitigation. In G.K. Mishra and G.C. Mathur (eds). *Natural Disaster Reduction.* New Delhi: Reliance Publishing House. 156–179.

3254 • 1997 Samanta, A. Cyclone hazards and community response in coastal West Bengal: An anthropo-historical perspective. *Economic and Political Weekly*, 32 (38): 2424–2428.

3255 ▲ 1997 Stenchion, P. Cyclones and disaster management. *Mausam*, 48 (4): 609–620.

3256 • 1997 World Meteorological Organisation. *Tropical Cyclone Operational Plan for the Bay of Bengal and the Arabian Sea.* Geneva: World Meteorological Organisation.

3257 ■ 1999 Kamala, P. Orissa cyclone and disaster management. *Mainstream*, 37: 19–24.

3258 • 2000 Doctor, H. *et al.* Disaster to Development: *Grassroots Experiences after the Orissa Super Cyclone.* London: Christian Aid.

3259 • 2000 Gupta, M. Orissa cyclone and after, managing of public health. *Economic and Political Weekly*, 35 (20): 1716–17.

3260 ■ 2000 Oxford Committee for Famine Relief. *Village Contingency Plan for Cyclones.* Hyderabad: Bookline.

3261 • 2000 Pandey, V.N. Traditional systems of medicine. In S. Parasuraman and P.V. Unnikrishnan (eds). *India Disasters Report: Towards a Policy Initiative.* New Delhi: Oxford University Press. 49–50.

3262 ■ 2000 Rajawat, A.S. and S. Nayak. Remote sensing applications in coastal zone hazard. In P.S. Roy, *et al* (eds). *Natural Disasters and their Mitigation*: A Remote Sensing and Geographical Information Systems Perspective. Dehradun: Indian Institute of Remote Sensing. 158–174.

3263 ▲ 2000 Reddy, A.V.S. Cyclone mitigation measures. In S. Parasuraman and P.V. Unnikrishnan (eds). *India Disasters Report: Towards a Policy Initiative.* New Delhi: Oxford University Press. 195–197.

3264 • 2000 United Nations Development Programme. *Crisis Prevention and Recovery in Orissa.* Bhubaneswar.

3265 • 2000 Winchester, P. Cyclone mitigation, resource allocation and post-disaster reconstruction in South India: Lessons from two decades of research. *Disasters: The Journal of Disaster Studies, Policy and Management*: 18–37.

3266 ▲ 2001 Choudhury, R. C. *et al. Disaster Management: Orissa Cyclone, Floods and Tidal Wave Disaster (October 1999)* Hyderabad: National Institute of Rural Development.

3267 ■ 2001 Das, K. C. Role of geomorphology in land use planning: A study from Super Cyclone affected areas of Jagatsinghpur district (Orissa). In *Proceedings of the Role of Earth Sciences in Integrated Development and Related Societal Issues.* 2001 Special Publication No. 65 (2) Kolkata: Geological Survey of India. 130–137.

3268 ● 2001 International Federation of Red Cross and Red Crescent Society. Shelters save lives and livelihoods. In *World Disasters Report 2001*. Geneva. 24–25.

3269 ● 2001 Mahapatra, S. *Post-Disaster Management of Cyclone Affected Areas.* New Delhi: School of Planning and Architecture (Unpublished Postgraduate Dissertation).

3270 ■ 2001 Naidu, R. Disaster mitigation measures for cyclone-prone areas of Andhra Pradesh. In P. Sahni, *et al* (eds). *Disaster Mitigation: Experiences and Reflections.* New Delhi: Prentice-Hall of India Private Limited. 67–75.

3271 ▲ *2001* National Disaster Management Division. *Guidelines for Preparing Cyclone Risk Mitigation in States and Union Territories.* New Delhi.

3272 ■ 2001 Ramesh, K. S. Cyclone disaster management in a coastal district of Andhra Pradesh–A case study. In V. K. Sharma (ed). *Disaster Management.* New Delhi: Indian Institute of Public Administration. 189–207.

3273 ● 2002 Lorna, V. *Impact Assessment Study of the Orissa Disaster Management Project.* Bangkok: Asian Disaster Preparedness Centre.

3274 ● 2003 Harichandan, H. Community participation in disaster mitigation and reduction (with special reference to cyclone). In P. Sahni, *et al* (eds). *Disaster Risk Reduction in South Asia.* New Delhi: Prentice-Hall of India Private Limited. 210–216.

3275 ● 2003 Panigrahi, N. Disaster management and the need for convergence of services of welfare agencies: A case study of Super Cyclone of Orissa. *Social Change*, 33 (1): 1–25.

3276 ● 2003 Samal, K. C. *et al. Beyond Relief: Disaster Mitigation, Livelihood Rehabilitation and the Post-Cyclone Recovery of Orissa* (*Village Level Studies in the Three Most Cyclone Affected Districts).* Bhubaneswar: National Centre for Development Studies.

3277 ● 2004 International Federation of Red Cross and Red Crescent Society. Harnessing local capabilities in rural India. In *World Disasters Report 2004*. Geneva. 56–77.

3278 ■ 2004 Rajawat, A.S., and S. Nayak. Applications of satellite survey in coastal disasters. In K.S. Valdiya (ed). *Coping with Natural Hazards: Indian Context.* Hyderabad: Orient Longman. 204–215.

3279 ▲ 2004 Sahu, B.P., and S. Mahapatra. *Post-Disaster Management of Super Cyclone: A Case Study of Orissa.* New Delhi: Mittal Publication.

3280 ● 2004 Thomalla, F., and H. Schmuck. We all knew that a cyclone was coming: Disaster preparedness and the cyclone of 1999 in Orissa, India. *Disasters: The Journal of Disaster Studies, Policy and Management,* 28(4): 373–387.

3281 ● 2006 Appa Rao, T.V.S.R. Research and development for cyclone disaster mitigation. In H.N. Srivastava and G.D. Gupta (eds). *Management of Natural Disasters in Developing Countries.* New Delhi: Daya Publishing House. 50–63.

3282 ■ 2006 Choudhary, J.R. Mitigating cyclone and storm surge disasters. In H.N. Srivastava and G.D. Gupta (eds). *Management of Natural Disasters in Developing Countries.* New Delhi: Daya Publishing House. 1–9.

3283 ■ 2006 Mahajan, P.N. Satellite inputs for better disaster management of tropical cyclones, In N.C. Mahanti, *et al* (eds). *Disaster Management.* New Delhi: Narosa Publishing House.

3284 ■ 2006 Pal, P.K. Use of satellite observation for tropical cyclone monitoring and prediction for disaster management. In N.C. Mahanti, *et al* (eds). *Disaster Management.* New Delhi, Narosa Publishing House.

Management: Drought

3285 ▲ 1973 Planning Commission. *Integrated Agricultural Development in Drought Prone Areas.* New Delhi: Government of India.

3286 ● 1973 Revenue and Forest Department. *Sukhatankar Committee Report.* Mumbai: Government of Maharashtra.

3287 ● 1976 Subramanian, K.K. *Public Works Policy during Drought and Famines and its Lessons for an Employment Policy.* New Delhi: Indian Council of Social Science, (Project Report).

3288 ▲ 1978 Jaiswal, N.K. and N.V. Kolte. *Drought Prone Area Programme in Kurnool. Hyderabad*: National Institute of Rural Development.

3289 ▲ 1978 Srivastava, U.K. *Management of Drought Prone Area Programme: Analysis and Case Studies from Jhabua District.* Ahmedabad: Indian Institute of Management

3290 ● 1979 Arunachalam, B. Socio-economic and demographic implications in planning for drought prone areas of Maharashtra. In N.B.K Reddy (ed). *Proceedings of the All India Symposium on Drought Prone*

Areas of India. Tirupati: Rayalseema Geographical Society. 269–276.

3291 ● 1979 Arunachalam, B. The problems of drought: What can the geographer do? In N. B. K Reddy (ed). *Proceedings of the All India Symposium on Drought Prone Areas of India.* Tirupati: Rayalseema Geographical Society. 61–68.

3292 ● 1979 Bali, J. S. Watershed management to fight droughts. In *Proceedings of the International Symposium on Hydrological Aspects of Drought.* New Delhi: Indian Institute of Technology. 61–66.

3293 ▲ 1979 Bhandi, B. L. and B. V. Rao. Interdisciplinary approach to the Integrated Development of Drought Prone Area Programme. In N.B.K Reddy (ed). *Pr oceedings of the All India Symposium on Drought Prone Areas of India.* Tirupati: Rayalseema Geographical Society. 281–285.

3294 ● 1979 Bhat, L. S. Development strategy for drought prone areas. In N. B. K Reddy (ed). *Proceedings of the All India Symposium on Drought Prone Areas of India.* Tirupati: Rayalseema Geographical Society. 249–252.

3295 ▲ 1979 Choudry, R. K. Some aspects of strategy and planning for agricultural development in drought prone areas. In N. B. K Reddy (ed). *Proceedings of the All India Symposium on Drought Prone Areas of India.* Tirupati: Rayalseema Geographical Society. 303–311.

3296 ■ 1979 Jaiswal, N. K. Measures suggested for the development of drought prone areas. In N. B. K Reddy (ed). *Proceedings of the All India Symposium on Drought Prone Areas of India.* Tirupati: Rayalseema Geographical Society. 253–268.

3297 ■ 1979 Rakshit, D. Drought: A consequence of short-term climatic fluctuation and its possible mitigation. *Bhagirath*, 26 (October-December): 159–163.

3298 ■ 1979 Ramamurthy, V. Drought mitigation strategy through weather modification. In *Proceedings of the International Symposium on Hydrological Aspects of Drought.* New Delhi: Indian Institute of Technology. 67–75.

3299 ● 1979 Reddy, G. N. A strategy for the diffusion of agricultural innovations in Drought Prone Areas Programme. In N. B. K Reddy (ed). *Proceedings of the All India Symposium on Drought Prone Areas of India.* Tirupati: Rayalseema Geographical Society. 295–297.

3300 ■ 1979 Reddy, G. N. and Hemavathi. Some aspects of the development of drought prone areas. In N. B. K Reddy (ed). *Proceedings of the All India Symposium on Drought Prone Areas of India.* Tirupati: Rayalseema Geographical Society. 277–279.

3301 ▲ 1981 Jaiswal, N. K. and N. V. Kolte. *Development of Drought Prone Areas.* Hyderabad: National Institute of Rural Development Hyderabad.

3302 ▲ 1981 National Committee on the Development of Backward Areas. *Report on Development of Drought Prone Areas.* New Delhi: Government of India.

3303 ▲ 1982 Ministry of Rural Development. *Report of the Task Force on the Drought Prone Areas Programme and the Desert Development Programme.* New Delhi: Government of India.

3304 ■ 1982 Naganna, C. and D. Barai. Drought mitigation strategy: A case study in Kolar region (Karnataka). *Transactions, Institute of Indian Geographers*, 4 (2): 125–131.

3305 ■ 1982 Tapeshwar, S. Drought disaster management: An Indian perspective. *Geographical Review of India*, 44 (1 and 4): 206–215.

3306 ▲ 1983 Babu, M. *Planning in Drought Prone Area.* New Delhi: School of Planning and Architecture (Unpublished Postgraduate Dissertation).

3307 ■ 1983 Brahme, S. *Drought in Maharashtra, 1972: A Case for Irrigation Planning.* Mumbai: Orient Longman and Gokhale Institute of Politics and Economics.

3308 ▲ 1983 Government of Rajasthan. *Drought Prone Area Programme.* Jaipur: Government of Rajasthan.

3309 ▲ 1983 Jayachandran, G. *Agricultural Development in the Drought Prone Area of Bharmapuri Region.* Mysore: Mysore University (Unpublished Ph.D Thesis).

3310 • 1983 Mathur, K. and A. Gupta. *Establishment of Planning Cells in Selected Drought Prone Area Programme Districts.* New Delhi: Report of the Action Research Project, Indian Institute of Public Administration.

3311 ■ 1984 Baba, G. S. Disaster management: A case study of droughts in India. In B. R. Naidu (ed). *Disaster Management.* Tirupati: Sri Venkateswara University, 35–48.

3312 ▲ 1985 Chaudhary, S. Management of mini watersheds: An alternative for developing drought prone areas of Puruliya. *In Proceedings of the Sixth Indian Geographical Congress,* Department of Architecture and Regional Planning. Kharakpur: Indian Institute of Technology.

3313 ▲ 1985 Ministry of Agriculture. *A Model Manual for Drought Management.* New Delhi: Government of India.

3314 ▲ 1985 Sharma, I. *Water Management in Drought Prone Areas.* New Delhi: Criterion Publisher .

3315 ▲ 1986 Dandekar, V. A national policy for distribution of irrigated water in drought prone areas. In *Proceedings of the Seminar on Control of*

Drought, Desertification and Famine. New Delhi: India International Centre.

3316 ▲ 1986 Koshy, C. K. *Drought Management: The Gujarat Experience*. Gandhinagar: Office of the Commissioner Relief, Government of Gujarat.

3317 ▲ 1986 Mitra, A. K. Under utilization revisited: Surface irrigation in drought prone areas of Western Maharashtra. *Economic and Political Weekly*, 21 (17): 767–768.

3318 ● 1986 Mohandas, P. *Integrated Area Development Planning under the Drought Prone Area Programme in Karnataka: A Critical Appraisal*. New Delhi: Indian Institute of Public Administration.

3319 ● 1987 Bandhopadhyay, J. Political ecology of drought and water scarcity: Need for an ecological water resources policy. *Economic and Political Weekly*, 22 (50): 2159–2169.

3320 ▲ 1987 Revenue and Forests Department. *Drought Prone Area Review Committee*. Mumbai: Government of Maharashtra.

3321 ▲ 1987 Sarma, S. Managing drought. *Frontline*, 20 (5): 4–8.

3322 ▲ 1987 Sathe, M. D. Drought Prone Area Programme in Ahmednagar. *Economic and Political Weekly*, 22 (25): 1533–1537.

3323 ● 1988 Mohandas, P. *Integrated Area Development Planning under the Drought Prone Area Programme in Karnataka: A Critical Appraisal*. New Delhi: Advanced Professional Programme in Public Administration, Indian Institute of Public Administration (Unpublished Master of Philosophy Dissertation).

3324 ● 1988 Rao, C. H. H. *et al. Unstable Agriculture and Droughts: Implications for Policy*. New Delhi: Vikas Publishing House.

3325 ● 1988 Sen, A. Union Budget 1988–89: On management of drought and of public opinion. *Economic and Political Weekly*, 23 (14 and 15): 733–738.

3326 ▲ 1989 Bhatia, B. M. 1987–Drought in retrospect: Some policy issues. In *Proceedings of the National Workshop on Management of Drought*, Vigyan Bhawan New Delhi: Department of Science and Technology.

3327 ● 1989 Chopra, R. Drought 1987: The role of voluntary organizations in management of droughts. In *Proceedings of the National Workshop on Management of Drought*, Vigyan Bhawan. New Delhi: Department of Science and Technology.

3328 ● 1989 Mathur, K. Politics and management of droughts in India. In *Proceedings of the National Workshop on Management of Drought*. New Delhi: Department of Science and Technology.

3329 ■ 1989 Puttaswamiah, K. *Cost-Benefit Analysis of Irrigation and Drought Proofing Management in Government Management*. New Delhi: Oxford and IBH Publishing Co. Pvt. Ltd.

3330 ▲1989 Quaraishi, M.A. *Drought Strategy*. Delhi: B.R Publishing Corporation.

3331 • 1989 Rangasami, A. Financing the management of drought: Policy perspectives. In *Proceedings of the National Workshop on Management of Drought,* Vigyan Bhawan. New Delhi: Department of Science and Technology.

3332 ▲1990 Ministry of Agriculture. *The Drought of 1987: Response and Management in India (Volumes I and II)*. New Delhi: Government of India.

3333 • 1990 Misra, R.P. Planning for a drought prone district: The case of Chitradurga district (Karnataka). In R.P. Misra and R.N. Achcyutha (eds). *Micro-level Rural Planning: Principles, Methods and Case Studies*. New Delhi: Concept Publishing Company.

3334 ■ 1991 Ahluwalia, D. Drought proofing in the Indian foodgrain economy. *Indian Journal of Agricultural Economics,* 46 (2): 111–120.

3335 ▲1991 Jodha, N.S. Drought management: Farmers' strategies and their policy implications. *Economic and Political Weekly*, 26 (13): A–98 A–104.

3336 ▲1991 Saptarshi, P.G. *Geoenvironmental Management in Drought Prone Areas, with Special Reference to Paschim Zila Parishad, District Ahmednagar (Maharashtra)*. New Delhi: Indian Council of Social Science, (Project Report).

3337 ■ 1992 Agarwal, R.P. Tackling drought through soil and water management: A study of Northwest India. *Kurukshetra,* 41(3): 25–28.

3338 • 1992 Bandopadhyay, J. and V. Shiva. Ecology of drought and water scarcity: Need for an ecological water resource policy. In S. Bahuguna (ed). *Environmental Crisis and Sustainable Development*. Dehradun: Natraj Publication. 522–550.

3339 • 1992 Dubhashi, P.R. Drought and development. *Economic and Political Weekly*, 27 (13): A27–A36.

3340 ▲1992 Mathur, K. and N.G. Jayal. Drought management in India: The long term perspective. *Disasters: The Journal of Disaster Studies*, Policy and Management, 16 (1): 60–65.

3341 ▲1992 Ministry of Rural Development. *Manual for Drought Prone Area Programme and Desert Development Programme*. New Delhi: Government of India.

3342 ■ 1992 Patil, N.P. Measures to contain drought: A case study of Bijapur in Karnataka. *Southern Economy*, 31 (6): 20–23.

3343 ■ 1992 Rakshit, D.K. On some aspects of drought mitigation and management. *Science and Culture*, 58 (5–8): 112–114.

3344 • 1992 Sen, S.R. Drought, starvation, unemployment: Some corrective measures. *Economic and Political Weekly*, 27 (19): 989–991.

3345 ▲1992 Subbiah, A.R. Drought Management through Anticipatory Multidimensional Approaches: A Case Study-Disaster Management. New Delhi: Indian Institute of Public Administration.

3346 ▲1993 Ganesh, S.P. *Resource Appraisal and Planning Strategy for the Drought Prone areas: A Study of the Karjat Tehsil, Ahmadnagar district (Maharashtra)*. Pune: Pune University (Unpublished Ph.D Thesis).

3347 • 1993 Katiyar, V.S. *Human Dimensions and Drought Management in India.* Delhi: Pointer Publishers.

3348 ▲1993 Mathur, K. *et al. Drought, Policy and Politics in India*: The Need for a Long-Term Perspective. New Delhi: Sage Publications.

3349 • 1993 Menon, M. Carving oasis in drought prone Kutch. *Down to Earth*, 2 (7): 22–23.

3350 • 1993 Padmanabhan, B.S. Dealing with drought: Managing rain water holds the key. *Frontline*, 26 (June 4): 33–34.

3351 • 1993 Sastri, A.S.R.S. Agricultural drought management strategies to alleviate impacts: Examples from arid and subhumid regions of the Indian subcontinent. In D.A. Wilhite (ed). *Drought Assessment, Management, and Planning: Theory and Case Studies.* Boston: Kluwer Academic Publishers. 65–86.

3352 ■ 1993 Shrivastava, M.P. Drought prone area management. *Journal of Rural Development*, 12 (2): 165–181.

3353 • 1993 Subbiah, A.R. Indian drought management: From vulnerability to resilience. In D.A. Wilhite (ed). *Drought Assessment, Management, and Planning: Theory and Case Studies.* Boston: Kluwer Academic Publishers. 157–179.

3354 • 1994 Council of Professional Social Workers. Drought. In *State of Orissa's Environment: A Citizen's Report.* Bhubaneswar. 159–187.

3355 ■ 1994 Dubhashi, P.R. Strategies to counter drought. *Economic and Political Weekly*, 29 (40): 2602.

3356 • 1994 Gaugi, R.Y. *The Study of Rural Industrialization in Drought Prone Region with Special Reference to Anantpur District.* Anantapur: Sri Krishna Devaraya University (Unpublished Ph.D Thesis).

3357 ▲1994 Gurjar, R.K. *Drought Planning in India.* Jaipur: Printwell Publishers.

3358 ▲1994 Ministry of Rural Development. *Report of the Technical Committee on Drought Prone Area Programme and Desert Development Programme.* New Delhi: Government of India.

3359 ▲1994 Special Correspondent. Report of the Technical Committee on Drought Prone Areas Programme and Desert Development Programme. *Indian Journal of Agricultural Economics,* 49 (4): 624.

3360 ▲ 1995 Planning Commission. *Evaluation Report on Drought Prone Area Programme, 1992–93.* New Delhi Government of India.

3361 • 1996 Basavaidh, M.C. Andhra Pradesh: People's action to protect water sources. *Economic and Political Weekly*, 31 (19): 1609–1623.

3362 ▲ 1997 Hooda, R.S. Lift irrigation and commercial crops in the drought prone areas of Southwest Haryana. *Geographical Review of India*, 59 (2): 151– 158.

3363 ▲ 1997 Planning Commission. Drought Prone Area Programme and Desert Development Programme. In *Ninth Five Year Plan.* New Delhi: Government of India. 618–620.

3364 • 1999 Bharara, L.P. *Man in the Desert: Drought, Desertification and Indigenous Knowledge for Sustainable Development.* Jodhpur: Scientific Publishers.

3365 ▲ 1999 Hooja, R. and R. Joshi (eds). *Desert, Drought and Development: Studies in Resource Management and Sustainability.* Jaipur: Rawat Publication.

3366 • 2000 Bhatnagar, B.P. Long term drought proofing of the people of the arid areas of Rajasthan. In *Proceedings of the National Seminar on Water Conservation and Drought Proofing of the Arid Areas and Its People.* Jaipur: Social Policy Research Institute.

3367 ■ 2000 Bokil, M. Drought in Rajasthan: In search of a perspective. *Economic and Political Weekly*, 35 (48): 4171–4175.

3368 • 2000 Narian, P. *et al. Strategy to Combat Drought and Famine in the Indian Arid Zone.* Jodhpur: Centre for Arid Zone Research Institute.

3369 • 2000 White, D.H. Drought policy monitoring and management in arid lands. *Annals of Arid Zone,* 39 (2): 105–330.

3370 ▲ 2001 Bavadan, L. Dealing with drought *Frontline,*18 (13): 51–54.

3371 ■ 2001 Bhanwar, D. and B.D. Bithu. Drought protection through water management survival strategy. *Bhagirath*, 48 (January-March): 17–20.

3372 • 2001 Dhameja, A. Droughts and floods: A case for dying wisdom. In P. Sahni, *et al* (eds). *Disaster Mitigation: Experiences and Reflections.* New Delhi: Prentice-Hall of India Private Limited. 76–91.

3373 • 2001 Gopal, K. Neemi's water conference: *Jal biradari* launches movement in drought-proof villages. *Down to Earth*, 9: 39.

3374 • 2001 Khurana, I. Rainwater rights. *Down to Earth*, 10 (6): 36–39.

3375 ▲ 2001 Mohnot, S.M. A way out of drought. *Down to Earth*, 9 (17): 52.

3376 • 2001 Prasad, E. Drought of 1999–2000: Emulating civil society. In A. Agarwal, *et al* (eds). *Making Water Everybody's Business: Practice and Policy of Water Harvesting.* New Delhi: Centre for Science and Environment. 363–381.

3377 ■ 2001 Rajendran, S. Drought in Karnataka: Need for long term perspective. *Economic and Political Weekly*, 36 (36): 3423–3426.

3378 ▲ 2001 Sharma, A.K. and R.K. Goyal. Drought mitigation versus management: Vaccination for recurring problem of drylands. *Bhagirath*, 48 (July–September): 121–123.

3379 ● 2001 Singh, R. Drought proofing by adopting people's methods. *Kurukshetra*, 50 (4): 41–43.

3380 ● 2001 Special Correspondent. Think institution. *Down to Earth*, 10 (2): 42.

3381 ● 2001 Subbiah, A.R. Drought management through participatory multidimensional approaches: A case study. In V.K. Sharma (ed). *Disaster Management*. New Delhi: Indian Institute of Public Administration. 91–107.

3382 ▲ 2002 Faroda, A.S. Some considerations for drought management in Indian arid zone. In P. Narain (ed). *Drought Management in Indian Arid Zone*. Jodhpur: Centre Arid Zone Research Institute, Jodhpur.

3383 ▲ 2002 Gambheer, S.S. *Strategy for Stabilising Farm Income and Employment in Drought Prone Areas: A Case Study of Sericulture in Karnataka*. Dharwad: Karnataka University (Unpublished Ph.D Thesis).

3384 ▲ 2002 Gupta, S. *Water Policy for Drought Proofing Chhattisgarh*. New Delhi: Institute for Human Development.

3385 ■ 2002 Indian Water Resource Society. Drought management. *Bhagirath*, 49 (January–March): 17–49.

3386 ■ 2002 Jana, R. Analysis of drought management over land: A case study. In R. Nagarajan (ed). *National Workshop Proceedings on Recent Trends in Drought Assessment Monitoring and Management*. Mumbai: Centre of Studies in Resources, Engineering Indian Institute of Technology. 71–80.

3387 ▲ 2002 Narain, P. *Drought Management in Indian Arid Zone*. Jodhpur: Centre Arid Zone Research Institute, Jodhpur.

3388 ■ 2002 Osman, M. and P.K. Mishra. An approach to contingent planning for drought mitigation in Andhra Pradesh. In R. Nagarajan (ed). *National Workshop Proceedings on Recent Trends in Drought Assessment Monitoring and Management*. Mumbai: Centre of Studies in Resources, Engineering Indian Institute of Technology. 66–70.

3389 ■ 2002 Parmar, P.K.S. Drought management in Madhya Pradesh. In R. Nagarajan (ed). *National Workshop Proceedings on Recent Trends in Drought Assessment Monitoring and Management*. Mumbai: Centre of Studies in Resources, Engineering Indian Institute of Technology. 58–65.

3390 ■ 2002 Ramakrishna, Y.S. *et al.* Climatological aspects of droughts and their management. In R. Nagarajan (ed). *National Workshop Proceedings on Recent Trends in Drought Assessment Monitoring and Management.* Mumbai: Centre of Studies in Resources, Engineering Indian Institute of Technology. 8–16.

3391 ■ 2002 Singh, D.K. *et al.* Prioritization of artificial recharge structures for maximising the impact based on remote sensing techniques in drought hit part of Rajgarh district, Madhya Pradesh (India). In R. Nagarajan (ed). *National Workshop Proceedings on Recent Trends in Drought Assessment Monitoring and Management.* Mumbai: Centre of Studies in Resources, Engineering Indian Institute of Technology. 93–97.

3392 ■ 2002 Singh, H.N. and K.D. Sharma. Drought mitigation and farming system research in India. In R. Nagarajan (ed). *National Workshop Proceedings on Recent Trends in Drought Assessment Monitoring and Management.* Mumbai: Centre of Studies in Resources, Engineering Indian Institute of Technology. 30–37.

3393 • 2002 Singh, V.K. Contingency planning for livestock during drought in Rajasthan. In P. Narain. *Drought Management in Indian Arid Zone.* Jodhpur: Centre Arid Zone Research Institute, Jodhpur.

3394 • 2002 Sinha, A.K. and D. Dhar. Disaster management information system for Maharashtra state with special reference to drought. In R. Nagarajan (ed). *National Workshop Proceedings on Recent Trends in Drought Assessment Monitoring and Management.* Mumbai: Centre of Studies in Resources, Engineering Indian Institute of Technology. 85–92.

3395 • 2002 Special Correspondent. Battle lines are drawn as drought proofing remains a distant dream. *Down to Earth*, 11 (1): 6.

3396 • 2002 Srinivasulu, K. *et al.* Drought management techniques for dryland development. *Kurukshetra,* 51 (11): 29–33.

3397 ▲ 2002 Swami, S.K. (eds). *Drought Management in India: Best Practices, Lessons Learnt.* New Delhi: Government of India.

3398 • 2003 Dash, J. and Behura, N.K. Crisis management: Lessons from the past, a case study of drought situation in Kalahandi (undivided) of Orissa. *Man in India,* 83 (1–2): 121–134.

3399 ■ 2003 Desai, B.M. Drought impact and vision for proofing. *Economic and Political Weekly*, 37 (21): 2023–2024.

3400 • 2003 Joshi, P.K. *et al.* Socio-economic considerations in effective implementation of drought management strategies. In N.P. Saxena (ed). *Management of Agricultural Drought: Agroeconomic and*

Genetic Options. New Delhi: Oxford and I B H Publishing Co. Pvt. Ltd. 1–18.

3401 ▲ 2003 Nagarajan, R. (eds). Drought: *Assessment, Monitoring, Management and Resources Conservation*. New Delhi: Capital Publication.

3402 ■ 2003 Rao, D. P. Droughts. In H.K. Gupta (ed). *Disaster Management*. Hyderabad: University Press (India) Private Limited. 128–142.

3403 ▲ 2003 Sahani, P. Drought profile, management and risk reduction in India. In P. Sahni, *et al* (eds). *Disaster Risk Reduction in South Asia*. New Delhi: Prentice-Hall of India Private Limited. 299–326.

3404 ▲ 2003 Singh, K. Pilot demonstration project for drought risk reduction. In P. Sahni, *et al* (eds). *Disaster Risk Reduction in South Asia*. New Delhi: Prentice-Hall of India Private Limited. 236–249.

3405 ● 2004 Das, H.P. Agricultural drought mitigation and management of sustained agricultural development in India. In M. V. K. Sivakumar, *et al* (eds). *Natural Disasters and Extreme Events in Agriculture: Impacts and Mitigation*. New York: Springer. 277–304.

3406 ■ 2004 Rathore, M. S. *Adaptive Strategies to Droughts in Rajasthan*. Jaipur: Institute of Development Studies.

3407 ● 2005 Jana, B. L. *Appropriate Technologies for Drought Prone Areas*. Udaipur: Agrotech Publications.

3408 ● 2005 Misra, R. P. Managing droughts in India: Experiences and prospects. In B. Thakur *et al* (eds). *Urban and Regional Development in India: Essays in Honour of Professor L. N. Ram*. New Delhi. Concept Publishing Company. 581–597.

3409 ▲ 2005 Reddy, A. R. (ed). *Watershed Management for Sustainable Development, with Reference to Drought Prone Areas*. New Delhi: Mittal Publication.

3410 ● 2005 Singh, R. Drought *johad a*nd self reliance. *Yojana*, 46: 9.

3411 ▲ 2005 Singh, T. Drought disaster management for agricultural security in India. In B. Thakur (ed). *Urban and Regional Development in India: Essays in Honour of Professor L. N. Ram*. New Delhi: Concept Publishing Company. 598–620.

Management: Earthquake

3412 ● 1963 Behrman, D. Earthquake disasters need not happen. *Science and Culture*, 29 (3): 238–239.

3413 ● 1970 Mazumdar, A. K. *Urban Structure in a Seismic Zone in Assam*. New Delhi: School of Planning and Architecture (Unpublished Postgraduate Dissertation).

3414 ● 1970 Patel, V. J. *et al*. Stability of earth dams for earthquake conditions. *ISET Journal of Earthquake Technology*, 7 (1): 41–46.

3415 • 1980 Rao, B.G. and A.K. Shama. *Skirted Granular Pile Foundations for Seismic and Flood-Prone Zones.* Chichester: John Wiley and Sons.

3416 ■ 1983 Gupta, H.K. Induced seismicity hazard mitigation through water level manipulation at Koyna (India). *Bulletin of the Seismological Society of America*, 73: 679.

3417 • 1984 Bureau of Indian Standards. *Indian Standards Criteria for Earthquake Resistant Design of Structures.* New Delhi.

3418 • 1985 Annamalai, G. Seismic design of bearing and shear wall reinforced masonry buildings. *Indian Concrete Journal*, 59 (1).

3419 • 1986 Bureau of Indian Standards. *Criteria for Earthquake Resistant Design of Structures:* 1893–1984. New Delhi.

3420 • 1986 International Association for Earthquake Engineering. *Guidelines for Earthquake Resistant Non-Engineered Construction.* Kanpur: National Information Center of Earthquake Engineering. Indian Institute of Technology Kanpur,

3421 • 1988 Gambhir, M.L. Earthquake resistance of high-rise buildings. *Indian Concrete Journal*, 62 (3).

3422 • 1989 Selvam, M. *et al.* A note on the design of reinforced concrete chimneys subjected to earthquake effects *Indian Concrete Journal*, 63 (8).

3423 • 1990 Jain, A.K. Non-deterministic seismic response of tall reinforced concrete towers. *Indian Concrete Journal*, 64 (3).

3424 • 1991 Jain, A.K. and R.A. Mir. Inelastic response of reinforced concrete frames under earthquakes. *Indian Concrete Journal*, 65 (11).

3425 ■ 1992 Bagla, P. Proper earthquake proofing to save life and property. *Kurukshetra*, 41(2): 37–38.

3426 • 1992 Jain, S.K. On better engineering preparedness: Lessons from the 1988 Bihar earthquake. *Earthquake Spectra*, 8 (3): 391–402.

3427 • 1992 Khatri, K.N. Planning for earthquake safe development. *Yojana*, 36 (17): 22–28.

3428 • 1992 Medhekar, M.S. *et al.* Shear reinforcement for a seismic design of flexural members. *Indian Concrete Journal*, 66 (6).

3429 • 1993 Arya, A.S. Relief and prevention issues in earthquake disaster mitigation in North Bihar. In P. Hall, *et al* (eds). *Multilateral Cooperation for Development in the Twenty-First Century: Training and Research for Regional Development.* Nagoya, Japan: United Nations Centre for Regional Development. 233–250.

3430 • 1993 Arya, A.S. Training and drills for the general public in emergency response to a major earthquake. In *Training and Education for Improving Earthquake Disaster Management in Developing Countries.* Proceedings of the Sixth International Research and Training

Seminar on Regional Development Planning for Disaster Prevention. Nagoya: Japan. 103–114.

3431 ■ 1993 Building Material Technology Promotion Council. *Base Line Socio-Economic and Participation Design Exercise in Earthquake affected Areas of Maharashtra: An Interim Report.* New Delhi: Housing and Urban Development Corporation.

3432 ● 1993 Bureau of Indian Standards.*Indian Standards Code of Practice for Earthquake Resistant Design and Construction of Buildings.* New Delhi.

3433 ● 1993 Bureau of Indian Standards. *Improving Earthquake Resistances of Earthen Buildings (Guidelines).* New Delhi.

3434 ● 1993 Bureau of Indian Standards. *Improving Earthquake Resistances of Low Strength Masonry Buildings (Guidelines).* New Delhi.

3435 ● 1993 Bureau of Indian Standards. *Ductile Derailing of Reinforced Concrete Structures Subjected to Seismic Forces (Code of Practice).* New Delhi.

3436 ● 1993 Bureau of Indian Standards. *Repair and Seismic Strengthening of Buildings (Guidelines).* New Delhi.

3437 ● 1993 Bureau of Indian Standards. *Earthquake resistant design and construction of buildings (Code of practice).* New Delhi.

3438 ■ 1993 Kurien, C.T. What needs to be done? *Frontline*, 26 (October 22): 16.

3439 ● 1993 Medhekar, M.S. and S.K. Jain. Seismic behaviour, design and detailing of reinforced concrete shear walls, Part II: Design and detailing. *Indian Concrete Journal*, 67 (9).

3440 ● 1993 Medhekar, M.S. and S.K. Jain. Seismic behaviour, design and detailing of reinforced concrete shear walls, Part I: Behaviour and strength. *Indian Concrete Journal,* 67 (7).

3441 ● 1993 Ravindra, A. and S. Das. Technology transition in Garhwal. *Economic and Political Weekly*, 28: 561–564.

3442 ● 1993 Valdiya, K.S. Latur earthquake of 30 September 1993: Implication and planning for hazard preparedness. *Current Science*, 65 (7): 515–517.

3443 ● 1994 Arya, A.S. Earthquake resistant construction for non-engineered buildings. *Indian Concrete Journal*, 68 (11): 64–67.

3444 ● 1994 Government of India. *Advisory Committee Report on Latur Earthquake.* New Delhi.

3445 ● 1994 Murthy, C.V.R. and S.K. Jain. A review of Indian Standard: 1893–1984 provisions on seismic design of buildings. *Indian Concrete Journal*, 68 (11).

3446 ● 1994 Muruzzuma, M.D. Housing *Development in Earthquake Prone Areas: Case Study of Uttarkashi.* New Delhi: School of Planning and Architecture (Unpublished Postgraduate Dissertation).

3447 • 1994 Tandon, M. Considerations in seismic design of bridges. *Indian Concrete Journal*, 68 (11).

3448 • 1994 Thakkar, S.K. Base isolation, supplemental dumping and active control for a seismic construction. *Indian Concrete Journal*, 68 (11).

3449 • 1995 Jain, S.K. Seismic strengthening of existing reinforced concreate buildings. *Indian Concrete Journal*, 69 (1).

3450 • 1996 Anand, S. A. Vulnerability reduction: Key to earthquake protection. *Himalayan Geology*, 17: 33–44.

3451 • 1996 Jain, A. K. Quality assurance and control for structures subjected to earthquakes *Indian Concrete Journal*, 70 (11).

3452 • 1996 Singh, R. B. Recent earthquake in Maharashtra (India): Uncertainty, existing reality, and task ahead in disaster management in metropolitan areas for the 21 century. In R.B. Singh (ed). *Research in Geography, Disasters and Environment: Monitoring and Forecasting, Volume II*. New Delhi: A P H Publishing Corporation. 83–86.

3453 • 1996 Sreedhar, R. Earthquake disaster mitigation: Lessons from recent experiences. *Himalayan Geology*, 17: 225–234.

3454 • 1997 Krishnaraj, M. Gender issues in disaster management: The Latur earthquake." *Gender, Technology and Development,* 1 (3): 395–411.

3455 ▲ 1998 Government of Maharashtra. *Disaster Action Plan of Latur District.* Mumbai.

3456 • 1998 Gupta, A. The strategy effectiveness chart: A tool for evaluating earthquake disasters mitigation strategies. *Applied Geography*, 18 (1): 55–67.

3457 • 1998 Jain, S.K. and C.V.R. Murthy. A state-of-the-art review on seismic design of bridge-Part–I: Historical development and Aasto Code. *Indian Concrete Journal*, 72 (2).

3458 • 1998 Jain, S.K. and C.V.R. Murthy. A state-of-the-art review on seismic design of bridge-Part–II: Caltrans, Tnz and Indian code. *Indian Concrete Journal*, 72 (3).

3459 • 1998 Jain, S.K. and S. Chaturvedi. *Standardisation Efforts in Mitigating Earthquakes.* New Delhi: Government of India.

3460 • 1998 Maharashtra Emergency Earthquake Rehabilitation Program. *Earthquake Resistant Construction and Seismic Strengthening of Non-Engineered Buildings in Rural Areas of Maharashtra.* Mumbai: Government of Maharashtra.

3461 • 1998 Paul, D.K. Lessons from Uttarkashi earthquake. *Indian Concrete Journal*, 72 (11).

3462 • 1998 Maharashtra Emergency Earthquake Rehabilitation Program. *Maharashtra Disaster Management Plan.* Mumbai: Government of Maharashtra.

3463 • 1998 Ray, D.C. Lessons from the Jabalpur earthquake. *Indian Concrete Journal*, 72 (11).

3464 • 1998 Sinha, R. and C.V.R. Murthy. The 1993 Killari earthquake: Engineering lessons and challenges. *Indian Concrete Journal*, 72 (11).

3465 • 1998 Thakkar, S.K. Lessons from Bihar earthquake. *Indian Concrete Journal*, 72 (11).

3466 • 1999 Center of Studies in Social Sciences. *Survey of Rehabilitated Households Affected by Killari Earthquake (Latur and Osmanabad Districts)*. Mumbai: Tata Institute of Social Sciences.

3467 • 1999 Government of Maharashtra. *Progress of Maharashtra Emergency Earthquake Rehabilitation Program*. Mumbai.

3468 • 1999 Jatana, B. L. Fail-safe large dams in earthquake prone Himalayan region. *ISET Journal of Earthquake Technology*, 36 (1): 1–14.

3469 • 1999 Singh, S. *Active Control of Buildings under Seismic Excitation*. New Delhi: Indian Institute of Technology (Unpublished Ph.D thesis).

3470 ▲ 2000 Acharyya, S.K. Natural hazards and disaster mitigation in India during 90's, with special reference to earthquakes.In *Annual Convention of the Indian Society of Remote Sensing and Symposium on Spatial Technologies for Natural Hazards and Management*. Kanpur: Indian Institute of Technology.

3471 • 2000 Arya, A.S. Recent developments towards earthquake risk reduction in India. *Current Science*, 79 (9): 1270–1277.

3472 ▲ 2000 Despande, B.G. Disaster management in earthquakes. In R.B. Singh (ed). *Disaster Management*. Jaipur: Rawat Publications, 159–164.

3473 • 2001 Arya, A.S. Action plan for earthquake disaster mitigation. In V.K. Sharma (ed). *Disaster Management*. New Delhi: Indian Institute of Public Administration. 60–73.

3474 • 2001 Arya, A.S. Case study in India: Diagnosis, repair and strengthening of damaged buildings. In V.K. Sharma (ed). *Disaster Management*. New Delhi: Indian Institute of Public Administration, 157–176.

3475 • 2001 Arya, A.S. Uttarkashi: Houses that hold on. *Down to Earth*, 9 (24): 28–29.

3476 • 2001 Brabant, K.V. *The Right to Information: Exploring the Challenges of Transparency, Responsiveness and Accountability in Post-earthquake Gujarat*. Geneva: International Federation of Red Cross and Red Crescent Society.

3477 • 2001 Chaudhuri, S. Disaster management beyond financial assistance (Bhuj). *Economic and Political Weekly*, 36 (11): 880–881.

3478 • 2001 Editorial. Northeast Assam type solutions. *Down to Earth*, 9 (24): 30.

3479 ▲ 2001 Kumar, P. Earthquake disaster impact mitigation made easy. In H.R. Wason, *et al* (eds). *Proceedings of the Workshop on Recent Earthquakes of Chamoli and Bhuj*. Roorkee. 70–80.

3480 ▲ 2001 Lal, A.K. Preparing for future earthquakes. *Yojana*, 45: 24–36.

3481 • 2001 Mishra, P.K. *Control on Quality of Construction in Earthquake Affected Areas of Gujarat*. Gandhinagar: Gujarat State Disaster Management Authority, Government of India.

3482 • 2001 Murthy, C.V.R. *et al.* Anchorage details and joint design in seismic reinforced concrete frames. *Indian Concrete Journal*, 75 (4).

3483 ■ 2001 Mutreja, S.K. *Disaster Management: A Case Study of Earthquake Disaster in Gujarat, From Response to Crisis.* New Delhi: Advanced Professional Programme in Public Administration, Indian Institute of Public Administration (Unpublished Master of Philosophy Dissertation).

3484 ■ 2001 Rajalakshmi, T.K. For preparedness and prevention. *Frontline*, 18 (March 17): 11.

3485 • 2001 Rawal, V. and T.S. Nair. Gujarat: Earthquake plan prepared in haste. *Economic and Political Weekly*, 36 (18): 820–822.

3486 • 2001 Ray, P.N. *Earthquake Relief and Rehabilitation in Gujarat: Issues in Disaster Management.* Working Paper 19. Ahmedabad: Centre for Environmental Planning.

3487 • 2001 Reddy, P.R. National disaster mitigation and reduction institute: An immediate necessity. *Current Science*, 80 (9): 1090–1091.

3488 • 2001 Sharma, C. Earthquake engineering. *New Age*, 49 (7): 13–14.

3489 • 2001 World Bank and Asian Development Bank. *Gujarat Earthquake Recovery Programme.* New Delhi: World Bank and Asian Development Bank.

3490 • 2002 Aggarwal, P. *et al.* Seismic performance of reinforced concrete buildings during Bhuj earthquake of 26 January 2001. *ISET Journal of Earthquake Technology*, 39 (3): 195.

3491 • 2002 Basu, P.C. Seismic upgradation of buildings: An overview. *Indian Concrete Journal*, 76 (8).

3492 • 2002 Bureau of Indian Standards. *Indian Standards Institute 1893–2002 (Part 1): Criteria for Earthquake Resistant Design of Strucures-Part 1 (General Provisions and Buildings).* New Delhi.

3493 ■ 2002 Chhangte, L. *Risk Assessment of Earthquake Prone Areas, Aizowl (Mizoram).* New Delhi: School of Planning and Architecture (Unpublished Postgraduate Dissertation).

3494 • 2002 Dhargalkar, U. Seismic assessment of buildings: A suggested methodology. *Indian Concrete Journal*, 76 (8).

3495 ● 2002 Gupta, R. *Planning for a Post Disaster Area: A Case Study of Anjar (Gujarat)*. New Delhi: School of Planning and Architecture (Unpublished Postgraduate Dissertation).

3496 ● 2002 International Federation of Red Cross and Red Crescent Society. A livelihood context for disaster management. In *World Disasters Report 2003*. Geneva: 32–33.

3497 ● 2002 Jain, S.K. and T. Srikant. Analysis for seismic retrofitting of buildings. *Indian Concrete Journal*, 76 (8).

3498 ● 2002 Jha, Y. *Role of Institutions in Disaster Management: A Case Study of Anjar (Gujarat)*. New Delhi: School of Planning and Architecture (Unpublished Postgraduate Dissertation).

3499 ● 2002 Kalra, G. *Significance of Indigenous Technology in Earthquake Affected Areas of Chamoli (Block Dasoli)*. New Delhi: School of Planning and Architecture (Unpublished Postgraduate Dissertation).

3500 ● 2002 Mehta, S. Gujarat: From epicentre of quake to epicentre of development. *India Today*, (January 21): 55–56.

3501 ● 2002 Mukherjee, A. and M.V. Joshi. Seismic retrofitting technique using fibre composites. *Indian Concrete Journal*, 76 (8).

3502 ■ 2002 Mutreja, S.K. *Disaster Management: A Case Study of Earthquake Disaster in Gujarat, From Response to Crisis*. New Delhi: Indian Institute of Public Administration.

3503 ● 2002 Sheth, A. Seismic retrofitting by conventional methods. *Indian Concrete Journal*, 76 (8).

3504 ▲ 2002 Singh, H. *Planning for Disaster Mitigation in Earthquake Prone Urban Areas: A Case Study of Earthquake Prone City (Ahmedabad)*. New Delhi: School of Planning and Architecture (Unpublished Postgraduate Dissertation).

3505 ● 2002 Sinha, S. and L. Roy. Estimation of plastic energy absorbed in reinforced concrete columns under seismic loading. *Indian Concrete Journal*, 76 (7).

3506 ▲ 2002 Vatsa, K.S. Reducing earthquake losses: Towards a national perspective. *Economic and Political Weekly*, 37 (16):1503–1507.

3507 ● 2003 Dasgupta, K. *et al.* Seismic shear design of reinforced concrete structural walls-Part I: Review of code provisions. *Indian Concrete Journal*, 77 (11).

3508 ● 2003 Dasgupta, K. *et al.* Seismic shear design of reinforced concrete structural walls-Part–II: Proposed improvements in IS 13920:1993 provisions. *Indian Concrete Journal*, 77 (11).

3509 ● 2003 Goswami, R. and C.V.R. Murthy. Seismic shear design of reinforced concrete bridge piers-Part I: Review of code provision. *Indian Concrete Journal*, 77 (6).

3510 • 2003 Goswami, R. and C.V.R. Murthy. Seismic shear design of reinforced concrete bridge piers, Part–II: Numerical investigation of irc provisions. *Indian Concrete Journal*, 77 (7).

3511 • 2003 Jain, S.K. Review of Indian Seismic Code, IS 1893 (Part I), 2002. *Indian Concrete Journal*, 77 (11).

3512 • 2003 Kewalramani, M. and R. Gupta. Re-engineering of earthquake prone buildings. In *Proceedings of the Second Conference of Disaster Management Case Histories*. Pilani: Birla Institute of Technology. 117–124.

3513 • 2003 Kumar, M. and R. Gupta. *Proceedings of the Workshop on ReEngineering of Earthquake Prone Buildings*. Pilani: Birla Institute of Technology.

3514 • 2003 Mathur, S. and S.K. Deb. Seismic response control of Reinforced concrete setback building with friction dampers. *Indian Concrete Journal*, 77 (11).

3515 • 2003 Mokha, B. *Intervening in a Devasted Urban Precinct: Case of Old City Core (Bhuj)*. New Delhi: School of Planning and Architecture (Unpublished Postgraduate Dissertation).

3516 • 2003 Narasimhan, B. Lessons from Latur: A decade after the earthquake. *Economic and Political Weekly*, 38 (45): 4730–4732.

3517 • 2003 Rai, D.C. Seismic retrofitting of reinforced concrete shaft support of elevated tanks. *Indian Concrete Journal*, 77 (11).

3518 ■ 2003 Saraf, A.K. Remote sensing and geographic information systems applications in earthquake studies. In M. Mohanty and S. Aggarwal (eds). *Applications of Geographic Information Systems in Earthquake and Landslide Hazard Assessment*. New Delhi: Department of Science and Technology, Government of India. 17–19.

3519 • 2003 Sheth, A. Use of intermediate reinforced concrete moment frames in moderate seismic zones. *Indian Concrete Journal*, 77 (11).

3520 • 2003 Subramanian, N. and D.S. Prakasa Rao. Seismic design of joints in reinforced concrete structures – A review. *Indian Concrete Journal*, 77 (2).

3521 • 2004 Das, D. and C.V.R. Murthy. Brick masonry infills in seismic design of reinforced concrete framed buildings Part-I: Cost implications. *Indian Concrete Journal*, 78 (8).

3522 • 2004 Das, D. and C.V.R. Murthy. Brick masonry infills in seismic design of reinforced concrete framed buildings Part-II: Behaviour. *Indian Concrete Journal*, 78 (8).

3523 • 2004 Kanitkar, R. and V. Kanitkar. Seismic performance of conventional multistoryed buildings with open ground floors for vehicular parking. *Indian Concrete Journal*, 78 (2).

3524 • 2004 Nakagawa, Y. and S. Rajib. Social capital: A missing link to disaster recovery. *International Journal of Mass Emergencies and Disasters*, 22 (1): 5–34.

3525 • 2004 Srivastava, H.N. and G.D. Gupta. Disaster mitigation vis-a-vis time of occurrence and magnitude of earthquake in India. *Natural Hazard Journal*, 31 (2): 343–356.

3526 • 2005 Atif Syed. Disaster preparedness. *Indian Journal for Practising Doctors*, 2 (4):

3527 • 2005 Bureau of Indian Standards. *National Building Code of India*. New Delhi.

3528 • 2005 Chugh, R. and D. Menon. Design aids for estimation of shear in seismic design of reinforced concrete beams. *Indian Concrete Journal*, 79 (3).

3529 • 2005 Dasgupta, K. and C.V.R. Murthy. Seismic design of reinforced concrete columns and wall sections Part I: Consistent limit state design philosophy. *Indian Concrete Journal*, 79 (3).

3530 • 2005 Dasgupta, K. and C.V.R. Murthy. Seismic design reinforced concrete column wall sections Part II: Proposal for limiting strain in steel. *Indian Concrete Journal*, 79 (4).

3531 ▲ 2005 Department of Science and Technology. *National Seismicity Programme*. New Delhi: Seismology Division, Department of Science and Technology, Governtment of India.

3532 ▲ 2005 Matoo, T.K. Disaster and immunization. *Indian Journal for Practising Doctors*, 2 (4).

3533 • 2005 Singh, P. A strategy for mitigation of earthquake hazards in an urban centre. In R.K. Sharma and G. Sharma (eds). *Natural Disaster: Institutional Building for Mitigating the Impact on Vulnerable Sections*. New Delhi: APH Publishing Corporation. 119–124.

3534 • 2006 Asha, P. and R. Sundararajan. Evaluation of seismic resistance of exterior beam-column joint with detailing as per IS 13920: 1993. *Indian Concrete Journal*, 80 (2).

3535 • 2006 Comerio, M.C. Can buildings be made earthquake-safe? *Science*, 312 (5771): 204–206.

3536 • 2006 Durgesh, C. *et al.* Behaviour of seismic and non-seismic reinforced concrete frames under cyclic loads. *Indian Concrete Journal*, 80 (7).

3537 • 2006 Gupta, I. *et al.* Himalayan population at earthquake risk: Strategies for preparedness. *Disaster Prevention and Management*, 15 (4): 608–620.

3538 • 2006 Jain, S.K. Proposed codal provisions for design and detailing of beam-column joints in seismic regions. *Indian Concrete Journal*, 80 (8).

3539 • 2006 Pande, R.K. Participation in practice and disaster management: Experience of Uttaranchal (India). *Disaster Prevention and Management*, 15 (3): 425–428.

3540 ■ 2006 Ranade, P.S. Natural disasters management: A study of Kutch (Gujarat) 2001 Earthquake. In P.S. Ranade (ed). *Natural Disasters: Management and Preparedness.* Hyderabad: Institute of Charted Financial Analysts of India University Press. 234–247.

3541 ▲ 2006 Thakur, V.C. Earthquake hazard assessment and mitigation strategy: A case study of Himachal Pradesh. In K.K. Sharma, *et al* (eds). *Environmental Geohazards: Science and Society.* New Delhi: Research India Press. 53–63.

3542 ● 2006 Uma, S.R. and A.M. Prasad. Seismic behaviour of beam-column joints in reinforced concrete moment resisting frames: A review. *Indian Concrete Journal,* 80 (1).

Management: Flood

3543 ■ 1932 Saha, M. Catastrophic floods in Bengal and how they can be combated. *Modern Review:* 51 (2): 163–166.

3544 ■ 1939 Special Correspondent. Prevention of flood in the Indo-Gangetic plain. *Science and Culture,* 5: 166–167.

3545 ■ 1952 Indian Bureau of Flood Control. *Report on Extension of Irrigation and Further Flood Control of the Mahanadi Delta (Orissa).* New Delhi.

3546 ■ 1953 Sinha, B.N. Flood control in the Mahanadi and its bearing on the delta. *Geographical Review of India,* 15(4) 21–27.

3547 ■ 1954 Bose, N.K. Floods and measures for flood control. *Science and Culture,* 20 (3): 117–118.

3548 ■ 1955 Gairola, S.S. Flood control of the Ganga drainage basin. *National Geographical Journal of India,* 1 (1): 26–34.

3549 ■ 1955 Sen, A. Soil conservation and flood control. *Indian Journal of Power and River Valley Development,* 5 (2): 18–20.

3550 ▲ 1956 Bose, J.N. Solution of the Kosi problem. *Modern Review,* 99 (6): 479–481.

3551 ■ 1957 Banerjee, E.A.R. Problem of flood control in Kalighat river valley. *Indian Journal of Power and River Valley Development,* 7 (9): 66–68.

3552 ▲ 1957 Ministry of Irrigation and Power. *Report of the High Level Committee on Floods (Volume 1): General Assessment, Principles and Policies.* New Delhi: Government of India.

3553 ■ 1961 Chakravarti, S.N. Soil conservation and flood control in West Bengal. *Modern Review:* 110 (3): 191–194.

3554 ● 1961 Sachdev, M.R. A campaign against floods. *Bhagirath,* 8 (July): 3–5.

3555 ■ 1962 Department of Public Works and Irrigation. *Report of the Flood Inquiry Committee (Volume 1).* Bhubaneswar: Government of Orissa.

3556 ■ 1962 Mani, S. R. S. Flood control II: Controlled flooding. *Indian Journal of Power and River Valley Development*, 12 (11): 24–26.

3557 ▲ 1964 Khosla, A. N. *Orissa Decade of Destiny, 1963–1973: A Plan for the Integrated Development of the River Basin of Orissa.* Cuttack: Government of Orissa.

3558 ▲ 1965 Burton, I. Towards development of a comprehensive programme for flood damage reduction. *Indian Journal of Power and River Valley Development*, 15 (4): 1–3.

3559 ■ 1965 Special Correspondent.Flood control progress in perspective. *Bhagirath*, 12 (October): 106–107.

3560 ■ 1967 Editorial. Flood control measures in India. *Bhagirath*, 24 (January): 77.

3561 ■ 1967 Kumra, P. N. Flood control in the Fourth Plan. *Bhagirath*, 14 (October): 111–113.

3562 ▲ 1967 Singhota, J. S. Flood control. *Indian Journal of Power and River Valley Development*, 17 (8): 31–34.

3563 ■ 1968 Kumara, P. N. Flood protection methods in India. *Geographical Review of India*, 30 (4): 45–54.

3564 ■ 1968 Special Correspondent. Assam: Taming the Brahmaputra. *Economic and Political Weekly*, 3 (33): 1166.

3565 ▲ 1969 Bose, N. K. Should the Damodar Valley Corporation flood control potential be increased. *Indian Journal of Power and River Valley Development*, 19 (12): 501–504.

3566 ■ 1969 Planning Commission. Flood control and management. In *Fourth Five year Plan*. New Delhi: Government of India. 258–265.

3567 ■ 1970 Barua, B. B. Flood and erosion control in Brahmaputra Valley by making use of structural features. *Indian Journal of Power and River Valley Development*, 20 (2).

3568 ■ 1972 Banerji, S. and V. B. Lal. Flood control in India: Basin approach to the problem. *Indian Journal of Power and River Valley Development*, 22 (10): 410–416.

3569 ▲ 1973 Bhattacharya, A. K. Flood control in Damodar Valley. *Geographical Review of India*, 35 (2): 136–146.

3570 ● 1975 Khale, M.D. *Planning Approach to Flood Prone Settlements.* New Delhi: School of Planning and Architecture (Unpublished Postgraduate Dissertation).

3571 ▲ 1975 Sinha, A. The Bihar floods: Need for joint Indo-Nepal measures. *Economic and Political Weekly*, 10 (34): 1421–1423.

3572 ■ 1976 Tripathi, J. Flood control in India: A review. *Bhagirath*, 23 (April–September): 61–66.

3573 ■ 1977 Burman, S. G. Flood and its control. In D. Bandhu and E. Chauhan (eds). *Trends in Indian Environment.* New Delhi: Today and Tomorrow's Printers and Publishers. 199–206.

3574 ■ 1977 Ministry of Agriculture and Irrigation. *Floods and their Control in India.* New Delhi: Government of India.

3575 ■ 1977 Special Correspondent. Flood control: An *ad hoc* approach continues. *Economic and Political Weekly*, 12: 1520–1521.

3576 ■ 1978 Chitale, S.V. Some thoughts on flood cotrol problems in India. *Journal of Irrigation and Power,* 35: 493–496.

3577 ■ 1978 Ganga Flood Control Commission. *An Outline Plan of Flood Control for the Entire Ganga Basin.* Patna

3578 ■ 1978 Ministry of Agriculture. *Flood Fighting.* New Delhi: Report of the Working Group on Integrated Action Plan for Flood Control in Indo-Gangetic Basin.

3579 ■ 1978 Rout, S.P. Operation of Hirakud reservoir for flood control in delta. *Indian Journal of Power and River Valley Development*, 28 (11): 322–230.

3580 ■ 1978 Shah, R.B. Master plan for flood control and water utilisation in the Sahibi Basin. *Bhagirath*, 25 (October–December): 180–185.

3581 ■ 1978 Sinha, S. West Bengal flood control: Care for development by conservation. *Mainstream*, 16 (December 2).

3582 ■ 1978 Subramanyam, K.V. Damodar Valley Corporation and flood control. *Economic and Political Weekly*, 13 (47): 1847–1848.

3583 ■ 1978 Varma, C.J.V. Fighting the floods. *Indian Journal of Power and River Valley Development*, 28 (4).

3584 ■ 1979 Singh, C. Design flood for design of spillways. *Indian Journal of Power and River Valley Development*, 29 (9 and 10): 222–223.

3585 ■ 1979 Singh, P.N. Flood control in India. *Bhagirath*, 26 (October–December): 139–143.

3586 ■ 1980 Central Water and Power Commission. *Survey of Flood and Earthquake Affected Areas of Brahmaputra Valley in Assam: Role of Flood Proofing in Flood Damages.* New Delhi: Government of India.

3587 ■ 1980 Planning Commission. Flood control and its management. In *Sixth Five Year Plan.* New Delhi: Government of India. 158–175.

3588 ■ 1980 Rashtriya Barh Ayog. *Report of the National Flood Commission.* New Delhi: Government of India.

3589 ■ 1980 Sanyal, A. Flood: Making the waters flow. *Economic and Political Weekly*, 15 (17): 1387.

3590 • 1981 Bhadra, G. Technocratic view of flood control. *Economic and Political Weekly*, 16 (4): 107–108.

3591 • 1981 Naidu, M.D.N. Disaster management: Preventive, rescue, relief, rehabilitation and reconstruction operations. In *Proceedings of the International Conference on Flood Disaster.* New Delhi: Indian National Science Academy. 19–29.

3592 ■ 1981 Patnaik, M. Role of soil conservation and afforestation for flood moderation. In *Proceedings of the International Conference on Flood Disaster*. New Delhi: Indian National Science Academy. 1–7.

3593 ▲ 1981 Thakur, D.N. Strategy of agricultural production in flood-prone areas, with special reference to the Brahmaputra Valley. In *Proceedings of the International Conference on Flood Disaster*. New Delhi: Indian National Science Academy.

3594 ■ 1982 Hazarika, R.N. *A Study on Optimal Trans Basin Transfer from Brahmaputra River Basin: A Systems Approach to Flood Moderation in the Basin in Context of the Indian Continental Water Balance*. Mumbai: Indian Institute of Technology (Unpublished Ph.D Thesis).

3595 ■ 1982 Patel, C.C. Flood problem and flood control programme. *Bhagirath*, 29 (October): 148–151.

3596 ■ 1982 Singh, N.P. *Theory of Extreme Values and its Application to Flood Control*. Hissar: Haryana Agricultural University (Unpublished Ph.D Thesis).

3597 ■ 1985 Basu, S. Floods and water management. *Yojana*, 29 (6 and 7): 13–17.

3598 ▲ 1985 Irrigation and Power Department. *Hydrology and Flood Delta Development Plan*. Bhubaneswar: Government of Orissa.

3599 ■ 1985 Jha, B. *River Management of Flood, Drought and Power Shortage in Patna (Bihar)*. Patna: Janashakti Printing Press.

3600 ■ 1985 Planning Commission. Flood control and management. In *Seventh Five Year Plan*. New Delhi: Government of India.

3601 ■ 1985 Sharma, N.K. Flood control in India. *Bhagirath*, 32 (January): 1–9.

3602 ■ 1986 Roy, S.R. *Morphology of the Taraphini Basin through Remote Sensing Techniques and Flood Mitigation*. Kharagpur: Indian Institute of Technology, Kharagpur (Unpublished Ph.D thesis).

3603 ■ 1987 Prasad, K. Flood control achievements and challenges. *Yojana*, 31 (14 and 15): 81–83.

3604 ■ 1987 Special Correspondent.Role of Central Water Commission in the recent floods in Godavari. *Bhagirath*, 34 (January): 42–44.

3605 ■ 1988 Department of Public Works and Irrigation. *Interim Reports of the Flood Advisory Committee, 1940 and 1942*. Bhubaneswar: Government of Orissa.

3606 ▲ 1988 Lyngdoh, J.M. Disaster management: A case study of Kosi security system in Northeast Bihar. *Journal of Rural Development*, 7 (5): 519–39.

3607 ■ 1989 Darbari, H.N. and S. Singh. Flood control problems in Jammu region. *Bhagirath*, 36 (October–December): 220–225.

3608 ■ 1989 Ministry of Water Resources. *Report of the Working Group on Flood Management for the Eighth Five Year Plan*. New Delhi: Government of India.

3609 ▲1989 Rogers, P. *et al. Eastern Water Study: Strategies to Manage Flood and Drought in the Ganges-Brahmaputra Basin.* Washnigtion D C: United States Agency for International Development (Irrigation Support Project for Asia and the Near East).

3610 ■ 1990 Bhattacharya, A.K. *A Socio-Economic Analysis of Alternate Schemes of Flood Control.* Kharagpur: Indian Institute of Technology, Kharagpur (Unpublished Postgraduate Dissertation).

3611 ■ 1991 Bhattacharya, A.K. Variation in designing floods depending on inclusion or non-inclusion of upstream factors in economic analysis of flood control reservoir schemes. *Indian Journal of Power and River Valley Development*, 41 (9): 171–172.

3612 ▲1991 Central Water Commission. *Brahmaputra Board Master Plan, Assam.* New Delhi: Government of India.

3613 ■ 1991 Department of Water Resources. *Flood Control Order.* Patna: Government of Bihar.

3614 ■ 1991 Ganga Flood Control Commission. *Comprehensive Plan of Flood Management for the Ganga Sub-Basin, Part 11/9, The Bagmati River System.* New Delhi: Ministry of Water Resources, Government of India.

3615 ▲1991 Singh, S.K. *Disaster Managemnt in Flood Prone Areas: A Case Study of Muzaffarpur District.* New Delhi: Deparment of Housing, School of Planning and Architecture (Unpublished Postgraduate Dissertation).

3616 ■ 1992 Dhillon, G.S. and V.V. Sastry. Non-structural measures to reduce flood losses. *Indian Journal of Power and River Valley Development*, 42 (7): 222–226.

3617 ■ 1992 Varadarajan, P.K. Comprehensive management of floods. *Kurukshetra*, 41: 21–24.

3618 ■ 1993 Kunwar, U. Problems of flood control: A case study of Bihar. *Southern Economy*, 32 (15–16): 24–26.

3619 • 1993 Navalawala, B.N. Flood insurance: Concept and operational problems. *Yojana*, 37 (13 and 14): 14–16.

3620 ■ 1993 Special Correspondent. Non-structural aspects of flood management in India. *Bhagirath*, 40 (4) October–December): 139–147.

3621 ▲1994 Mangat, H.S. Patiala floods: Where lies the solution? *Transactions, Institute of Indian Geographers*, 16 (1): 29–33.

3622 ■ 1994 Rao, R.S. Role of remote sensing in flood management: The May 1990 cyclone. *International Journal of Remote Sensing*, 15 (8): 1557–1558.

3623 ■ 1995 Brihan Mumbai Mahanagar Palika. *Anti-Flooding Measures during Monsoons*. Mumbai: Brihan Mumbai Mahanagar Palika.

3624 ■ 1995 Editorial. Exercise in upgradation. *Down to Earth*, 4 (9): 12.

3625 ■ 1995 Editorial. All party proposals for flood control in Assam. *New Age*, 43 (37): 9.

3626 ● 1995 Lieten, G.K. and P. Datta. Panchayats in dealing with natural disasters: 1993 floods in Jalpaiguri. *Economic and Political Weekly*, 30 (15): 828–832.

3627 ● 1995 Sarkar, S. *Housing for Flood Affected Areas: The Case Study of Ghatal (West Bengal)*. New Delhi: School of Planning and Architecture (Unpublished Postgraduate Dissertation).

3628 ▲ 1995 Singh, P. *Colonised Rivers of North Bihar: Colonial Intervention in Irrigation and Flood Control, 1880–1940*. New Delhi: Center for Historical Studies, Jawahar Lal Nehru University (Unpublished Master of Philosophy Dissertation).

3629 ■ 1995 Singh, R. *et al.* Flood control in the Brahmaputra valley. *Southern Economy*, 33 (18): 19–20.

3630 ▲ 1995 Sud, S. Tackling the flood menace: Water for development. *Yojana*, 39 (1 and 2): 59–61.

3631 ▲ 1996 Central Water Commission. *Report of the Working Group on Flood Management for the Ninth Five Year Plan (1997–2002)*. New Delhi: Government of India.

3632 ■ 1996 Chopra, R. *et al.* Flood management using remote sensing technology: The Punjab (India) experience. *International Journal of Remote Sensing*, 17 (17): 3511–3521.

3633 ■ 1996 Gauriar, S. *Natural Disaster Management with Emphasis on Flood Management*. New Delhi: Advanced Professional Programme in Public Administration, Indian Institute of Public Administration (Unpublished Master of Philosophy Dissertation).

3634 ■ 1996 Kumar, A. *Flood Control and Rural Settlement Planning*. New Delhi: Mohit Publications.

3635 ▲ 1996 Kumar, P. *Disaster Mitigation and Management for Flood Prone Areas*. New Delhi: School of Planning and Architecture (Unpublished Postgraduate Dissertation).

3636 ■ 1996 Sharma, K.K. *An Appraisal of Structural Measures of Flood Control in Delhi*. New Delhi: Advanced Professional Programme in Public Administration, Indian Institute of Public Administration (Unpublished Master of Philosophy Dissertation).

3637 ▲ 1996 Shekhar, S. *Development Strategy for a Flood Prone Area: A Case Study in Community Development Block Salkhua (Bihar)*. New Delhi: School of Planning and Architecture (Unpublished Postgraduate Dissertation).

3638 ■ 1997 Bakshit, A.M. and K. Sarkar. Control of basin parameters in flood recurrence in the Lower Barak valley, Cachhar . In *Proceedings of the Symposium on the Earth Sciences in Environmental Assessment and Management.* Special Publication No. 48. Kolkata: Geological Survey of India. 85–94.

3639 ■ 1997 Keskar, G.R. *Report of the Flood Management for the Ninth Five Year Plan.* New Delhi: Central Water Commission, New Delhi.

3640 ■ 1997 Planning Commission. Flood control and management. In *Ninth Five Year Plan.* New Delhi: Government of India.

3641 ● 1998 Anand, V. and T. Prasad. Socio-economics of flood control: A case of study in North Bihar. In B.N. Verma (ed). *Flood Devastation and Agricultural Development in Eastern India.* Delhi: B.R. Publishing Corporation. 13–24.

3642 ■ 1998 Editorial. Flood decontrol. *Down to Earth*, 7 (10): 3.

3643 ■ 1998 Gorde, R.J. Floods and flood control: Engineering approach. In V.S. Kale (ed). *Flood Studies in India.* Bangalore: Memoirs of the Geological Society of India, No. 41. 173–194.

3644 ▲ 1998 Government of India. *Majuli Master Plan.* New Delhi.

3645 ■ 1998 Gupta, M. and A. Sharma. Disaster mapping as a tool for development management: An approach to flood damage control. In B.N. Verma (ed). *Flood Devastation and Agricultural Development in Eastern India.* Delhi: B.R. Publishing Corporation. 63–68.

3646 ■ 1998 Jha, U.M. and R.K. Sinha. Flood devastation and control measures in North Bihar. In B.N. Verma (ed). *Flood Devastation and Agricultural Development in Eastern India.* Delhi: B.R. Publishing Corporation. 219–226.

3647 ▲ 1998 Munsi, S.K. Problems of flood management in India: Policies and programmes. *Geographical Review of India*, 60 (3): 239–247.

3648 ■ 1998 Pandey, R.K. Flood problem and water resource management in Eastern Region. In B.N. Verma (ed). *Flood Devastation and Agricultural Development in Eastern India.* Delhi: B.R. Publishing Corporation. 41–58.

3649 ■ 1998 Rao, D.P *et al.* Remote sensing and geographic information systems in flood management in India. In V.S. Kale (ed). *Flood Studies in India.* Bangalore: Memoirs of the Geological Society of India No. 41. 195–256.

3650 ■ 1998 Seth, S.H. Flood hydrology and flood management in India. In V.S. Kale (ed). *Flood Studies in India.* Bangalore: Memoirs of the Geological Society of India, No. 41. 155–172.

3651 ■ 1998 Singh, K.M. and N. Sharma. Flood control: What not to learn from Bihar. In B.N. Verma (ed). *Flood Devastation and Agricultural Development in Eastern India.* Delhi: B.R. Publishing Corporation. 199–220.

3652 ▲ 1998 Thakur, R. *et al.* Development of sustainable systems in flood prone areas of Bihar. In B.N. Verma (ed). *Flood Devastation and Agricultural Development in Eastern India.* Delhi: B.R. Publishing Corporation. 341–357.

3653 ■ 1999 Ahuja, B.S. Non-structural flood management-cost effective options. *Shelter,* special issue, *October.* 51–53.

3654 ■ 1999 Biswas, R. *et al.* A study of intersectoral co-ordination in disaster management in flood-prone districts of West Bengal. *Indian Journal of Public Health,* 43 (3): 106–111.

3655 ■ 1999 Central Water Commission. *Flood Management and River Configuration Study of River Yamuna from Tajewala to Okhla.* New Delhi.

3656 ■ 1999 Planning Commission. Flood control and management. In *Annual Plan: 1999–2000.* New Delhi: Government of India. 336–354.

3657 ▲ 1999 Vamshikrishna, K.M. *Comprehensive Flood Hazard Reduction Plan, Bhadrachalam (Andhra Pradesh).* New Delhi: School of Planning and Architecture (Unpublished Postgraduate Dissertation).

3658 • 2000 Awasthi, S. Uttar Pradesh seeds of the future. In S. Parasuraman and P.V. Unnikrishnan (eds). *India Disasters Report: Towards a Policy Initiative.* New Delhi: Oxford University Press. 163.

3659 ■ 2000 Jain, B.K. Flood management in Uttar Pradesh. In P.S. Roy, *et al* (eds). *Natural Disasters and their Mitigation: A Remote Sensing and Geographic Information Systems Perspective.* Dehradun: Indian Institute of Remote Sensing. 155–157.

3660 ■ 2000 Jana, M.M. Flood and their control in North Bengal: Application of remote sensing techniques. In R.B. Singh (ed). *Disaster Management.* Jaipur: Rawat Publications. 94–108.

3661 ■ 2000 Jena, D. *Flood and Flood Control Measures: A Study with Special Reference to Mahanadi Delta in Orissa.* Dehradun: Research Foundation for Science, Technology and Ecology.

3662 ■ 2000 Kumar, B. and A.K. Srivastava. Use of remote sensing and geographic information systems in Ganga Flood Control Commission: Past experience and future approach. In P.S. Roy, *et al.* (eds). *Natural Disasters and their Mitigation: A Remote Sensing and Geographic Information Systems Perspective.* Dehradun: Indian Institute of Remote Sensing. 151–154.

3663 ▲ 2000 Kumar, P.K. *Disaster Mitigation and Management for Flood: Hajo Circle.* New Delhi: Deparment of Housing, School of Planning and Architecture (Unpublished Postgraduate Dissertation).

3664 • 2000 Muralidharan, S. Proactive monsoon management plan is needed. *Frontline,* 17 (11): 77–81.

3665 ■ 2000 Orissa State Disaster Management Authority and United Nations Organization. *Community Contingency Plan for Floods and Cyclone: Orissa.* Bhubaneswar.

3666 ■ 2000 Planning Commission. Flood control and management. In *Tenth Five Year Plan* New Delhi: Government of India. 65–105.

3667 ▲ 2000 Rath, B. Perspective remedial action plan for the flood and cyclone disaster prone areas of coastal Orissa. In *Proceedings of the International Conference on Sustainable Development of Water Resources.* New Delhi: Institute of Resource Management and Economic Development.

3668 ▲ 2000 Shafi, M. The hazards of floods in India: Measures for prevention. In R. B.Singh (ed). *Disaster Management.* Jaipur: Rawat Publications. 77–83.

3669 ▲ 2000 Tanu, P. *Urban Development Strategies for Flood Prone Areas, Patna.* New Delhi: School of Planning and Architecture (Unpublished Postgraduate Dissertation).

3670 ■ 2000 Valunjkar, S.S. *Operating Policies for Flood Control Management and Development of Models for Optimal Utilization of Available Water Resources.* Nagpur: Nagpur University (Unpublished Ph.D Thesis).

3671 ■ 2001 Indian Water Resource Society. Management of floods and droughts. *Bhagirath*, 48 (January–March): 1–11.

3672 ■ 2001 Jain, P.C. Flood mitigation practices in India. In V.K. Sharma (ed). *Disaster Management.* New Delhi: Indian Institute of Public Administration. 38–45.

3673 ▲ 2001 Jena, A.C. *Flood Disaster Management: A Study.* Hyderabad and New Delhi: National Institute of Rural Development and A.P.H. Publishing Corporation.

3674 ■ 2001 Rangachari, R. Disaster preparedness and mitigation in flood management: Cooperation and networking among countries. In P. Sahni, *et al* (eds) *Disaster Mitigation: Experiences and Reflections.* New Delhi: Prentice-Hall of India Private Limited. 55–66.

3675 ▲ 2001 Rani, N. *Flooding and Sustainable Water Management in the Lower Bagmati River Basin.* Delhi: Department of Geography, Delhi School of Economics, University of Delhi (Unpublished Master of Philosophy Dissertation).

3676 ■ 2001 Sandhu, K. Flood management. *Bhagirath*, 48 (October): 169–199.

3677 ▲ 2001 Sharma, V.K. and T. Priya. Development strategies for flood prone areas: Case study, Patna .*Disaster Prevention and Management*, 10 (2): 101–110.

3678 ▲ 2002 Central Water Commission. *Appraisal of Flood Management in India: 2002*. New Delhi: Government of India.

3679 ■ 2002 Das, A. K. L. Different development phases of flood control and drainage system. *Bhagirath*, 49 (April–June): 83–86.

3680 ● 2002 Mishra, D. K. *Living with the Politics of Floods: The Mystery of Flood Control.* Dehradun: Peoples Science Institute.

3681 ● 2003 Chapman, G. P. and K. Rudra. *Bengal's Millennium Flood and the Question of Appropriate Development.* Lancaster: Department of Geography, Lancaster University, United Kingdom.

3682 ● 2003 Gupta, S. *et al.* Economics of flood protection in India. *Natural Hazard Journal*, 28 (1): 199–210.

3683 ■ 2003 Jain, V. and R. Sinha. Evaluation of geomorphic control on flood hazard through geomorphic instantaneous unit hydrograph. *Current Science*, 85 (11): 1596.

3684 ■ 2003 Mahapatra, P. K. and R. D. Singh.Flood management in India. *Natural Hazard Journal*, 28 (1): 131–143.

3685 ▲ 2003 Mirza, M. *et al* (eds). *Flood Problem and Management in South Asia.* Netherlands: Kluwer Academic Publishers.

3686 ▲ 2003 Rath, B. Perspective remedial action plan for the flood and cyclone disaster prone areas of Coastal Orissa. In K. Prasad (ed). *Water Resources and Sustainable Development.* New Delhi: Shipra Publication. 201–222.

3687 ● 2003 Rural Volunteers Centre. *Some Best Practices in Flood Preparedness in Brahmaputra River Basin: A Case Study of Dhemaji District (Assam).* Dhemaji: Rural Volunteers Centre.

3688 ■ 2004 Chittibabu, P. *et al.* Mitigation of flooding and cyclone hazard in Orissa. *Natural Hazard Journal*, 31 (2): 455–485.

3689 ● 2004 Krishna, K. S. Brahmaputra flooding: Problems for administration and prospects. In V. P. Singh *et al.* (eds). *The Brahmaputra Basin Water Resources.* Netherlands: Springer. 309–320.

3690 ■ 2004 Mitra, A. K. Flood management.In V. P. Singh *et al.* (eds). *The Brahmaputra Basin Water Resources.* Netherlands: Springer. 535–559.

3691 ■ 2004 Murty, V.B. and V.H. Prasad. Remote sensing and geographic information systems in flood management. In V. P. Singh *et al.* (eds). *The Brahmaputra Basin Water Resources.* Netherlands: Springer. 560–577.

3692 ■ 2005 Dhillon, G. S. Natural disaster-floods: Non-structural measures for mitigation of impact on vulnerable sections. In R. K. Sharma and G. Sharma (eds). *Natural Disaster: Institutional Building for Mitigating the Impact on Vulnerable Sections.* New Delhi: APH Publishing Corporation. 17–24.

3693 ▲2005 Dossal, M. A master plan for the city: Looking at the past. *Economic and Political Weekly*, 40 (36), 3897–3900.

3694 • 2005 Mittal, S.P. Peoples participation in disaster management, with special reference to flood management in Shivalik region. In R.K. Sharma and G. Sharma (eds). *Natural Disaster: Institutional Building for Mitigating the Impact on Vulnerable Sections.* New Delhi: APH Publishing Corporation. 217–222.

3695 • 2005 Special Correspondent. Facing the flood: How prepared is Bihar? *Down to Earth*, 14 (7): 56.

3696 • 2005 Special Correspondent. Mumbai flood, the deluge: Causes and lessons for the future. *Economic and Political Weekly*, 40 (36), 3896– 3897.

3697 • 2006 Gupta, S.C. and R. Kumar. Flood management and monitoring in Himachal Pradesh. In K.K. Sharma, *et al* (eds). *Environmental Geohazards: Science and Society.* New Delhi: Research India Press. 185–194.

3698 ■ 2006 Sood, R.K. and S.S. Randhawa. Application of remote sensing and geographic information systems for flood control management in Himachal Pradesh: An overview. In K.K. Sharma, *et al* (eds). *Environmental Geohazards: Science and Society.* New Delhi: Research India Press. 195–209.

Management: Landslide

3699 ▲1959 Srivastava, J.P. *Report on the Investigation of Foundation Conditions and Stability of Hillside of Snowdown Hospital of Himachal Government, Shimla and Landslides at Jutogh and Narkanda.* Kolkata: Geological Survey of India (Unpublished Report).

3700 ▲1969 Srivastava, A.K. *Third Progress Report on the Geological Investigations of the Stability of the Hill Slopes around Nainital.* Kolkata: Geological Survey of India (Unpublished Report).

3701 ■ 1980 Datye, K.R. and S.S. Nagaraju. Flexible piles for landslide control. In *Proceedings of the International Symposium on Landslides.* New Delhi: Sarita Pradhan. 439–441.

3702 ■ 1980 Koshy, M.G. Avalanche control measures. In *Proceedings of the International Symposium on Landslides.* New Delhi: Sarita Pradhan. 419–424.

3703 ■ 1984 Dhir, M.P. *Stabilisation of Landslides on National Highways.* New Delhi: Central Road Research Institute, (Unpublished Report).

3704 ■ 1986 Bhandari, R.K. and C. Gupta. Problems of landslides in the Himalaya and future directions. In J.S. Singh (ed). *Environmental Regeneration in Himalaya, Concepts and Strategies.* Nainital: Central Himalayan Environment Association and Gyanodaya Prakashan. 39–57.

3705 ■ 1988 Didwal, R.S. Remedial methods employed in Jammu Province in India to control landslides and their efficacy. In C. Bonnard (ed).

Proceedings of the Fifth International Symposium on Landslides. Rotterdam: Balkema. 897–902.

3706 • 1988 Viladhar, M.N. *et al.* Slope stabilisation for building safety. In C. Bonnard (ed). *Proceedings of the Fifth International Symposium on Landslides.* Rotterdam: Balkema. 1015–1022.

3707 • 1990 Haigh, M.J. Landslides evaluation for the highway maintenance in the Himalaya. In T.V. Singh (ed). *Geography of the Mountains: Countributions to Indian Geography,* (Series 12). New Delhi: Heritage Publishers. 80–94.

3708 ■ 1993 Kuberan, R. *Control Measures of Landslides.* New Delhi: Department of Science and Technology, Government of India.

3709 ■ 1996 Central Road Research Institute. *Geotechnical Report on Landslide and its Corrective Measures on National Highway–21, Bilaspur Division of Himachal Pradesh.* New Delhi.

3710 • 1996 Malhothra, G.S. *et al.* Landslides hazard assessment: A need for environmental management of Himalaya. In K. Senneset (ed). *Proceedings of Seventh International Symposium on Landslides.* Rotterdam: Balkema. 315–320.

3711 ■ 1996 Thakur, V.C. *Landslides Hazard Management and Control in India.* Kathmandu: International Center for Integrated Mountain Development.

3712 ▲ 1999 Pilgrim, N.K. Landslides, risk and decision-making in Kinnaur district: Bridging the gap between science and public opinion. *Disasters: The Journal of Disaster Studies, Policy and Management,* 23 (1), 45–65.

3713 • 2001 Navalli, S. Safe as houses. *Down to Earth,* 9 (24), 26–27.

3714 ■ 2002 Rajinder, B. *et al.* Landslides hazard and mitigation measures at Gangtok (Sikkim Himalaya). *Engineering Geology,* 64 (4), 351–368.

3715 • 2003 Arora, B.R. and M.P. Sah. *Establishment of Geotechnical Laboratory for Landslide Investigation and Training of Manpower at Dehradun.* New Delhi: Ministry of Science and Technology, Government of India.

3716 ▲ 2004 Ganju, A. and A. P. Dimri. Prevention and mitigation of avalanche disaster in Western Himalayan region. *Natural Hazard Journal,* 31 (2), 357–371.

3717 ■ 2004 Khire, M.V. Terrain evaluation and mitigation for landslides: A case study on Varandh Ghat. In R. Nagarajan (ed). *Lanshde Disaster: Assessment and Monitoring.* New Delhi: Anmol Publications. 37–56.

3718 ■ 2004 Nagarajan, R. Strategy for landslides hazard mitigation. In R. Nagarajan (ed). *Landslide Disaster: Assessment and Monitoring.* New Delhi: Anmol Publication. 284–293.

3719 ▲ 2004 Rajaram, B. *et al.* Management of slides in Konkan railway alignment. In R. Nagarajan (ed). *Landslide Disaster: Assessment and Monitoring.* New Delhi: Anmol Publications. 294–302.

3720 ▲ 2004 Sah, M. P. and S. K. Bartarya. Landslides hazards in the Himalaya: Strategy for their management. In K. S. Valdiya (ed). *Coping with Natural Hazards: Indian Context.* Hyderabad: Orient Longman. 165–179.

3721 ■ 2004 Sharma, S. S. Mitigation of avalanche disasters in Indian Himalaya. In K. S. Valdiya (ed). *Coping with Natural Hazards: Indian Context.* Hyderabad: Orient Longman. 124–133.

3722 ■ 2004 Sinha, U. N. and P. Kumar. Control measures at Kaliasaur landslides, Garhwal Himalaya (Uttaranchal). In R. Nagarajan (ed). *Lanslide Disaster: Assessment and Monitoring.* New Delhi: Anmol Publications. 264–277.

3723 ● 2005 Aniruddha, U. and P. Rautela. Disaster management strategy for avoiding landslide induced losses to the villages in the vicinity of the Himalayan township of Mussoorie in Uttaranchal . *Disaster Prevention and Management,* 14 (3), 378–387.

3724 ● 2006 Aniruddha, U. and C. Prasad. Disaster management strategy for mass wasting hazard prone Naitwar Bazaar and surrounding areas in Upper Tons valley in Uttarkashi district . *Disaster Prevention and Management,* 15 (5), 821–837.

3725 ■ 2006 Rautela, P. and R.K. Pande. Non-monsoonal landslides in Uttaranchal Himalaya (India): Implications upon disaster mitigation strategy. *Disaster Prevention and Management,* 15 (3), 448–460.

3726 ■ 2006 Shukla, A. K. *et al.* Layout of avalanche control structures at microsphere plate 10: A case study. In K.K. Sharma, *et al* (eds). *Environmental Geohazards: Science and Society.* New Delhi: Research India Press. 17–127.

3727 ■ 2006 Singh, B. Landslides hazard mitigation: Department of Science and Tecnology's initatives. In K.K. Sharma, *et al* (eds). *Environmental Geohazards: Science and Society.* New Delhi: Research India Press. 171–184.

Management: Storm

3728 ▲ 2004 Roy, S. *Natural Disaster Management: A Case Study of Tornado.* New Delhi: Abhijeet Publications.

Management: Tsunami

3729 ■ 2005 Bagla, P. Indian Ocean tsunami: A dead spot for the tsunami network? *Science,* 310: 1604.

3730 ▲ 2005 Bird, J. *et al.* Managing tsunami risk. *The Lancet*, 366: 271–273.

3731 ● 2005 Koutitas, C. *et al.* A computational approach to design codes for tsunami resisting coastal structures. *ISET Journal of Earthquake Technology*, 42 (4): 137.

3732 ▲ 2005 Kurien, J. Securing the future against tsunamis. *Economic and Political Weekly*, 40 (2): 98–99.

3733 ▲ 2005 Ministry of Home Affairs. *Prevention, Protection and Mitigation from Risk of Tsunami Disasters: A Concept Note*. New Delhi: Government of India.

3734 ▲ 2005 Mukhopadhyay, R. and S. De. Dealing with 26/12. *Current Science*, 88 (11): 1713.

3735 ▲ 2005 Roy, S. C. Eluding nature's wrath. *Science and Culture*, 71 (1–2):1–2.

3736 ● 2005 Standing, G. Tsunami recovery grants: Looking at long term sustainable plans. *Economic and Political Weekly*, 40 (6): 510–514.

3737 ● 2005 Stone, R.and R. A. Kerr. Gearing for the next killer wave. *Science*, 310: 1602.

3738 ● 2005 Subbarao, K. V. An action plan for preparedness and mitigation of damage and destruction due to tsunamis. *Journal of Geological Society of India*, 65 (3): 367

3739 ● 2005 Subrahmanyam, K. R. Tsunami and the coastal regulation zone. *Journal of Geological Society of India*, 65 (6): 769.

3740 ● 2006 Chandrasekar, N. *et al.* Classification of tsunami hazard along the Southern Coast of India: An initiative to safeguard the coastal environment from similar debacle. *Science of Tsunami Hazard*, 24 (1): 5–23.

3741 ▲ 2006 Gupta, H. K. Indian initiative to mitigate tsunami and storm surge. In *Proceedings of the Brain Storming Session on the Great Tsunami Event of 26 December 2004*. New Delhi: Department of Science and Technology, Government of India: 47–52.

3742 ■ 2006 Hegde, V. S. and B. Manikiam. Disaster management support with space inputs. In *Proceedings of the Brain Storming Session on the Great Tsunami Event of 26 December 2004* Science Academy. New Delhi: Department of Science and Technology, Government of India. 85–90.

3743 ● 2006 Kumar, B. Recent Asian tsunami hazard and disaster management programs over India. In N. C. Mahanti, *et al* (eds). *Disaster Management*. Delhi: Narosa Publishing House.

3744 ● 2006 Kuriakose, B. *Experiences in the Study of the Villages Chinnangudi and Tarangambadi in Connection with the Housing Reconstruction Project for the Tsunami Victims*. Mumbai: Department of Humanities and Social Sciences, Indian Institute of Technology.

3745 ■ 2006 Rao, U.R. Role of space in disaster management with special emphasis on tsunami. In *Proceedings of the Brain Storming Session on the Great Tsunami Event of 26 December 2004*. New Delhi: Department of Science and Technology, Government of India. 79–84.

3746 ● 2006 Shetye, S.R. Public awareness, education and international co-operative endeavours. In *Proceedings of the Brain Storming Session on the Great Tsunami Event of 26 December 2004*. Government of India. 65–70.

Management: Assorted

3747 ● 1973 Banerjee, N. Role of Indian Medical Association during natural calamities and national emergencies. *Journal of Indian Medical Association*, 61(11): 477–481.

3748 ● 1978 Rao, V. *Planning for Disaster Prone Area*. New Delhi: Department of Urban and Regional Planning, School of Planning and Architecture (Unpublished Postgraduate Dissertation).

3749 ● 1980 Joint Assistance Centre. *Natural Disaster: A Guide for Relief Workers*. New Delhi: Government of India.

3750 ● 1980 Venugopal, R. *Housing Development Strategies for Disaster Prone Area*. New Delhi: School of Planning and Architecture (Unpublished Postgraduate Dissertation).

3751 ● 1981 State Planning Board. *Report of the Group of Disaster Preparedness and Management*. Mumbai: Government of Maharashtra.

3752 ● 1981 Winchester, P. From disaster to development: Notes from India. *Disasters: The Journal of Disaster Studies, Policy and Management,* 5 (2): 154–163.

3753 ■ 1983 Jain, N.K. Training for disaster preparedness in India. *Disasters: The Journal of Disaster Studies, Policy and Management*, 7 (1): 1–78.

3754 ● 1984 Chandrasekhar, S. Disaster management in the Indian economy. In B. R. Naidu (ed). *Disaster Management*. Tirupati: Sri Venkateswara University. 15–18.

3755 ▲ 1984 Naidu, B.R. Disaster management. In B.R. Naidu (ed). *Disaster Management*. Tiru-pati: Sri Venkateswara University. 1–14.

3756 ▲ 1984 Naidu, B.R. (eds). *Disaster Management*. Tirupati: Sri Venkateswara University.

3757 ▲ 1984 Raghavulu, C. V. Disaster management. In B.R. Naidu (ed). *Disaster Management*. Tirupati: Sri Venkateswara University. 19–26.

3758 ● 1984 Raju, B.Y. Disaster management: Role of financial institutions. In B. R. Naidu (ed). *Disaster Management*. Tirupati: Sri Venkateswara University. 95.

3759 ▲ 1984 Rao, K.N. Disaster management. In B.R. Naidu (ed). *Disaster Management*. Tirupati: Sri Venkateswara University. 91–94.

3760 ● 1984 Reddy, V.M. Role of government and voluntary organizations in disaster management. In B.R. Naidu (ed). *Disaster Management.* Tirupati: Sri Venkateswara University. 101–104.

3761 ▲ 1985 Maddox, J. Avoiding recurrent catastrophes. *Nature*, 313 (June 6), 453–453.

3762 ● 1987 Dhatwala, H.R. Natural calamaties and preventive measures. *Journal of Himachal Pradesh Institute of Public Administration,* 11 (2), 29–33.

3763 ▲ 1988 Special Correspondent. Disaster policies. *India Today*, (May 8): 56.

3764 ● 1989 Kumar, V.S. and S.G. Krishnan. Disaster management: A psychological crisis. *Social Welfare*, 36 (5), 33–34.

3765 ● 1989 National Building Organisation. *Housing in Disaster Prone Areas: Report of the Development Group* . New Delhi: Indian Meterological Department.

3766 ▲ 1990 Desai, B. Managing ecological upheavals: A third world perspective. *Social Science and Medicine,* 30 (10), 1065–1072.

3767 ▲ 1990 Sinha, D.K. (eds). *Readings on Natural Disaster Reduction.* Kolkata: University of Calcutta.

3768 ● 1991 Kayastha, S.L. Population, disaster and disaster reduction strategies. *Geographical Review of India*, 53 (1 and 4): 1–14.

3769 ▲ 1991 Kulshrestha, S.M. Natural disaster management. *Social Welfare*,38 (1): 14–15.

3770 ● 1991 Ministry of Agriculture. *Discussion Papers of the National Advisory Council on International Decade for Natural Disaster Reduction.* New Delhi: Natural Disaster Management Division, Government of India.

3771 ● 1991 Saha, S. *Shelter and Planning for Disaster Prone Areas.* New Delhi: School of Planning and Architecture (Unpublished Postgraduate Dissertation).

3772 ■ 1992 Basu, D. Management of natural disasters in West Bengal. In D.K. Sinha (ed). *Natural Disaster Reduction for Nineties: Perspectives, Aspects and Strategies.* Kolkata: International Journal Services. 533–550.

3773 ● 1992 Chaudhri, A.K. Preparing disaster relief and management plans. *Kurukshetra*, 41(3): 8–10.

3774 ● 1992 Elo, O. Preparedness programme for natural and man-made disasters. *Indian Journal of Public Health,* 36 (1): 6–14.

3775 ● 1992 Guha, M.K. Role of media in natural disaster reduction before the commencement of incident. *Science and Cu lture*, 58 (5–8): 95–97.

3776 ● 1992 Haldar, K.K. Nutritional measures in the management of disasters. *Indian Journal of Public Health,* 36 (1): 21–26.

3777 • 1992 Jain, N.K. and R.N. Panda. *National Disaster Prediction: Second National Convention on Role of Voluntary Agencies.* New Delhi: Jac Trust.

3778 ■ 1992 Karale, R.L. *Remote Sensing for Control and Mitigation of Disaster.* Nagpur: State Institute of Science Education.

3779 • 1992 Moni, M. *et al.* Natural calamites relief management information system. In *Proceedings of the World Congress on Natural Hazards Reduction.* 317–333.

3780 ▲ 1992 Narasimhan, B. Perspective for natural disaster reduction in India. *Indian Journal of Public Health,* 36 (1): 27–32.

3781 ▲ 1992 Padmanabhan, B.S. Disaster management and development. *Yojana,* 36 (22): 15–16.

3782 • 1992 Raghavulu, C.V. and B.P.C. Bose. Centre-State relations in disaster mitigation and preparedness. *Indian Journal of Political Science,* 53 (3): 314–331.

3783 • 1992 Seshagiri, N. National Informatics Centre network based natural hazards relief management. In D.K. Sinha (ed). *Natural Disaster Reduction for Nineties: Perspectives, Aspects and Strategies.* Kolkata: International Journal Services. 251–269.

3784 ▲ 1992 Sinha, D.K (eds). *Natural Disaster Reduction for Nineties: Perspectives, Aspects and Strategies.* Kolkata: International Journal Services.

3785 ▲ 1992 Special Correspondent. On natural disaster reduction. *Science and Culture,* 58 (5–8): 105.

3786 • 1992 Venkateshwarlu, J. Disaster management: A national perspective. *Kurukshetra,* 41(3): 15–20.

3787 ▲ 1993 Mishra, G.K. and G.C. Mathur (eds). *Natural Disaster Reduction.* New Delhi: Reliance Publishing House and Indian Institute of Public Administration.

3788 ■ 1993 Murty, R. *et al.* Geomorphological studies for disaster mitigation: A case study of the Krishna delta. *International Journal of Remote Sensing,* 14: 3269–3274.

3789 ■ 1993 Rao, V.R. and L. Lalitha. *Disasters in India: Can Remote Sensing do Something?* Bangalore: Indian Space Research Organisation.

3790 • 1993 Supe, A.N. Disaster management: Are we ready? *Journal of Postgraduate Medicine,* 39 (1).

3791 • 1993 Voluntary Health Association of India. *Turning Disasters into Development: A Glimpse of Indian Voluntary Action.* Bangalore: Government of India.

3792 ▲ 1994 Gaur, V.K. Managing disaster precipitated by natural hazard. *Sadhana: Academy Proceedings in Engineering Sciences,* 19 (4): 543–550.

3793 • 1994 Mishra, V.S. *Environment, Disasters and the Law.* New Delhi: Ashish Publishing House.

3794 ● 1994 United Nations. *National Disaster Prevention, Preparedness and Mitigation*. New York: United Nations.

3795 ▲1995 Mathur, G.C. Planning and management for the prevention and mitigation of natural disasters in metropolises of India. In G.K. Mishra and G.C. Mathur (eds). *Natural Disaster Reduction*. Delhi: Reliance Publishing House. 1–27.

3796 ● 1995 Mathur, G.C. Housing in areas prone to natural disasters. In G.K. Mishra and G.C. Mathur (eds). *Natural Disaster Reduction*. New Delhi: Reliance Publishing House. 67–78.

3797 ● 1995 Mishra, G.K and G.C. Mathur (eds). *Natural Disaster Reduction*. New Delhi: Reliance Publishing House.

3798 ● 1995 Venkateswara, R. Risk management: Time for innovative approaches. *Environmental Management*, 3: 313–321.

3799 ● 1996 Army Training Command. *Policy Approach and Role of Various Organisations for Disaster Management in India*. New Delhi: Government of India.

3800 ■ 1996 Council for Advancement of Peoples Action and Rural Technology. *Disaster Preparedness: A Handbook for Trainers (Volume I and II)*. New Delhi: Government of India.

3801 ● 1996 Gupta, G. Action Plan role of universities in International Decade for the Natural Disaster Reduction. In A.P. Santhakumar, *et al* (eds). *Proceedings of the International Conference on Disasters and Mitigation*, (A–4). Chennai: Anna University. 46.

3802 ● 1996 Jacob, D. *Mobilising Community Participation in Disaster Management (Mimeo)*. New Delhi: APH Publishing Corporation.

3803 ● 1996 Kayastha, S.L. Disasters in India: Certain characteristics, mitigation and management. In R.B. Singh (ed). *Disasters, Environment and Development*. New Delhi: Oxford and IBH Publishing Co. Pvt. Ltd. 63–76.

3804 ● 1996 Khan, A.A. *et al*. *Role of Non-Government Organisations in Disaster Management in India*. New Delhi: Indian Institute of Public Administration (Mimeo).

3805 ▲1996 Narasimhan, B. Natural disaster management 'current concerns'. In R.B. Singh (ed). *Disasters, Environment and Development*. New Delhi: Oxford and IBH Publishing Co. Pvt. Ltd. 19–30.

3806 ● 1996 National Institute of Rural Development. *Course Material on the Role of Panchayats in the Management of Natural Disasters*. Hyderabad:

3807 ● 1996 Reddy, A.V.S. *et al*. Disaster mitigation through people's participation: Role of local self government institutions. *Asia-Pacific Journal of Rural Development*, 6 (2), 15–34.

3808 ● 1996 Sharma, V.K. Natural disaster preparedness, mitigation and reduction: Efforts in India during International Decade for the

Natural Disaster Reduction. In R.B. Singh (ed). *Disasters, Environment and Development.* New Delhi: Oxford and IBH Publishing Co. Pvt. Ltd. 93–106.

3809 ● 1997 Awaradi, S.A. *Disaster Management Plan for Andaman and Nicobar Islands.* Nicobar: District Administration.

3810 ▲ 1997 Bandyopadhyay, S. Natural environmental hazards and their management: A case study of Sagar Island (India). *Singapore Journal of Tropical Geography*, 18 (1): 20–45.

3811 ▲ 1997 Disaster Mitigation Institute and Oxford Centre for Disaster Studies. *Urban Risk Reduction: Action Planning in Ahmedabad.* Oxford: Oxford Centre for Disaster Studies.

3812 ▲ 1997 Jayaraman, V. *et al.* Managing the natural disaster from space technology inputs. *Acta Astronautica*, 40 (2–8): 291–325.

3813 ▲ 1997 Oxford Centre for Disaster Studies. *Reducing Urban Risk, India.* London: Technology Development Research Project for Department of International Development of the British Government.

3814 ■ 1997 Reddy, A.V.S. Utilization of remote sensing techniques in disaster management: An administrative experience. *Asia-Pacific Journal on Rural Development*, 7 (2): 97–100.

3815 ▲ 1997 Sharma, V.K. Natural disaster management in India. *Indian Journal of Public Administration (Special Number)*, 43 (3): 764–774.

3816 ■ 1998 Goswami, S. Disaster Management-Assam's dire need. In *National Workshop on Geodynamics and Natural Resources of Northeastern India.* Dibrugarh: Dibrugarh University. 36–41.

3817 ● 1998 Menon, V.C. Role of non-governmental organisations in disaster management. *Indian Journal of Adult Education*, 53 (9): 33–39.

3818 ● 1998 Mohanty, B.B. Role of media in disaster management. *Indian Journal of Adult Education*, 53 (9): 43–54.

3819 ● 1998 Pratinidhi, A.K. and R.M. Gokhale. Disaster management: Knowledge attitude and practice of urban slum dwellers. *Indian Journal of Medical Science*, 52 (4).

3820 ▲ 1998 Roy, M. *Disaster Mamagement: An Indian Perspective.* Australia: University of Wollongong.

3821 ● 1998 Sachdeva, J.L. Role of adult education in creating awareness of natural disaster management. *Indian Journal of Adult Education,* 53 (9): 41–42.

3822 ▲ 1998 Sharma, V.K. Natural disaster management: An overview. *Indian Journal of Adult Education*, 53 (9): 5–17.

3823 ▲ 1998 Singh, S.K. *et al. Disaster Management.* New Delhi: Mittal Publication.

3824 • 1998 Upadhyay, J. N. Role of education and training in disaster mitigation. *Indian Journal of Adult Education*, 53 (9): 27–32.

3825 ▲ 1999 Aich, B. N. Natural calamities: Mitigation and management. In A. D. Mukhopadhaya (ed). *Disaster Management.* Kolkata: Jadavpur University.

3826 ▲ 1999 Basu, P. K. Natural disaster preparedness. In A. D. Mukhopadhaya (ed). *Disaster Management.* Kolkata: Jadavpur University.

3827 • 1999 Bhandari, R. K. Less is more in natural disaster prevention, but many times more is less in natural disaster correction. *Shelter,* special issue, October: 57–60.

3828 ▲ 1999 Bose, A. Urban landuse planning for preparedness of built environment in the wake of natural disasters. *Shelter,* special issue, October: 49–50.

3829 ■ 1999 Celly, R. K. and T. N. Gupta. Dimensions of natural disaster management in India. *Shelter,* special issue, October: 1–3.

3830 ▲ 1999 Gadhok, T. K. Risk assessment: A key to prevention.*Shelter,* special issue, October: 18–20.

3831 ▲ 1999 Gupta, A. and A. Sharma. Landuse planning and urban risk reduction. *Shelter,* special issue, October: 32–34.

3832 ■ 1999 Gupta, R. Geographic information systems and remote sensing for natural disaster prevention. *Shelter,* special issue, October: 21–23.

3833 • 1999 Jain, N. K. Role of non-government organisations in community based disaster preparedness. *Shelter,* special issue, October: 30–31.

3834 • 1999 Krishnan, M. Disaster Management Action Plan for the State of Maharashtra: A review of its unique components. *Shelter,* special issue, October: 27–29.

3835 ▲ 1999 Kumar, J. Disaster management: A case study. *Yojana*, 43 (2): 36–38.

3836 • 1999 Misra, V. K. *Role of Armed Forces in Natural Disaster Management.* New Delhi: Advanced Professional Programme in Public Administration, Indian Institute of Public Administration (Unpublished Master of Philosophy Dissertation).

3837 ▲ 1999 Mukhopadhaya, A. D. (eds). *Disaster Management.* Kolkata: Jadavpur University.

3838 • 1999 Sharma, V. K. Status of preparedness planning in India for disaster mitigation. *Shelter,* special issue, October: 13–15.

3839 • 1999 Singh, K. and T. K. Gadhok. Capacity building for national disaster mitigation. *Shelter,* special issue, October: 81–82.

3840 ▲ 1999 Sinha, A. and V. K. Sharma. *Culture of Prevention: Natural Disaster Management, India.* New Delhi: National Centre for Disaster Management, Indian Institute of Public Administration.

3841 • 1999 Suresh, V. Housing and Urban Development Corporation Limited's role and contribution in disaster preparedness and mitigation. *Shelter,* special issue, October: 68–76.

3842 ▲ 1999 Suresh, V. and R.K. Safaya. Hill towns: Environmental planning as a pre-requisite for disaster mitigation. *Shelter,* special issue, October: 35–42.

3843 • 1999 Suresh, V. and S. Rinki. Traditional technologies for disaster mitigation. *Shelter,* special issue, October: 100–104.

3844 • 1999 Suresh, V. and T.K. Gadhok. Disasters to development. *Shelter,* special issue, October: 105–110.

3845 • 1999 Swarup, V. *et al.* Strategies to mitigate impacts of disasters: Capacity building of communities. *Shelter,* special issue, October: 83–85.

3846 • 2000 Ayyanag, R.S. *et al.* Disaster management planning and experience in Maharashtra. In P.S. Roy, *et al* (eds). *Natural Disasters and their Mitigation: A Remote Sensing and Geographic Information Systems Perspective.* Dehradun: Indian Institute of Remote Sensing, 175–180.

3847 • 2000 Das, P.B. How to avert disasters? *Down to Earth*, 9 (24): 54.

3848 ■ 2000 Disaster Mitigation and Management Centre. *Work Plan of the Disaster Mitigation and Management Centre for Uttaranchal.* Nainital: Disaster Management Cell.

3849 ■ 2000 Disaster Mitigation and Management Centre. *Draft of District Disaster Management Action Plan.* Nainital: Disaster Management Cell.

3850 ▲ 2000 Gaur, V.K. Disaster mitigation. In P.S. Roy, *et al* (eds). *Natural Disasters and their Mitigation: A Remote Sensing and Geographic Information Systems Perspective.* Dehradun: Indian Institute of Remote Sensing. 7–16.

3851 ■ 2000 Gupta, A. *Information Technology and Natural Disaster Management in India.* New Delhi: National Centre for Disaster Management, Indian Institute of Public Administration.

3852 ■ 2000 Indian Institute of Remote Sensing. *Proceedings of Indian Institute of Remote Sensing National Symposium on Spatial Technologies for Natural Hazards and Management.* Dehradun: Indian Institute of Remote Sensing.

3853 • 2000 Jain, N.K. Joint assistance center-An all India agency for assistance in disaster situations. In P.S. Roy, *et al* (eds). *Natural Disasters and their Mitigation: A Remote Sensing and Geographic Information Systems Perspective.* Dehradun: Indian Institute of Remote Sensing. 181–185.

3854 • 2000 Nair, C.N. The role of the internet. In S. Parasuraman and P.V. Unnikrishnan (eds). *India Disasters Report: Towards a Policy Initiative.* New Delhi: Oxford University Press. 93–95.

3855 • 2000 Narayan, S. (eds). *Anthropology of Disaster Management.* New Delhi: Gyan Publishing House.

3856 • 2000 Narayanan, A. Media can play preventive role in all disasters. *Mainstream*, 38 (November 25): 7–8.

3857 ■ 2000 Padiyath, B. Maharashtra state government disaster management plan. In S. Parasuraman and P.V. Unnikrishnan (eds). *India Disasters Report: Towards a Policy Initiative.* New Delhi: Oxford University Press. 219–221.

3858 • 2000 Parasuraman, S. and P.V. Unnikrishnan (eds). *India Disasters Report: Towards a Policy Initiative.* New Delhi: Oxford University Press.

3859 • 2000 Periera, M.S. Teaching disaster management: Addressing complexities of social work practice. *Indian Journal of Social Work*, 61(4): 693–701.

3860 ▲ 2000 Pilgrim, M.K. Disaster reduction and the limits of expectation: A case study from the Indian Himalayas. In R.B. Singh (ed). *Disaster Management.* Jaipur: Rawat Publications. 32–58.

3861 • 2000 Planning Commission. Disaster management: The development perspective. In *Tenth Five Year Plan.* New Delhi Government of India.

3862 • 2000 Rajarathnam, S. Role of universities in disaster reductions. In P.S. Roy, *et al* (eds). *Natural Disasters and their Mitigation: A Remote Sensing and Geographic Information Systems Perspective.* Dehradun: Indian Institute of Remote Sensing. 186–190.

3863 • 2000 Ramanathan, U. Compensation and insurance. In S. Parasuraman and P.V. Unnikrishnan (eds). *India Disasters Report: Towards a Policy Initiative.* New Delhi: Oxford University Press. 107–114.

3864 ■ 2000 Roy, P.S *et al* (eds). *Natural Disasters and their Mitigation: A Remote Sensing and Geographic Information Systems Perspective.* Dehradun: Indian Institute of Remote Sensing.

3865 ▲ 2000 Sahoo, M. Andhra Pradesh Government policy on disaster management. In S. Parasuraman and P.V. Unnikrishnan (eds). *India Disasters Report: Towards a Policy Initiative.* New Delhi: Oxford University Press. 104–106.

3866 • 2000 Sharma, V.K. Nongovernment-Government collaboration. In S. Parasuraman and P.V. Unnikrishnan (eds). *India Disasters Report: Towards a Policy Initiative.* New Delhi: Oxford University Press. 124–125.

3867 • 2000 Singh, B. Role of the government: Responsibilities and performance. In S. Parasuraman and P.V. Unnikrishnan (eds). *India Disasters Report: Towards a Policy Initiative.* New Delhi:Oxford University Press. 96–101.

3868 ■ 2000 Singh, R. B. Remote sensing and geographic information systems for disaster management. In R. B. Singh (ed). *Disaster Management.* Jaipur: Rawat Publications. 11–24.

3869 ▲ 2000 Singh, R. B (ed). *Disaster Management.* Jaipur: Rawat Publication.

3870 ■ 2000 Tandon, O. P. Disaster management: An integrated approach for disaster preparedness, mitigation and awareness. In R. B. Singh (ed) *Disaster Management.* Jaipur: Rawat Publications. 25–31.

3871 ■ 2000 Vankatachary, K. V. Space technology for disaster management. In P. S. Roy, *et al* (eds). *Natural Disasters and their Mitigation: A Remote Sensing and Geographic Information Systems Perspective.* Dehradun: Indian Institute of Remote Sensing. 17–30.

3872 ■ 2000 Westen, C. J. V. and R. Soeters. Remote sensing and geographic information systems for natural disaster management. In P. S. Roy, *et al.* (eds). *Natural Disasters and their Mitigation: A Remote Sensing and Geographic Information Systems Perspective.* Dehradun: Indian Institute of Remote Sensing. 31–76.

3873 ● 2001 Agarwal, A. *et al. Making Water Everybody's Business: Practice and Policy of Water Harvesting.* New Delhi: Centre for Science and Environment.

3874 ■ 2001 Gardner, J. S. and R. B. Singh. Management of water related disasters in the context of the Himalayan mountain region of India. *Annals of National Association of Geographers, India,* 21 (2): 41–46.

3875 ▲ 2001 Goel, S. L. and R. Kumar. Issues in disaster management. In S. L. Goel and R. Kumar (eds). *Disaster Management.* Delhi: Deep and Deep Publications Pvt. Ltd. 3–31.

3876 ● 2001 Goel, S. L. Organisation structure for disaster management. In S. L. Goel and R. Kumar (eds). *Disaster Management.* Delhi: Deep and Deep Publications Pvt. Ltd. 71–94.

3877 ▲ 2001 Goel, S. L. and R. Kumar (eds). *Disaster Management.* New Delhi: Deep and Deep Publications Pvt. Ltd.

3878 ▲ 2001 Government of Gujarat. *Disaster Management Action Plan for Kachachh District.* Gandhinagar, Gujarat.

3879 ▲ 2001 Grover, A. R. Planning for disaster mitigation. In S. L. Goel and R. Kumar (eds). *Disaster Management.* Delhi: Deep and Deep publications Pvt. Ltd. 95–105.

3880 ■ 2001 Gupta, M. C. *et al* (eds). *Manual on Natural Disaster Management in India.* New Delhi: National Centre for Disaster Management, Government of India.

3881 ● 2001 Kumar, S. Disaster management: Analyzing vulnerability. In P. Sahni, *et al* (eds). *Disaster Mitigation: Experiences and Reflections.* New-Delhi: Prentice-Hall of India Private Limited.

3882 ▲ 2001 Ministry of Agriculture. *High Powered Committee on Disaster Management Report 2001*. New Delhi: Government of India.

3883 ▲ 2001 Ministry of Agriculture. *High Powered Committee for Preparation of Disaster Management Plans, Report II*. New Delhi: Government of India.

3884 • 2001 Ministry of Agriculture. *Natural Disaster Mitigation: Do's and Don'ts*. New Delhi: Government of India.

3885 • 2001 Ministry of Agriculture. *National Disaster Response Plan: Building a Culture of Prevention*. New Delhi: Government of India.

3886 ▲ 2001 Ministry of Agriculture and Ministry of Urban Development and Poverty Alleviation. *Natural Disaster Management in India*. New Delhi: Government of India.

3887 ▲ 2001 Narayan, B. *Disaster Management*. New Delhi: Ashish Publishing House.

3888 • 2001 National Disaster Management Division. *Disaster Risk Management Programme: Community Based Disaster Reduction and Recovery Through Participation of Communities and Local Self Governments*. New Delhi.

3889 ■ 2001 Pant, J.C. *Disaster Management: Towards a National Response System*. Lucknow: Indian Literacy Board.

3890 • 2001 Rajan, K. Natural disaster management in national development: An Indian perspective. In V.K. Sharma (ed). *Disaster Management*. New Delhi: Indian Institute of Public Administration. 26–37.

3891 ▲ 2001 Sahni, P. and A. Dhameja. Managing disasters in the new millenium. *Man and Development*, 23 (4): 127–137.

3892 • 2001 Sahni, P. *et al.* (eds). *Disaster Mitigation: Experiences and Reflections*. New Delhi: Prentice-Hall of India Private Limited.

3893 • 2001 Sarkar, A.K. *Disaster Mitigation Plan for Rural Settlements of Coastal Orissa*. New Delhi: Deparment of Housing, School of Planning and Architecture (Unpublished Postgraduate Dissertation).

3894 ▲ 2001 Sharma, V.K. Managing disasters in the new millennium. In P. Sahni, *et al* (eds). *Disaster Mitigation: Experiences and Reflections*. New Delhi: Prentice-Hall of India Private Limited. 15–21.

3895 • 2001 Sharma, V.K. Functioning of disaster management organization and methodology to cope with disaster in India. In V.K. Sharma (ed). *Disaster Management*. New Delhi: Indian Institute of Public Administration. 79–88.

3896 ▲ 2001 Sharma, V.K (ed). *Disaster Management*. New Delhi: National Centre for Disaster Management, Indian Institute of Public Administration.

3897 • 2001 Sikka, D.R. Role of media in disaster preparedness. In P. Sahni, *et al* (eds). *Disaster Mitigation: Experiences and Reflections*. New Delhi: Prentice-Hall of India Private Limited. 124–131.

3898 ▲ 2001 Singh, R. P. *et al.* Management of water related disasters in the context of the Himalayan mountain region of India. *Annals of the National Association of Geographers, India,* 21 (2): 41–46.

3899 ▲ 2001 Sinha, A (eds). *Disaster Management: Lessons Drawn and Strategies for Future.* New Delhi: National Centre for Disaster Management, New.

3900 ▲ 2001 Valdiya, K. S. Public policy for natural hazard management. *Current Science,* 80 (4): 486–487.

3901 • 2001 Verma, B. K. and B. Bhushan. Disaster management in India: A community perspective. In V. K. Sharma (ed). *Disaster Management.* New Delhi: Indian Institute of Public Administration. 46–59.

3902 ▲ 2001 Virk, H. S. Natural hazards mitigation and management. *Current Science,* 80 (10): 1257–1258.

3903 ▲ 2002 Agarwal, P. *Disaster Preparedness and Mitigation Plan for Hill Regions of Uttaranchal, Uttarkashi District.* New Delhi: School of Planning and Architecture (Unpublished Postgraduate Dissertation).

3904 • 2002 Andharia, J. Institutionalising community participation: The challenges in disaster management practice. *Indian Journal of Social Work,* 63 (2): 236–242.

3905 • 2002 Asian Disaster Preparedness Centre. *Technical Assistance for Strengthening Disaster Mitigation and Management at State Level in India.* Bangkok.

3906 • 2002 Brahma, S. *Planning and Disaster Management in Kamrup District (Assam).* New Delhi: School of Planning and Architecture (Unpublished Postgraduate Dissertation).

3907 • 2002 Kropec, M. *Urban Development and Disaster Mitigation.* Ahmedabad: Disaster Mitigation Institute.

3908 • 2002 Mishra, P. K. Maps and disaster management. *Economic and Political Weekly,* 37 (49): 4676–4677.

3909 • 2002 Prasad, J. K. *Role of Non-Government Organisations in Natural Disaster Management.* New Delhi: Advanced Professional Programme in Public Administration, Indian Institute of Public Administration (Unpublished Master of Philosophy Dissertation).

3910 • 2002 Sharma, A. K. *Disaster Mitigation by Regulating Engineering Profession: A Legislative Framework.* New Delhi: Advanced Professional Programme in Public Administration, Indian Institute of Public Administration (Unpublished Master of Philosophy Dissertation).

3911 • 2002 Thapliyal, B. K. and A. C. Jena. *Community Based Disaster Management (Mimeo).* New Delhi: National Institute of Rural Dvelopment, Hyderabad and A P H Publishing Corporation.

3912 ■ 2003 Ahmed, A. *Disaster Management: Through the New Millennium*. New Delhi: Anmol Publications.

3913 ▲ 2003 Ahmed, N. *Managing Disasters*. New Delhi: Kilaso Books.

3914 ● 2003 Andharia, J. Disaster risk reduction through development orientation in disaster management: The significance of peoples participation in India. In P. Sahni, *et al* (eds). *Disaster Risk Reduction in South Asia*. New Delhi: Prentice-Hall of India Private Limited. 327–351.

3915 ● 2003 Ariyabandu, M. M. Bringing together disaster and development concepts and practices: Some experiences from South Asia. In P. Sahni, *et al* (eds). *Disaster Risk Reduction in South Asia*. New Delhi: Prentice-Hall of India Private Limited. 26–36.

3916 ■ 2003 Arora, M. K. Role of remote sensing in disaster risk reduction in South Asia. In P. Sahni, *et al* (eds). *Disaster Risk Reduction in South Asia*. New Delhi: Prentice-Hall of India Private Limited. 84–100.

3917 ● 2003 Atmanand. Insurance and disaster management: The Indian context. *Disaster Prevention and Management*, 12 (4): 286–304.

3918 ● 2003 Dhameja, A. Disaster risk reduction through disaster-resistant construction techniques. In P. Sahni, *et al* (eds). *Disaster Risk Reduction in South Asia*. New Delhi: Prentice-Hall of India Private Limited. 197–209.

3919 ▲ 2003 Gardner, J. S and R. B Singh. Management of water related disasters in the context of the Himalayan region of India. In K. Prasad (ed). *Water Resources and Sustainable Development*. New Delhi: Shipra Publication. 223–228.

3920 ▲ 2003 Gupta, H. K. (eds). *Disaster Management*. Hyderabad: Universities Press.

3921 ■ 2003 Indian National Academy of Engineering. *Proceedings of the National Seminar on Disaster Management and Mitigation 2003*. Phoenix Publishers.

3922 ● 2003 Kohli, R. Armed forces in disaster risk reduction in South Asia. In P. Sahni, *et al* (eds). *Disaster Risk Reduction in South Asia*. New Delhi: Prentice-Hall of India Private Limited. 217–235.

3923 ■ 2003 Manikiam, B. Remote sensing application in disaster management. *Mausam*, 54 (1): 173–182.

3924 ● 2003 Medury, U. Disaster risk reduction: A preparedness approach in South Asia. In P. Sahni, *et al* (eds). *Disaster Risk Reduction in South Asia*. New Delhi: Prentice-Hall of India Private Limited. 132–144.

3925 ▲ 2003 Menon, K. *et al*. *Together, Towards a Safer India: An Introduction to Disaster Management for Class VIII*. New Delhi: Central Board for Secondary Education.

3926 ■ 2003 Misra, J. P. *Forestry in Disaster Management: A Case Study of Himachal Pradesh*. New Delhi: Advanced Professional Programme in Public

Administration, Indian Institute of Public Administration (Unpublished Master of Philosophy Dissertation).

3927 ● 2003 Palakudiyil, T. and Todd, M. *Facing Upto the Storm, How Local Communities Can Cope with Disasters (Lessons from Orissa and Gujarat).* London: Christian Aid.

3928 ■ 2003 Ramakrishnan, P. S. Vegetal cover and natural disaster management in Himalaya. In H. K. Gupta (ed). *Disaster Management.* Hyderabad: University Press (India) Private Limited. 1–21.

3929 ● 2003 Sahani, P. and A. Dhameja. Disaster risk reduction through capacity building of the community and *panchayati raj* institutions. In P. Sahni, *et al* (eds). *Disaster Risk Reduction in South Asia.* New Delhi: Prentice-Hall of India Private Limited. 162–184.

3930 ▲ 2003 Sahni, P. and M. M. Ariyabandu (eds). *Disaster Risk Reduction in South Asia.* New Delhi: Prentice-Hall of India Private Limited.

3931 ● 2003 Shah, H. Assessing abilities for disaster risk reduction: Discovering possibilities and exploring options. In P. Sahni, *et al* (eds). *Disaster Risk Reduction in South Asia.* New Delhi: Prentice-Hall of India Private Limited. 185–196.

3932 ● 2003 Sharma, V. K. *Sustainable Development and Disaster Management.* Allahabad: R. N. Dubey Foundation.

3933 ● 2003 Swami, S. K. Trigger mechanism: The emergency response plan for disaster risk reduction. In P. Sahni, *et al* (eds). *Disaster Risk Reduction in South Asia.* New Delhi: Prentice-Hall of India Private Limited. 77–83.

3934 ● 2003 Vatsa, K. S. and J. Joseph. Disaster management plan for the state of Maharashtra: Evolutionary process. *Natural Hazard Review,* 4 (4): 206–212.

3935 ● 2003 Vayunandan, E. Stakeholder's response in disaster risk reduction. In P. Sahni, *et al* (eds). *Disaster Risk Reduction in South Asia.* New Delhi: Prentice-Hall of India Private Limited. 352–368.

3936 ▲ 2004 Disaster Management Unit. *Disaster Management: A Regional Perspective for Development.* Shillong: North-Eastern Council.

3937 ▲ 2004 Kumar, P. *Disaster Management in India: Need for an Integrated Approach.* New Delhi: Advanced Professional Programme in Public Administration, Indian Institute of Public Administration (Unpublished Master of Philosophy Dissertation).

3938 ▲ 2004 Lalmuanawma, C. *Disaster Mitigation and Preparedness Plan for Aizwal City (Mizoram).* New Delhi: School of Planning and Architecture (Unpublished Postgraduate Dissertation).

3939 ● 2004 Menon, M. G. K. Preparing and planning: The key to hazard management. In K. S. Valdiya (ed). *Coping with Natural Hazards: Indian Context.* Hyderabad: Orient Longman. 1–6.

3940 • 2004 Ministry of Home Affairs. *Disaster Management in India: A Status Report.* New Delhi: Government of India.

3941 ▲ 2004 Murthy, K.R. *Disaster Management.* New Delhi: Dominant Publishers.

3942 ▲ 2004 Paul, D.K. Public policies for hazard management: Appropriate structural designs and building codes. In K.S. Valdiya (ed). *Coping with Natural Hazards: Indian Context.* The National Academy of Sciences and Orient Longman. 71–90.

3943 ▲ 2004 Satendra, and V.K. Sharma. *Sustainable Rural Development for Disaster Mitigation.* New Delhi: Concept Publishing Company.

3944 • 2004 Sharma, V.K. Disaster management strategies in India. In K.S. Valdiya (ed). *Coping with Natural Hazards: Indian Context.* Hyderabad: Orient Longman. 250–256.

3945 • 2004 Singh, R.B. Current curriculum initiatives and perspectives in education for natural disaster reduction in India. In J.P. Stoltman, *et al* (eds). *International Perspectives on Natural Disasters: Occurrence, Mitigation and Consequence.* Dordrecht: Kluwer. 409–416.

3946 ▲ 2004 Subbiah, A.R. *State of the Indian Farmer: A Millennium Study, Volume 21*: Natural Disaster Management. New Delhi: Ministry of Agriculture, Government of India.

3947 • 2004 Valdiya, K.S. and V.S. Kalra. Recommendations on the strategy for coping with natural hazards in India. In K.S. Valdiya (ed). *Coping with Natural Hazards: Indian Context.* Hyderabad: Orient Longman. 267.

3948 • 2004 Venton, C.C. and P. Venton. *Disaster Preparedness Programmes in India: A Cost-Benefit Analysis.* London: Overseas Development Institute, Commissioned and Published by the Humanitarian Practice Network.

3949 • 2005 Awasthi, K. Disaster Management Bill 2005 ignores local self-government. *Down to Earth*, 14 (14): 15.

3950 ■ 2005 Bandyopadhyay, C. and A.D. Kaushik. Strengthening the existing capacity of institutions of higher learning in India for disaster education and public awareness. In R.K. Sharma and G. Sharma (eds). *Natural Disaster: Institutional Building for Mitigating the Impact on Vulnerable Sections.* New Delhi: A P H Publishing Corporation. 107–118.

3951 • 2005 Bhattacharjee, Y. In wake of disaster, scientists seek out clues to prevention. *Science*, 307 (22): 22–23.

3952 ■ 2005 Champati Ray, P.K. Capacity building in remote sensing and geographic information systems for disaster management. In R.K. Sharma and G. Sharma (eds). *Natural Disaster: Institutional*

Building for Mitigating the Impact on Vulnerable Sections. New Delhi: APH Publishing Corporation. 143–156.

3953 • 2005 D' Souza, J.B. Disastrous management. *Economic and Political Weekly*, 40 (36): 3900–3901.

3954 • 2005 Dhaliwal, S.S. Disaster management: Role of urban local bodies. In R.K. Sharma and G. Sharma (eds). *Natural Disaster: Institutional Building for Mitigating the Impact on Vulnerable Sections.* New Delhi: APH Publishing Corporation. 87–98.

3955 • 2005 Editorial. Preparing for disaster. *Nature*, 438 (December 15): 889.

3956 ▲ 2005 Gawde, S. Preparedness, mitigation and management of natural disasters. In R.K. Sharma and G. Sharma (eds). *Natural Disaster: Institutional Building for Mitigating the Impact on Vulnerable Sections.* New Delhi: APH Publishing Corporation. 73–86.

3957 ▲ 2005 Gupta, J.K. Preparedness, mitigation and management of natural disasters. In R.K. Sharma and G. Sharma (eds). *Natural Disaster: Institutional Building for Mitigating the Impact on Vulnerable Sections.* New Delhi: APH Publishing Corporation. 45–58.

3958 • 2005 Jena, A.C. Empowering *panchayats* and community for disaster preparedness and mitigation: An institution building perspective. In R.K. Sharma and G. Sharma (eds). *Natural Disaster: Institutional Building for Mitigating the Impact on Vulnerable Sections.* New Delhi: APH Publishing Corporation. 189–216.

3959 ▲ 2005 Khanna, B.K. *All You Wanted to Know About Disasters.* New Delhi: New India Publishers.

3960 • 2005 Khanna, B.K. Natural disasters: Exposure and vulnerability. In R.K. Sharma and G. Sharma (eds). *Natural Disaster: Institutional Building for Mitigating the Impact on Vulnerable Sections.* Delhi: APH Publishing Corporation. 1–16.

3961 • 2005 Kulkarni, G.K. Disaster management: Issues for action. *Indian Journal of Occupational and Environmental Medicine*, 9.

3962 ▲ 2005 Malhotra, S. *Natural Disaster Management.* Jaipur: Aavishkar Publication.

3963 • 2005 Mathur, M. Government of India and United Nations Development Programme's Mandate on disaster management with a focus on national and Punjab state initiatives. In R.K. Sharma and G. Sharma (eds). *Natural Disaster: Institutional Building for Mitigating the Impact on Vulnerable Sections.* New Delhi: APH Publishing Corporation. 231–254.

3964 • 2005 Mittal, P. Natural disaster: Health sector planning. In R.K. Sharma and G. Sharma (eds). *Natural Disaster: Institutional Building for Mitigating the Impact on Vulnerable Sections.* New Delhi: APH Publishing Corporation. 177–188.

3965 • 2005 Palli, M. Natural disaster: Institutional building for mitigating the impact on vulnerable section. In R.K. Sharma and G. Sharma (eds). *Natural Disaster: Institutional Building for Mitigating the Impact on Vulnerable Sections.* New Delhi: A P H Publishing Corporation. 125–142.

3966 • 2005 Puri, I. Self help: A step towards disaster preparedness. *Indian Journal of Adult Education*, 18 (12): 9–15.

3967 • 2005 Rautela, P. Indigenous technical knowledge inputs for effective disaster management in the fragile Himalayan ecosystem. *Disaster Prevention and Management*, 14 (2): 233–241.

3968 • 2005 Samal, K.C. *et al. State, Non Government Organisations and Disaster Management.* Jaipur: Rawat Publication

3969 ▲ 2005 Sen, U.K. Disaster management. *Economic and Political Weekly*, 40(8): 694.

3970 • 2005 Sharma, R.K. and G. Sharma (eds). *Natural Disaster: Institutional Building for Mitigating the Impact on Vulnerable Sections.* New Delhi: A P H Publishing Corporation.

3971 • 2005 Soman, K. Disaster management efforts: Some thoughts on where to begin. *Journal of Geological Society of India*, 65 (6): 770–771.

3972 • 2005 Taori, K. *Disaster Management through Panchayati Raj Institutions.* New Delhi: Concept Publishing Company.

3973 ▲ 2005 Unnithan, S. Disaster management. *India Today*, (November 7): 19–21.

3974 ■ 2005 Vasudeva, P.K. Post disaster management. In R.K. Sharma and G. Sharma (eds). *Natural Disaster: Institutional Building for Mitigating the Impact on Vulnerable Sections.* New Delhi: A P H Publishing Corporation. 59–72.

3975 ▲ 2006 Aggarwal, G. Government initiatives towards disaster management. In K.K. Sharma, *et al* (eds). *Environmental Geohazards: Science and Society.* New Delhi: Research India Press. 392–407.

3976 • 2006 Arnold, M. Disaster reconstruction and risk management for poverty reduction. *Journal of International Affairs*, 59 (2): 269–279.

3977 ■ 2006 Banger, S.K. Remote sensing and geographical information systems for natural disaster management. In P.S. Ranade (ed). *Natural Disasters: Management and Preparedness.* Hyderabad: Institute of Charted Finanacial Analysts of India University Press. 220–226.

3978 • 2006 Basu, A. Human scarcity and disaster management. In P.S. Ranade (ed). *Natural Disasters: Management and Preparedness.* Hyderabad: Institute of Charted Finanacial Analysts of India University Press. 123–130.

3979 ▲ 2006 Ghosh, G. K. *Disaster Management.* New Delhi: A P H Publishing Corporation.

3980 ▲ 2006 Gill, H. S. Development strategy for multi-disaster prone state of Himachal Pradesh. In K. K. Sharma, *et al* (eds). *Environmental Geohazards: Science and Society.* New Delhi: Research India Press. 408–419.

3981 ■ 2006 Gupta, K. R. and R. K. Midha. Science and technology initiatives for natural hazard mitigation. In H. N. Srivastava and G. D. Gupta (eds). *Management of Natural Disasters in Developing Countries.* New Delhi: Daya Publishing House. 20–40.

3982 • 2006 Joshi, M. and C. Bhakuni. Community based learnings in disaster resistant construction in Himachal Pradesh. In K. K. Sharma, *et al* (eds). *Environmental Geohazards: Science and Society.* New Delhi: Research India Press. 367–379.

3983 ■ 2006 Kessler, E. Mitigation of geologic hazards. In K. K. Sharma, *et al* (eds). *Environmental Geohazards: Science and Society.* New Delhi: Research India Press. 285–303.

3984 ■ 2006 Khan, A. K. Management of natural disasters.In H. N. Srivastava and G. D. Gupta (eds). *Management of Natural Disasters in Developing Countries.* New Delhi: Daya Publishing House. 10–19.

3985 ▲ 2006 Kumar, N. (ed). *Disaster Management.* New Delhi: Alfa Publication.

3986 ■ 2006 Kumar, R. Role of earth system sciences in disaster management. In P. S. Ranade (ed). *Natural Disasters: Management and Preparedness.* Hyderabad: Institute of Charted Finanacial Analysts of India University Press. 210–219.

3987 ▲ 2006 Mahanti, N. C. *et al* (eds). *Disaster Management.* New Delhi: Narosa Publishing House.

3988 • 2006 Metha, S. Disaster and mass casualty management in a hospital: How well are we prepared? *Journal of Postgraduate Medicine*, 52 (2).

3989 ■ 2006 Metri, B. A. Disaster mitigation framework for India using quality circle approach. *Disaster Prevention and Management*, 15 (4): 621–635.

3990 ▲ 2006 Murthy, D. B. N. *Disaster Management: Text and Case Studies.* New Delhi: Deep and Deep Publications Pvt. Ltd.

3991 • 2006 Nanda, N. K. National Insurance Company: Plans and polices for post geohazard management. In K. K. Sharma, *et al* (eds). *Environmental Geohazards: Science and Society.* New Delhi: Research India Press. 380–391.

3992 • 2006 Ranade, P. S. Maps and atlases for management of natural disasters. In P. S. Ranade (ed). *Natural Disasters: Management and Preparedness.* Hyderabad: Institute of Charted Finanacial Analysts of India University Press. 248–265.

3993 ▲ 2006 Ranade, P.S. (ed). *Natural Disasters: Management and Preparedness*. Hyderabad: Institute of Charted Finanacial Analysts of India University Press.

3994 ■ 2006 Rupesh K. G. Space technology and natural hazards: An overview. In K.K. Sharma, *et al* (eds). *Environmental Geohazards: Science and Society*. New Delhi: Research India Press. 353–366.

3995 ▲ 2006 Sethi, V.K. *Disaster Management*. New Delhi: Maxford Books.

3996 ● 2006 Singh, A. and R.B. Singh. Community participation in disaster mitigation in India. In R. B. Singh (ed). *Natural Hazards and Disaster Management Vulnerability and Mitigation*. Jaipur: Rawat Publications. 306–313.

3997 ▲ 2006 Singh, R.B. Introduction to natural hazards and disaster: Conceptual framework, vulnerability, spatial pattern, impact and mitigation strategy: Indian experience. In R. B. Singh (ed). *Natural Hazards and Disaster Management, Vulnerability and Mitigation*. Jaipur: Rawat Publications. 3.

3998 ● 2006 Singh, R.B (ed). *Natural Hazards and Disaster Management, Vulnerability and Mitigation*. Jaipur: Rawat Publication.

3999 ▲ 2006 Singh, T. *et al. Disaster Management: Approaches and Strategies*. New Delhi: Akansha Publications.

4000 ● 2006 Sinha, P.C. *Disaster Management Process: Law Policy and Strategy*. Kolkata: S B S Publishers.

4001 ● 2006 Srivastava, A.K. Community-based disaster management: India experience. In R.B. Singh (ed). *Natural Hazards and Disaster Management Vulnerability and Mitigation*. Jaipur: Rawat Publications. 314–321.

4002 ▲ 2006 Srivastava, H.N. and G.D. Gupta. *Management of Natural Disasters in Developing Countries*. New Delhi: Daya Publishing House.

4003 ▲ 2006 Sur, D. and S.K. Bhattacharya. Natural disasters: Management and preparedness. In P.S. Ranade (ed). *Natural Disasters: Management and Preparedness*. Hyderabad: Institute of Charted Financial Analysts of India University Press. 182–197.

4004 ▲ 2006 Wisyanto. Natural disaster and its mitigation. In H.N. Srivastava and G.D. Gupta (eds). *Management of Natural Disasters in Developing Countries*. New Delhi: Daya Publishing House. 64–77.

Conclusion to the 4004

Therefore the 5017 research units lend themselves to various moulds. In their thematic form they fell into these eight themes Basics, Occurrence, Characteristics, Cause, Impact, Response, Forecasting and Management.

Even if a model design for the dimensions of research on disasters were spelt, it would have covered these eight aspects. Questioning and debating the fundamentals, inquiring into the spatio – temporal occurrence, describing the characteristics, analysing the cause and impact, moving ahead to understand the type of response and then predict the trends of the future. The design of research would lead on to identify the aspects of mitigation and management. This is exactly what the 5017 research units have revealed. Each of these eight has occupied a different share of the attention of research. Whereas works on the Basics formed 3 per cent of the total, those on Management fetched nearly 24 per cent (Table 6.28).

Table 6.28 Themes Researched for Disasters of India

Theme	Research units	Per cent of total
Basics	140	3
Occurrence	835	17
Characteristics	394	8
Cause	760	15
Impact	805	16
Response	579	11
Forecasting	280	6
Management	1224	24
Total	**5017**	**100**

Based on *4004*.

The works on Occurrences, Cause and Impact shared a near equal 15 to 17 per cent. On a reflective note even the proportions of research in different themes seem to be in order. Research on 'basics' cannot outnumber that on 'occurrences' while the inquiry of management has to be of greatest concern to the subject of disasters. Even when works on cause and effect are compared, in the case of works on disasters, it is to be expected that a research on 'impact' would outnumber that of 'cause'. Yet in spite of the strong upward surge in the number of research units in the last decade, recognition attracted by disasters, near equitable distribution in terms of the themes that were covered, it needs to be said that while recognition was accorded, realisation was missing. It can be argued that recognition itself is an outcome of realisation but if the premise of realisation falters, the outcome of recognition cannot be correct. When climbing a wall, the least one should know is that it is the right wall. What has been the overarching component that has been amiss in most of the 4004 research references? Where has realisation tripped? What is the future of the 4004?

VII

Back to the Future of 4004

Can the 4004 light the way towards building a disaster free India? Have they pursued a paradigm which is in sink with the reality of disasters in India? Does the spirit of the 4004 have strength to lever change in the attitude towards disasters? The need is to focus on the role of the 4004. This involves going beyond the thematic classification or that based on type and outlet of research. To move back to the future of 4004 one needs to capture the soul of each reference.

When the 5017 research units were read again with vigilance, this time not for time, type or theme, not even for Indian or foreign, single or multiple author, event or place but for the spirit that guided the work, it indeed was revealing that the 'natural' and not the disaster had captured almost the entire attention. A rejoinder to this could be that this research focused on natural disasters and therefore natural would not only belong in the scheme of adjectives but also be the objective. But this is the crux of the discourse.

It is this automatic connection forged between natural and disaster, which nourishes the problem. After all studying a cyclone, earthquake, landslide or storm is not studying a disaster. These are simply natural phenomena, and physical processes that operate on the surface of the Earth. When the world was uninhabited, even then storms, volcanoes and earthquakes occurred freely. Researching on the occurrence, cause, and characteristics of these naturals is interesting, important and contributes most significantly to the advancement of the knowledge of the physical sciences. But centring on them to understand and mitigate disasters would be simply looking at the wrong end.

Natural is Not Disaster

A disaster is only in the realm of humans. In fact a disaster is what happens to the people; all impact of disasters is on the people and all disaster management is for the people. Even the cause for the disaster are the people. Which are the people who cause the disasters is of course a different matter. Otherwise what can explain the reason why the 6.6 Richter scale earthquake in Los Angeles in 1994 led to the death of 20 people while the Bhuj earthquake of 7.7 Richter scale led to the death of 20,000? This also applies to cyclones. In the Hurricane Katrina, 2005, that had a velocity of 280 kilometres per hour, nearly 1836 people died. In the Orissa cyclone of 1999, where the winds had a speed of 260 kilometres per hour, the death toll was 9504. A near similar intensity of the naturals but a colossal difference in the number of deaths. Even within India it was the poor who constituted the profile of the victims.

In the Latur earthquake, among the total 7797 dead, 4307 were women compared to 3490 men. In the cyclone that hit Gujarat, it was the thousands of saltpan workers living in slums close to their place of work who perished. In Jamnagar as well, it was the labour colonies of Essar and Reliance that were heavily damaged and destroyed. 'The number of women killed by tsunami was 1.42 times the number of men, while the number of girls killed was 1.65 times the number of boys among children' (Krishnamoorthy, 2005). Women apparently were the more serious victims even in a post-disaster situation. 'Psychological morbidity was particularly high among women in the super-cyclone of Orissa 1999'. This was the conclusion drawn by a sample study conducted by the prestigious Bangalore-based National Institute of Health and Neuro Sciences. Many a report on drought relief programmes in Gujarat and Rajasthan confirms that women outnumbered men by a ratio of 3:1 of the total workforce at the camps.

In the Latur earthquake, over half the fatalities were children below the age of fifteen years. More than 3.3 million children were estimated to have been either orphaned or seriously injured or separated from their families in the wake of the 1999 Orissa cyclone (*The Times of India*). Even in the tsunami, children were the most affected. 'They constituted about 39 per cent of the total male victims, and 46 per cent of the females.' (Krishnamoorthy *et al*, 2005).

It would be worth an inquiry to also investigate the socio-economic profile of the people who die in a cold or heat wave. To search for the cause within the velocity of wind and magnitude of earthquake and not within the type of houses and kind of livelihoods that people pursue, would simply be a choosing to miss the point. The people at the sharp edge of disasters are not victims of chance or targets of an angry and violent nature. Nature is neutral. Why would nature choose to select women, or labour or salt workers or migrants or poor?

When the scientific understanding of the earth's physical system expanded, disasters became synonymous with the agents. In other words, disasters were equated with earthquake, flood, cyclone or landslide. This perspective is not immune to drawbacks. It downplayed the role of humans in creating or contributing to all types of catastrophic events. In India, there exists a paradigm of natural disasters where the disasters are suffixed to the preceding events. In reality, however, there 'may be no reason to infer the existence of one from the appearance of the other' (Hume, 1740). The fact that 'disasters are inherently social phenomena and all disasters are rooted in the social structure or social situation' (Quarantelli, 1995) gels the argument that just because an earthquake or cyclone precedes the deaths and damage, it does not automatically follow that it is the cause for the latter. Perpetuating such an outlook would not provide any solution for mitigation or management of disasters.

Therefore, there seems little cause to dispute that a disaster is from the people, about the people, on the people and is responded to by the people. Yet the focus of an over whelming majority of the 5017 research units has been the naturals and not the people. A purview of each and every relevant dimension of the 5017 research units confirmed that it is the naturals that dominate the inquiry of research. Even with a liberal selection of the total research units, 53 per cent were devoted exclusively to the study of the naturals. Be it the characteristic of a cyclone, measuring the aftershocks of an earthquake, gauging the levels of a flood, or classifying types of droughts, the inquiry is centred on the dimension of the natural. Even while forecasting a cyclone, monitoring the river level or studying earthquakes, the content is natural. The obsession with the natural can be seen across the different outlets of research, themes of research and even over the years of research.

Dominance of Natural in Research Outlets

A serious concern with the natural emerges when the outlets of research are checked. 66 per cent of the research in journals, 50 per cent of all reports and 59 per cent of all doctorates are anchored to the natural. It is mainly in features in magazines or books that the work veers on to the people. The more rigorous the outlet, the more serious it is about the natural. In a bid to sharpen the analysis, the inquiry was extended into another facet of research units. The conclusion did not change. Be it the atlases, data lists or the catalogues, all are exclusively devoted to the naturals. Take the group of atlases. There is one each on flood, earthquake, landslide and even one that plots the tracks of cyclone and storm. All the ninty-five plates in the *Vulnerability Atlas of India* are on the naturals. When tabulating the vulnerability of the type of house, the latter was weighed against a cyclone or flood and not in relation to programmes of human development. The focus is houses not homes. Till date, there is no atlas in India that maps deaths in disasters, range of people and area affected, the safety shelters and venues in the eventuality of a disaster or the response that disasters draw (Table 7.1).

All these and many other features are 'mapable' spatial entities. Though modern information technology is used, it still perpetuates age-old thinking.

Dominance of Natural in Technology and Models

Space technology has opened a vast vista of scientific enquiry. Of the total 157 research units that have used remote sensing imageries, radar technology, or geographic information systems for analytical purposes, 81 per cent arrived on the horizon of research during 1990–2006. Professionals with a background in physical sciences like physics, geology, engineering or hydrology, are adept at these tools. Using modern information technology, they join their work with study of the physical attributes. Be it measuring the scale and size of a landslide, the movement of a fault, the extent of a flood, the track of a cyclone or the changing coastline in a

Table 7.1 Dominance of Natural in Research Outlets

Outlet	**Per cent of total**			**Total**
	Natural	**Human**	**Ecological**	
Articles in journals	66	31	3	100
Ph.D. theses	59	35	6	100
Chapters in books and reports	56	37	7	100
Reports	50	36	14	100
M. Phil. dissertations	45	47	8	100
Books	22	52	26	100
Feature in magazines	17	82	1	100
Postgraduate dissertations	14	62	24	100

Based on *4004*.

Table 7.2 Use of Modern Information Technology in Research on Disasters

Decades	Research units[1]	Per cent of total
1960–69	5	3
1970–79	4	3
1980–89	20	13
1990–99	39	25
2000–06	89	56
Total	**157**	**100**

[1]*It includes research based on remote sensing, geographical information systems and radar observations.*

tsunami, these are attributes of the natural and not the disaster (Table 7.2). It indeed would be long before a remotely sensed image would be able to detect the emotional and mental agony of people in a disaster or capture the cunning and corrupt who have siphoned funds meant for development and thus exposed people to disasters.

Modelling is another favourite method among scientists who are keen to simulate reality. There are 67 research units that have tried to construct a model for the purpose of prediction. No less than 95 per cent of the works on models target the natural. Although human reality is more complex and difficult to model, the virtual absence of any effort in this direction is also a reflection of the dominating paradigm.

Dominance of Natural in Themes

In five out of the eight themes of classification of the research units, it again is the inquiry of the naturals that takes precedence. Their supremacy is by a large margin. All of the research on 'occurrences' was focused on the trend or distribution of the naturals. In the sphere of 'forecasting' again, the inquiry of the naturals took 91 per cent share. The 'characteristics' and 'basics' too give an 84 per cent weight to the naturals. The balance begins to tip in favour of the human in respect of the 'impact', but here too 26 per cent of the researchers have been keen to explore the impact of the natural on the natural like cyclone on storm surge or earthquake impact on landslide. Even within the theme of management, the concern of a quarter of the research units has been on how to manage the 'natural', be it in the form of technological solutions like dams and embankments in the case of floods or irrigation in face of drought. Management is based on the research on the durability of a house where its design is seen vis-à-vis the force of the shear wind or the movement of the land and not the fragility of the people and their priorities. It is a paradox indeed that while a large share of the research on cause, occurrence and characteristics floats within the natural, in the case of impact, response and management, the human aspect is explored. This impairment cannot resolve the problem of disasters in India. The existing paradigm and stock of literature refuses to even take a serious cognisance of humans (Table 7.3).

It is indeed shocking that in spite of the large number of disasters, a morsel of 34 research units took the trouble to inquire about the impact of disasters on human health; hardly a few were eager to investigate the impact on children and women. Surely the half a dozen research units like 'Children's reaction to disaster' or 'Impact of disasters on the school going child' cannot capture the pathos inflicted by disasters on children across India. Similarly a lone unit 'Disability issue' can hardly suffice as a research to respond to concerns of the physically challenged. India's demographic composition today has around 80 million old people. Yet impact or response of the elderly in and during a disaster finds no mention in the titles of the research units.

Disaster research in India, as reflected in the record of studies for a century and seven decades, provides evidence of wholesale bias towards the natural. While many new works have been added, they largely continue to portray the conventional, highly compartmentalised view of disasters, focusing on individual natural agents. Even when the content and syllabi

Table 7.3 Misleading Research – Natural in Cause, Human in Impact

Themes	**Per cent of total**			**Total**
	Natural	**Human**	**Ecological**	
Basics	83	17	0	100
Occurrence	100	0	0	100
Characteristics	84	5	11	100
Cause	79	21	0	100
Impact	26	73	1	100
Response	1	97	2	100
Forecasting	91	9	0	100
Management	25	42	33	100
Total	**53**	**38**	**9**	**100**

Based on *4004*.

Table 7.4 Lack of Concern for Humans in Research on Disasters

Concerns	Research units	Per cent of total (5017) research units
Health	34	0.68
Women	29	0.58
Children	10	0.20
Aged	0	0.00
Physically challenged	1	0.02
Vulnerability	44	0.88

Based on *4004*.

of the newly launched courses in the school and universities are carefully combed, they lean heavily towards explaining and holding the 'natural' as the culprit (Table 7.4). If this approach is not radically changed, many among the forthcoming 4004 research units will turn into a 404! The latter represents human beings that carry a confused, blank or uncertain expression on their face. If research on disasters continues to perpetuate the 404 mind set of blaming nature, it would be unable to evolve solutions to the problem of recurrent deaths and loss that India continues to face in disasters. Solutions after all arise only when the perspective is correct and truthful.

There are only two solutions to this vexing problem. Either the natural scientists turn sensitive to the critical role of humans or the social scientists rescue and nurture the research of disasters in their territory. The latter perhaps is a more durable, viable and sustainable option. After all the interest of the natural scientist would be on a particular physical event or class of phenomena, their ultimate professional concern is to understand and predict the natural. Their scientific affiliation and training is committed to the natural with scant reference to human. If disasters are going to spawn as a discipline in their own right in the coming years in India, the realisation of the need to research social dimensions has to stand stark.

It has taken many years, many national and international reminders, the death and loss of many before India could heed the call of disasters. Today, India is beginning to recognise the need to do something about disasters. India is willing to invest in research on disasters. Willingness is a space of limitless potential. It has to be kindled with consciousness that it is the doing and undoing of people which has created the disasters.

Chasing the path of blaming, fixing and researching on the natural is never going to build a disaster-less, leave alone a disaster-free, India. Many battles will be won if it is realised that a disaster is by, of, on and about the people. While it is a fact that India needs many times more 4004 units, it is also true that it urgently needs a different type of 4004.

Annexure 1 Libraries Consulted for Research

Name	**Place**
Agricultural Economics Research Centre	Delhi
All India Institute of Medical Sciences	Delhi
American Centre	Delhi
British Council	Delhi
Building Materials and Technology Promotion Council	Delhi
Central Road Research Institute	Delhi
Central Secretariat	Delhi
Centre for Policy Research	Delhi
Centre for Science and Environment	Delhi
Centre for the Study of Developing Societies	Delhi
Geological Survey of India	Lucknow
India International Centre	Delhi
India Meteorological Department	Delhi
Indian Agricultural Research Institute	Delhi
Indian Council of Agricultural Research	Delhi
Indian Council of Social Science Research	Delhi
Indian Habitat Centre Resource	Delhi
Indian Institute of Advanced Studies	Shimla
Indian Institute of Ecology and Environment	Delhi
Indian Institute of Mass Communication	Delhi
Indian Institute of Public Administration	Delhi
Indian Institute of Technology	Delhi
Indian National Science Academy	Delhi
Indira Gandhi National Open University	Delhi
Institute of Economic Growth	Delhi
Jawaharlal Nehru University	Delhi
Ministry of Environment and Forests	Delhi
Nehru Memorial Library	Delhi
Planning Commission of India	Delhi
Panjab University	Chandigarh
Delhi School of Economics	Delhi

Annexure 2 Bibliographies on Disaster

Ahearn, F.L. and R.E. Cohen. (1983). *Disasters and Mental Health: An Annotated Bibliography*. Rockville: National Institute of Mental Health.

Coppa and Avery Consultants. (1985). *Bibliographical Guide to Disaster Planning, Management, Insurance, the Case of Bhopal*. Vance Bibliographies.

Fitzsimmons, A.R. (1984). *Natural Hazards and Land Use Planning: An Annotated Bibliography*. Boulder: Natural Hazards Research and Applications Information Centre, University of Colorado.

Golant, S. (1969). *Human Behaviour Before the Disaster: A Selected Annotated Bibliography*. Boulder: Natural Hazards Research and Applications Information Centre, University of Colorado.

Mitchell, J.K. (1968). *A Selected Bibliography of Coastal Erosion, Protection and Related Human Activity in North America and the British Isles*. Boulder: Natural Hazards Research and Applications Information Centre, University of Colorado.

Morton, D.R. (1984). *A Selected, Partially Annotated Bibliography of Recent (1982–1983) Natural Hazards Publications*. Boulder: Natural Hazards Research and Applications Information Centre, University of Colorado.

Morton, D.R. (1986). *A Selected, Partially Annotated Bibliography of Recent (1984– 1985) Natural Hazards Publications*. Boulder: Natural Hazards Research and Applications Information Centre, University of Colorado.

Morton, D.R. (1979). *Bibliography on Natural Disaster Recovery and Reconstruction*. Boulder: Natural Hazards Research and Applications Information Centre, University of Colorado.

Morton, D.R. (1981). *A Selected Bibliography on Disaster Planning and Simulation*. Boulder: Natural Hazards Research and Applications Information Centre, University of Colorado.

Pope, T. and D. Wenger. (1984). Three Mile Island in the literature: A partially annotated bibliography. *International Journal of Mass Emergencies and Disasters*. 2(1): 197.

Quarantelli, E.L. (1970). A selected annotated bibliography of social science studies on disasters. *American Behavioural Scientist*. 13 (3): 452–456.

Quarantelli, E.L. (1980). *A 100 Item Annotated Bibliography on Disaster and Disaster Planning*. Newark: Disaster Research Centre, University of Delaware.

Quarantelli, E.L. (1982). *Inventory of Disaster Field Studies in the Social and Behavioral Sciences: 1919–1979*, Miscellaneous Report No. 32. Newark: Disaster Research Centre, University of Delaware.

Rayner, J. (1957). Annotated bibliography on disaster research. *Human Organization*. 16 (2): 30.

Relph, E.C. and S.B. Goodwillie. (1968). *Annotated Bibliography on Snow and Ice Problems*. Boulder: Natural Hazards Research and Applications Information Centre, University of Colorado.

Tierney, K.J. and B.Baisden. (1978). *An Annotated Bibliography on Disaster Mental Health and Crisis Intervention in Smaller Communities.* Miscellaneous Report No. 20. Delaware: Disaster Research Centre, University of Delaware.

Wilson, E. (1981). A selected annotated bibliography and guide to sources of information and planning for and responses to chemical emergencies. *Journal of Hazardous Materials.* 4 (4): 373.

Annexure 3 Doctorates on Disasters in the Universities of India

University/Institution	Number of doctorates	Per cent of total
Roorkee University, Roorkee[1]	16	14
Andhra Univerisity, Waltair, Vishakhapatnam	15	13
Benaras Hindu University, Varanasi	7	6
Indian Institute of Technology, Delhi	5	4
Sri Krishna Devaraya University, Anantpur	5	4
Guhati University, Guwahati	4	4
Indian Institute of Technology, Kharagpur	4	4
Calcutta University	4	4
Indian School of Mines, Dhanbad	4	4
Delhi University	3	3
Osmania University, Hyderabad	3	3
Burdwan University	3	3
Garhwal University, Sri Nagar	3	3
Kumaon University, Nainital	3	3
Other Universities/IITs and Institutions	34	30
Total	**113**	**100**

[1]*Now Indian Institute of Technology, Roorkee.*

Annexure 4 Research on Disasters of India Searched Online

Journal	Inception		Articles in a Journal	
	Place	**Year**[1]	**Total**	**Natural disasters, India**
Natural Hazard and Earth System Sciences	Paris, France	1970	470	2
Journal of Natural Disaster Science	Kyoto, Japan	1979	249	2
Science of Tsunami Hazards	Honolulu, USA	1982	266	9
International Journal of Mass Emergencies and Disasters	Oklahoma, USA	1983	398	2
Natural Hazard Journal	Netherlands	1988	964	22
Disaster Prevention and Management	Bradford, UK	1992	364	17
Indian Journal of Medical Ethics	Mumbai, India	1993	296	1
Environmental Hazards	New Jersey, USA	1999	91	1
Natural Hazard Review	Reston,USA	2000	162	1
Indian Journal for Practicing Doctors	Srinagar, India	2004	225	9
Total			**3485**	**66**

[1]*The journals were archived since their year of inspection and searched till 2006.*

Annexure 5 Journals Consulted Online[1] for References on Disasters in India

Journal	**Place of publication**	**Year of**		**Articles[2]**
		Inception	**Online**	
British Medical Journal	London, UK	1840	1994	4
Indian Concrete Journal	Mumbai, India	1927	1985	54
Journal of Indian Medical Association	Kolkata, India	1930	2005	3
Proceedings of Indian Academy of Sciences	Bangalore, India	1934	1978	29
Bulletin of Indian Society of Earthquake Technology	Roorkee, India	1964	1998	33
Journal of Indian Geophysical Union	Hyderabad, India	1964	2003	14
Disasters: The Journal of Disaster Studies, Policy and Management	Oxford, UK	1977	1997	18
Indian Cartographer	Hyderabad, India	1979	2001	3
International Journal of Remote Sensing	Nottingham, UK	1980	1997	19
Mountain Research and Development	Berne, Switzerland	1981	2000	5
Terra Nova	Oxford, UK	1989	1997	2
Gondwana Research-International Geoscience Journal	Thiruvananthapuram, India	1997	2004	2

[1]*From availability online to 2006.* [2]*On natural disasters in India.*

Annexure 6 Journals Searched Online with Keywords[1] for References on Disasters on India

Name	**Place of publication**	**Year of**		**Articles on natural disasters**
		Inception	**Online placement[2]**	
The Lancet	London, United Kingdom	1823	1823	18
Bulletin of the Seismological Society of America	Alexandria, United States of America	1911	1911	67
Indian Journal of Medical Research	Delhi, India	1913	2003	3
Current Science	Bangalore, India	1932	1932	115
Indian Journal of Medical Sciences	Mumbai, India	1947	1990	4
Tectonophysics	Amsterdam, Netherlands	1964	1964	33
Geomorphology	Amsterdam, Netherlands	1987	1987	7

[1]*Including hazard, disaster, natural disaster, earthquake, cyclone, flood, drought, landslide, storm, tsunami, heat and cold wave* [2]*Checked till 2006.*

Annexure 7 Journals/Magazines Searched in Print for References on Disasters of India

Journal/magazine[1]	Place of publication	Year of inception	Articles on natural disasters
Journal			
Mausam[2]	Delhi	1950	243
Economic and Political Weekly	Mumbai	1966	204
Journal of Geological Society of India	Bangalore	1958	78
Bhagirath	Delhi	1954	59
Indian Journal of Power and River Valley Development	Kolkata	1950	56
Vayumandal	Delhi	1971	19
Annals of Arid Zone	Jodhpur	1962	18
Transactions, Institute of Indian Geographers	Pune	1979	12
Annals of National Association of Geographers, India	Delhi	1981	9
Deccan Geographer	Pune	1962	7
Indian Geographical Journal	Chennai	1926	5
Magazine			
Frontline	Chennai	1984	247
Down to Earth	Delhi	1992	177
India Today	Delhi	1975	106

[1]*Volume of journals searched since their year of inception.* [2]*For change of name refer to Table 3.6.*

Annexure 8 Journals with Minimum of Ten Research Articles on Natural Disasters of India

Name of the Journal	No. of articles	Cumulative	
		Number	Percentage
Indian Journal of Meteorology and Geophysics[1]	243	243	13
Economic and Political Weekly	204	447	24
Current Science	120	567	31
Journal of Geological Society of India	78	645	35
Himalayan Geology[1]	68	713	39
Bulletin of the Seismological Society of America	67	780	42
Bhagirath	59	839	46
Indian Journal of Power and River Valley Development	56	895	49
Indian Concrete Journal	54	949	52
Geographical Review of India	46	995	54
ISET Journal of Earthquake Technology[1]	33	1028	56
Tectonophysics	33	1061	58
Shelter	32	1093	59
Proceedings of Indian Academy of Sciences[1]	30	1123	61
Proceedings of the Asiatic Society of Bengal[1]	27	1150	63
Natural Hazard Journal	22	1172	64
International Journal of Remote Sensing	19	1191	65
Vayumandal	19	1210	66
Annals of Arid Zone	18	1228	67
Disasters: The Journal of Disaster Studies, Policy and Management	18	1246	68
Indian Journal of Social Work	18	1264	69
The Lancet	18	1282	70
Disaster Prevention and Management	17	1299	71
Central Board of Geophysics, India	16	1315	72
Journal of Indian Geophysical Union	14	1329	72
Modern Review	14	1343	73
Physics of the Earth and Planetary Interiors	14	1357	74
Yojana	13	1370	75
Engineering Geology	12	1382	75
Kurukshetra	12	1394	76
Mainstream	12	1406	76
Transactions, Institute of Indian Geographers	12	1418	77
Photonirvachak, Journal of the Indian Society of Remote Sensing	10	1428	78
Others (with less than 10 articles each)	410	**1838**	**100**

[1]*For change in name of journal refer to Table 3.6.*

Annexure 9 Courses on Disaster Management in Institutes of India[1]

Course Title	Name of the Institute	Place	Year of		Course (Duration)
			Estb. of institute	Intro. of course	
PG Diploma in Disaster Management	Indira Gandhi National Open University	Delhi	1985	2006	1 to 4 years (distance learning)
PG Diploma in Disaster Preparedness and Rehabilitation	Disaster Management Center, Indian Red Cross Society	Delhi	1920*	2006	1 year (weekend)
MBA in Disaster Management	Guru Gobind Singh Indraprashtra University	Delhi	1998	2005	2 years (weekend)
MSc in Disaster Mitigation	Indian Institute of Ecology and Environment, Sikkim Manipal University	Delhi	1995	1999	2 years (distance learning)
PG Diploma Programme in Disaster Management	Bhavnagar University	Bhavnagar	1979	2005	n.a.
MTech in Disaster Management	SASTRA[2] Deemed University	Thanjavur	1984	2003	2 years (full-time)
Geoinformatics for Disaster Management & Environmental Assessment	Indian Institute of Remote Sensing	Dehradun	1966	n.a.	4 to 10 months
MSc in Geoinformatics for Disaster Management & Environmental Assessment	Indian Institute of Remote Sensing	Dehradun	1966	n.a.	17 months

[1]*Based on prospectus and websites of various institutes.* [2]*Shanmugha Arts, Commerce, Science, Technology and Research Academy*

Annexure 10 Professional/Educational Background of Candidates Enrolled for a Course on Disaster Management

Indira Gandhi National Open University	**Guru Gobind Singh Indraprastha University**	**Miranda House, University of Delhi**
PG Diploma in Disaster Management	Masters of Business Administration in Disaster Management	Applied Course in Disaster Management, Graduation Level
2005–2006	2006–2007	2006–2007
Science (Botany, Chemistry, Physics, Zoology, Environmental Studies, Forestry, Geology, Mathematics,) [37]	Engineers (govt. dept.) [29]	Environmental Courses [41]
Humanities (English, History, Political Science,) [26]	Officers (govt. dept.) [20]	History [26]
Commerce (Accounting, Commerce) [11]	Army [15]	Geography [19]
Social Science (Geography, Social Work, Economics) [10]	Police [15]	Hindi [7]
Engineering (Civil Engineering, Earthquake Engineering) [8]	Officials (fire dept.) [6]	Political Science [4]
Doctors [6]	Doctors [4]	Others [4]
Master in Business Administration [1]	Lawyers [4]	
Law [1]	Journalists [2]	
Mass Communication [1]	Teachers [2]	
Military Science [1]	Others [6]	
Public Administration [1]		
Total [100]	Total [100]	Total [100]
Total of 516 candidates in this course, of which details were not available for 321. These percentages are based on the remaining 195 candidates.	*Total candidates in the course is 54. The percentages are based on the total number of candidates, 2 did not furnish information.*	*Total candidates in the course is 27.The percentages are based on the total number of candidates*

Figures in brackets denote percentage to total students.

References

All India Association of Universities. *Bibliography of Doctoral Dissertations from 1857–2003*. Delhi: All India University Handbook.

Beck, U. (1998). *World Risk Society*. Cambridge: Polity Press.

Census of India. (1881, 1911, 1931, 1951). *Reports of the Census of India*. Kolkata and Delhi: Government Printing Press.

England P. and R. Bilham (2001). Plateau 'pop-up' in the great 1897 Assam earthquake. *Nature*, 410 (April 12). 806–809.

Hume, D. (1739). *A Treatise of Human Nature*. McMaster University Archive.

Iyengar, R.N. *et al.* (1999). Earthquake history of India in medieval times. *Indian Journal of History of Science*, 34 (3): 181–237.

Kapur, A. *et al.* (2005). *Disasters in India: Studies of Grim Reality*. Jaipur: Rawat Publication.

Kapur, A. (2005). Insensitive India: Attitudes towards disaster prevention and management. *Economic and Political Weekly*, 40 (42): 4551–4560

Kapoor, A. (2006). *2001 Bhuj Earthquake: A Geographical Study of Disaster and Response*. Delhi: Department of Geography, Delhi School of Economics, University of Delhi (Unpublished Ph.D. thesis).

Krishnamoorthy, K. (2005). Years of life lost and productivity loss due to tsunami in India. *Current Science,* 89 (5): 739–740.

Kuhn, T.S. (1962). *The Structure of Scientific Revolution*. Volume I–II. USA: Foundation of the Unity of Science.

Ministry of Finance (1952–2007). *Finance Commission Reports*. New Delhi: Government of India.

Parthasarathi, G. ed. (1986). *Jawaharlal Nehru: Letters to Chief Ministers, 1947–1964. Vol. I.* New Delhi: Oxford University Press. 165.

Quarantelli, E.L. (1998). *What is a Disaster? Perspectives on the Question*. London: Routledge.

Quarantelli, E.L. (1995). *Draft of a Sociological Disaster Research Agenda for the Future: Theoretical, Methodological and Empirical Issues*. Preliminary Paper No. 228. Delaware: Disaster Research Centre. University of Delaware.

Rajadhyaksha, A. and P. Willemen (1999). *Encyclopaedia of Indian Cinema.* Delhi: Oxford University Press.

Rodriquez, H. *et al.* (2006). *Handbook of Disaster Research.* New York: Springer.

Rusagi, N. K. (1988). *Crop Insurance in India: An Analysis.* Delhi: B. R. Publishing Corporation.

Sen, A. (1981). *Poverty and Famine: An Essay on Entitlement and Deprivation.* New York: Oxford University Press.

